Lecture Notes
in Computational Science
and Engineering

23

Editors

T. J. Barth, Moffett Field
M. Griebel, Bonn
D. E. Keyes, Norfolk
R. M. Nieminen, Espoo
D. Roose, Leuven
T. Schlick, New York

Springer
Berlin
Heidelberg
New York
Barcelona
Hong Kong
London
Milan
Paris
Tokyo

Luca F. Pavarino
Andrea Toselli
Editors

Recent Developments in Domain Decomposition Methods

With 69 Figures, 13 in Color, and 54 Tables

 Springer

Editors

Luca F. Pavarino
Department of Mathematics
University of Milano
Via C. Saldini 50
20133 Milano, Italy
e-mail: pavarino@mat.unimi.it

Andrea Toselli
Seminar for Applied Mathematics
ETH Zürich
Rämistraße 101
8092 Zürich, Switzerland
e-mail: toselli@sam.math.ethz.ch

Cover figure: Convective flow with an internal dissipative circular region. Discretization
with linear finite elements on non-matching grids and a discontinuous Galerkin approach
by A. Toselli

Cataloging-in-Publication Data applied for
Die Deutsche Bibliothek - CIP-Einheitsaufnahme
Recent developments in domain decomposition methods / Luca F. Pavarino ;
Andrea Toselli ed. - Berlin ; Heidelberg ; New York ; Barcelona ;
Hong Kong ; London ; Milan ; Paris ; Tokyo : Springer, 2002
(Lecture notes in computational science and engineering ; Vol. 23)
ISBN 3-540-43413-5

Mathematics Subject Classification (2000):
65N55, 65M55, 65N12, 65N22, 65N30, 65N35, 65F10, 65F50, 65Y05

ISSN 1439-7358
ISBN 3-540-43413-5 Springer-Verlag Berlin Heidelberg New York

Springer-Verlag Berlin Heidelberg New York
a member of BertelsmannSpringer Science + Business Media GmbH
http://www.springer.de
© Springer-Verlag Berlin Heidelberg 2002
Printed in Germany

Cover Design: Friedhelm Steinen-Broo, Estudio Calamar, Spain
Cover production: *design & production*
Typeset by the authors using a Springer TeX macro package

Printed on acid-free paper   SPIN: 10859419   46/3142/LK - 5 4 3 2 1 0

Preface

This volume collects some of the papers presented at the Workshop on Domain Decomposition held at ETH, Zürich, on June 7-8th 2001. The Workshop was organized by Luca F. Pavarino (University of Milan), Christoph Schwab (ETH Zürich), Andrea Toselli (ETH Zürich), and Olof B. Widlund (Courant Institute of Mathematical Sciences). Our sponsors were the University of Milan, Department of Mathematics (MURST projects: "Calcolo Scientifico: modelli e metodi numerici innovativi" and "Simmetrie, forme geometriche, evoluzione e memoria nelle equazioni alle derivate parziali"), the Seminar for Applied Mathematics, ETH Zürich, and the Program on Computational Science and Engineering at ETH Zürich.

The main goal of this meeting was to provide a forum for the exchange of ideas on the most recent developments in the field of Domain Decomposition Methods. We broadly understand Domain Decomposition as relating to the construction of preconditioners for the large algebraic systems of equations which often arise in applications, by solving smaller instances of the same problem. In our planning, we also wished to include studies of methods built from different discretizations in different subdomains such as in multi-physics models, mortar finite elements, wavelets, etc. Domain Decomposition methods are now fairly well understood for elliptic scalar and vector problems and are employed for the solution of large scale problems in computational sciences and engineering. However they remain less well understood for more general problems, such as scattering problems, mixed problems, wave propagation, and evolution problems. In addition, even for elliptic equations, some delicate important issues still need to be fully addressed, such as the improvement of some of the particular components of Domain Decomposition methods (coarse and local solvers) and their efficient application to a larger class of approximation methods (hp, spectral and wavelet approximations).

Among the most successful Domain Decomposition algorithms, we mention the Overlapping Schwarz, Neumann–Neumann and FETI (Finite Element Tearing and Interconnecting) methods. While the first class of methods is based on the solution of local problems on overlapping subdomains, Neumann–Neumann and FETI methods rely on a non-overlapping partition into subdomains (substructures). In a Neumann–Neumann method, a preconditioner is built with a low-dimensional coarse global problem, with a few degrees of freedom associated to each subdomain, and local Neumann problems on the subdomains. In a FETI method, the discrete problem is reformulated imposing the continuity of the finite element solution across the interface between the subdomains by introducing Lagrange multipliers. The primal variables are then implicitly eliminated, yielding an equation for the Lagrange multipliers. The continuity across the interface of the underlying finite element method is only fully satisfied at the convergence of the iteration. A FETI preconditioner is also built using local problems on the substructures. More recently, dual–primal FETI methods have been introduced. Here, Lagrange multipliers are still used but a few select interface continuity constraints are enforced in each iteration for particular sets of degrees of freedom on the interface. These new algorithms offer a number of advantages especially for very large and heterogeneous problems.

In mortar approximations, independent discretization methods can be employed in different subdomains. While the basic theory is quite well understood for elliptic problems, the application of mortar methods to more general problems is an ongoing field of research. Once the approximation properties of a mortar method are understood, one is left with the task of solving the corresponding linear system. If Lagrange multipliers are associated to the weak-continuity constraints across the interface between the subdomains, a mixed problem, formally the same as that employed in FETI methods, is obtained. It is then natural to generalize FETI preconditioners to linear systems arising from mortar approximations. Such generalizations are however far from straightforward.

The solution of a coarse problem is usually necessary in order to obtain scalable Domain Decomposition preconditioners, i.e., methods with a convergence rate that does not deteriorate with an increasing number of subdomains. The use of a coarse mesh is often quite non-trivial in particular when unstructured meshes are employed. An alternative is to construct coarse spaces that are not directly associated with a coarse mesh. In partition of unity and smoothed aggregation coarse spaces, the degrees of freedom can be associated instead with single subdomains of an overlapping partition and coarse basis functions can be properly constructed.

As a general introduction to Domain Decomposition methods we refer to the books by B. F. Smith, P.Bjørstad, and W. D. Gropp, *Domain Decomposition: Parallel Multilevel Methods for Elliptic Partial Differential Equations*, Cambridge University Press, 1996 and by A. Quarteroni and A. Valli, *Domain Decomposition Methods for Partial Differential Equations*, Oxford Science Publications, 1999. At the official Domain Decomposition web site http://www.ddm.org, interested readers can find information about Domain Decomposition meetings, proceedings and other related material.

About thirty scientists, from Europe and the United States, participated in the Workshop. This collection consists of fourteen of the twenty-one papers presented at the Workshop and their topics reflect some of the most active research areas in Domain Decomposition, such as:

- the development and analysis of novel FETI methods and Neumann-Neumann methods for the solution of systems arising from the approximations of partial differential equations (see Hetmaniuk and Farhat, Klawonn et al., Dryja and Widlund, and Goldfeld et al.);

- the construction and analysis of coarse solvers for two-level overlapping methods that do not require the introduction of a coarse triangulation (see Sarkis, and Lasser and Toselli);
- mortar methods for approximations on non-matching grids (see Maday et al., Bertoluzza et al., and Ben Belgacem et al.);
- preconditioners for scalar and vector scattering problems (see Lai et al., Hetmaniuk and Farhat, and Alonso and Valli);
- *hp* approximations and their efficient solution by iterative substructuring methods (see Bauer et al., Ben Belgacem et al.);
- block ILU preconditioners based on iterated filtering decompositions (see Achdou and Nataf);
- Domain Decomposition in time for evolution problems (see Bal and Maday).

We wish to thank Christoph Schwab and Olof Widlund for their help in organizing the Workshop.

Milan, Zürich,   *Luca F. Pavarino*
March 2002   *Andrea Toselli*

Contents

A Blended Fictitious/Real Domain Decomposition Method for Partially Axisymmetric Exterior Helmholtz Problems with Dirichlet Boundary Conditions

Ulrich Hetmaniuk and Charbel Farhat

Department of Aerospace Engineering Sciences and Center for Aerospace Structures, University of Colorado at Boulder, Boulder, CO 80309-0429, U.S.A.

Abstract. We blend a fictitious domain decomposition method and the FETI-H substructuring algorithm to construct a fast finite element based solver for exterior Helmholtz problems characterized by partially axisymmetric and sound-soft scatterers. We highlight the computational merits of this solver, and demonstrate between one and two orders of magnitude reduction of the CPU time associated with the straightforward solution of such exterior Helmholtz problems.

1 Introduction

It is well known that most partial differential equation problems defined over an axisymmetric domain can be efficiently solved by a Fourier based discretization method. However, for many engineering applications, the underlying computational domain is neither entirely axisymmetric, nor completely arbitrarily shaped, but has one or several major axisymmetric components (Figure 1). For such problems, an axisymmetric analysis method is not applicable, and a straightforward one can be inefficient because it does not exploit the properties of the axisymmetric regions. For example, a submarine can be represented as the assembly of a major cylindrical component — the tube — and a few minor "features" that are nevertheless essential for determining the submarine's acoustic signature, particularly in the medium- and high-frequency regimes. In these regimes, the finite element discretization of the exterior Helmholtz problem governing the acoustic scattered field can require hundreds of millions of grid points. However, this mesh size can be significantly reduced if the axisymmetry of the main tube is exploited by the solution methodology.

For elliptic problems as in thermal and elasticity applications, the gap in solution methods outlined above has been addressed by finite element based substructuring [1] and mortar [2] methods. For exterior Helmholtz problems with Dirichlet boundary conditions, a fictitious domain decomposition method [3] has been recently proposed in [4] for exploiting the axisymmetric regions. In this method, the original exterior Helmholtz problem is extended into an axisymmetric exterior problem that includes a fictitious region, and

Fig. 1. Partially axisymmetric bodies

parts of the Dirichlet boundary conditions are enforced by Lagrange multipliers. The axisymmetry of the enlarged domain is then exploited by expanding the solution into a Fourier series. The Fourier modes of the solution are obtained by solving a series of two-dimensional problems that are coupled by the Lagrange multipliers. Hence, this fictitious method transforms a three-dimensional problem into a suite of two-dimensional ones, reducing thereby the solution time by at least an order of magnitude [4].

In [4], a direct method was considered for solving the series of two-dimensional problems associated with the Fourier modes. In this paper, we investigate the solution of these problems by the FETI-H [5] substructuring algorithm, which leads us to the design of a blended fictitious/real domain decomposition method. We present this method in Section 4 after formulating the Helmholtz problem of interest in Section 2, and reviewing the basic fictitious method in Section 3. We demonstrate the superior computational performance of the blended fictitious/real solution algorithm in Section 5, and offer concluding remarks in Section 6.

2   The Sound-Soft Acoustic Scattering Problem

The scattering of time-harmonic acoustic waves by a three-dimensional sound soft obstacle embedded in a homogeneous medium can be formulated on a bounded domain Ω_b^{ex} as the following exterior boundary value problem (BVP)

$$\text{Find } u \in H^1(\Omega_b^{ex}) \text{ such that:}$$

$$-\Delta u - k^2 u = 0 \qquad \text{in } \Omega_b^{ex}, \tag{1}$$

$$u = -e^{ikd.x} \qquad \text{on } \Gamma, \tag{2}$$

$$\frac{\partial u}{\partial \nu} - Mu = 0 \qquad \text{on } \Sigma, \tag{3}$$

where H^1 is the standard Sobolev space, k denotes the wave number, i the imaginary number, x is a point in $\mathbb{R}^3$, d denotes the normalized direction of the incident wave, Γ is the surface of the scatterer and is assumed to be Lipschitzian, Σ is the artificial boundary of Ω_b^{ex} and is also assumed to be Lipschitzian, ν is the outward normal to Σ, M denotes a differential operator, and Eq. (3) is a general representation of absorbing boundary conditions. Different approaches for constructing an absorbing boundary condition are usually associated with different approaches for approximating the Dirichlet-to-Neumann (DtN) operator [6], and result in different expressions for M. However, all absorbing boundary conditions share the same objective, which is to reduce as much as possible the reflection of waves from the artificial boundary.

3   A Fictitious Domain Decomposition Method

For the sake of clarity, we assume throughout this paper that the scatterer Ω can be decomposed in two disjoint parts: a major axisymmetric component denoted by A, and a feature denoted by F (Figure 2). Hence, the open sets Ω, A and F satisfy

$$\overline{\Omega} = \overline{A} \cup \overline{F}. \tag{4}$$

The extension of the solution method proposed in this paper to multiple axisymmetric components and features is straightforward.

Given (4), the surface of the scatterer Γ is decomposed as follows

$$\Gamma = \Gamma_A \cup \Gamma_F,$$

where

$$\Gamma_A = \Gamma \cap \partial A,$$
$$\Gamma_F = \Gamma \cap \partial F. \tag{5}$$

Consider now the following exterior boundary value problem

$$\text{Find } (\tilde{u}, \mu) \in H^1(A_b^{ex}) \times (H_e^{1/2}(\Gamma_F))' \text{ such that:}$$

$$-\Delta\tilde{u} - k^2\tilde{u} + \tilde{\mu} = 0 \qquad \text{in } A_b^{ex}, \tag{6}$$
$$\tilde{u} = -e^{ikd.x} \qquad \text{on } \partial A, \tag{7}$$
$$\tilde{u} = -e^{ikd.x} \qquad \text{on } \Gamma_F, \tag{8}$$
$$\frac{\partial \tilde{u}}{\partial \nu} - M\tilde{u} = 0 \qquad \text{on } \Sigma, \tag{9}$$

where $H_e^{1/2}(\Gamma_F)$ is the space of functions belonging to $H^{1/2}(\Gamma_F)$ and whose extension on Γ belongs to $H^{1/2}(\Gamma)$ and satisfies (2) on Γ_A, $(H_e^{1/2}(\Gamma_F))'$ is

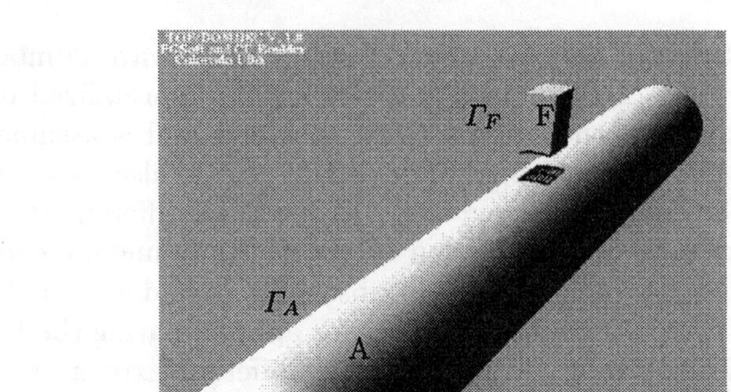

Fig. 2. Decomposition of a scatterer into a main axisymmetric component and a feature

the dual space of $H_e^{1/2}(\Gamma_F)$, A_b^{ex} is the exterior bounded domain associated with A (Figure 3), and $\tilde{\mu}$ is an extension of μ that satisfies

$$\forall v \in H^1(A_b^{ex}), \quad \int_{A_b^{ex}} \tilde{\mu} v \, dx \; = \; \int_{\Gamma_F} \mu v \, d\sigma. \tag{10}$$

We note that μ can be interpreted as a Lagrange multiplier for enforcing the original Dirichlet boundary condition on Γ_F (8).

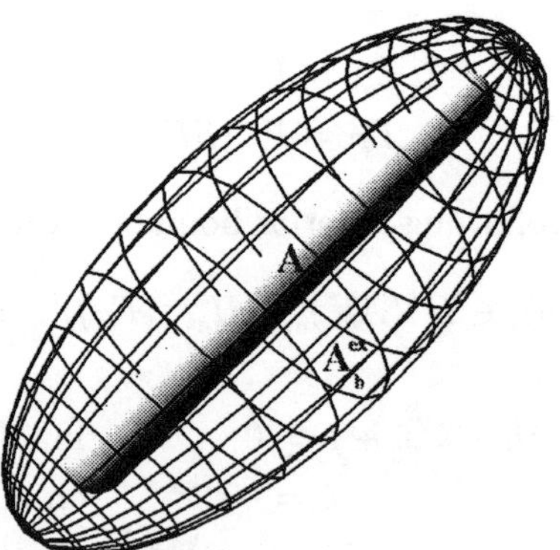

Fig. 3. Computational domain of the axisymmetric component

We refer to the above boundary value problem (6–9) as the *fictitious* boundary value problem (FBVP), because the region F, which is included in A_b^{ex}, is not part of Ω_b^{ex} and therefore is a fictitious region. The relation between this FBVP and the original BVP is given by the following theorem.

It is proved in [4] that solving the FBVP (6–9) is equivalent to solving the BVP (1–3) in the following sense:

- (Part 1) if $(\tilde{u}, \mu)$ is one solution of the FBVP (6–9), then the restriction of $\tilde{u}$ to Ω_b^{ex} is the solution of the BVP (1–3). Furthermore, the restriction of $\tilde{u}$ to F is *one* solution of an F-interior Helmholtz problem — that is, a Helmholtz problem defined inside F — with Dirichlet boundary conditions. The emphasis on *one* is because the F-interior problem can have multiple solutions if k^2 is an eigenvalue of the operator $-\Delta$.
- (Part 2) if u is the solution of the BVP (1–3) and u_F one solution of the F-interior Helmholtz problem with Dirichlet boundary conditions, then the couple $(\tilde{u}, \mu)$ defined by

$$
\begin{cases}
\tilde{u} = \begin{cases} u \text{ on } \Omega_b^{ex}, \\[1em] u_F \text{ on } F, \end{cases} \\[2em]
\mu = \dfrac{\partial u_F}{\partial \nu_F} - \dfrac{\partial u}{\partial \nu_F} \text{ on } \Gamma_F,
\end{cases}
\tag{11}
$$

where ν_F is the unitary normal to Γ_F and outward from Ω_b^{ex}, is solution of the FBVP (6–9).

Here, we note that the uniqueness of the solution of the FBVP (6–9) can be obtained by enforcing the uniqueness of the solution of the F-interior Helmholtz problem with Dirichlet boundary conditions. This can be achieved by a number of different regularization techniques (for example, see [5]). In all cases, the restriction of $\tilde{u}$ to Ω_b^{ex}, which provides the solution of the acoustic scattering problem of interest, is unique.

3.1 Variational Formulation

The FBVP problem described above has non-homogeneous Dirichlet boundary conditions. It can be reformulated as a FBVP with homogeneous Dirichlet boundary conditions as follows

$$\text{Find } (\tilde{u}, \mu) \in H^1(A_b^{ex}) \times (H_{00}^{1/2}(\Gamma_F))' \text{ such that:}$$

$$
\begin{aligned}
-\Delta \tilde{u} - k^2 \tilde{u} + \tilde{\mu} &= \tilde{f} && \text{in } A_b^{ex}, && (12) \\
\tilde{u} &= 0 && \text{on } \partial A, && (13) \\
\tilde{u} &= 0 && \text{on } \Gamma_F, && (14) \\
\frac{\partial \tilde{u}}{\partial \nu} - M\tilde{u} &= 0 && \text{on } \Sigma, && (15)
\end{aligned}
$$

where $\tilde{f}$ is a source function whose support is included in A_b^{ex} but does not intersect Σ, and which accounts for the influence of the non-homogeneous Dirichlet boundary conditions (7,8).

Let $H_{\partial A}^1(A_b^{ex})$ denote the following space

$$H_{\partial A}^1(A_b^{ex}) = \left\{ w \in H^1(A_b^{ex}) | w_{|\partial A} = 0 \right\}.$$

We introduce the following forms

$$a(v,w) = \int_{A_b^{ex}} (\nabla v . \nabla w - k^2 vw)dx - \int_{\Sigma} Mvwd\sigma, \quad \forall v, w \in H_{\partial A}^1(A_b^{ex}), \tag{16}$$

$$b(\mu,v) = \int_{\Gamma_F} \mu v \, d\sigma, \quad \forall(\mu,v) \in (H_{00}^{1/2}(\Gamma_F))' \times H_{\partial A}^1(A_b^{ex}), \tag{17}$$

$$L(v) = \int_{A_b^{ex}} fvdx, \quad \forall v \in H_{\partial A}^1(A_b^{ex}). \tag{18}$$

Using the above notation, the variational formulation associated with the homogeneous FBVP (12–15) is

$$\text{Find } (\tilde{u},\mu) \in H_{\partial A}^1(A_b^{ex}) \times (H_{00}^{1/2}(\Gamma_F))' \text{ such that:}$$

$$a(\tilde{u},v) + b(\mu,v) = L(v), \quad \forall v \in H_{\partial A}^1(A_b^{ex}), \tag{19}$$

$$b(\eta,\tilde{u}) = 0, \quad \forall \eta \in (H_{00}^{1/2}(\Gamma_F))'. \tag{20}$$

3.2   Discretization

Exploiting the axisymmetry of A requires defining an axisymmetric computational domain A_b^{ex}, which in turn calls for selecting an axisymmetric artificial boundary Σ. This is by no means a restrictive measure because most if not all artificial boundaries employed in acoustic scattering computations are surfaces of revolution. In that case, all functions defined on A_b^{ex} can be expressed as functions of the r, θ, z coordinates (Figure 4) and expanded in Fourier series with respect to the angle θ

$$\tilde{u}(r,\theta,z) = \sum_{n=-\infty}^{\infty} \tilde{u}_n(r,z)e^{in\theta}. \tag{21}$$

The Fourier coefficients $\tilde{u}_n(r,z)$ are defined on the meridian plane a_b^{ex} which generates A_b^{ex} by rotation around the z-axis (Figure 4).

Discretizing the two-dimensional domain a_b^{ex} by finite elements and truncating the Fourier expansion (21) leads to the following discrete expression of $\tilde{u}$ in a finite element e

$$\tilde{u}(r,\theta,z)^{(e)} = \sum_{n=-n_\theta}^{n_\theta} \sum_{j=1}^{n_j^{(e)}} X_j^{(e)}(r,z) \, \tilde{u}_{n,j} e^{in\theta}, \tag{22}$$

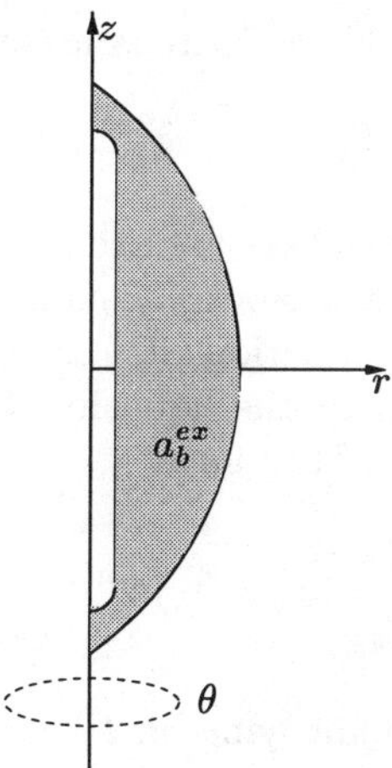

Fig. 4. Cylindrical coordinate system and generic meridian plane

where n_θ and $n_j^{(e)}$ denote respectively the selected highest consecutive Fourier mode number — in which case the number of Fourier modes is $2n_\theta + 1$ — and the number of nodes in the finite element e, $X_j^{(e)}$ is the shape function associated with node j of element e, and $\tilde{u}_{n,j}$ denotes the approximated value of $\tilde{u}_n(r,z)$ at node j.

Assuming that Γ_F and a_b^{ex} have matching discrete interfaces, the constraint equations implied by (20) can be enforced pointwise with discrete Lagrange multipliers.

Hence, the discretization of the variational formulation (19,20) described above leads to the following algebraic system of equations

$$\begin{cases} (\mathbf{K} - k^2\mathbf{M} - \mathbf{M}_\Sigma)\tilde{\mathbf{u}} + \mathbf{C}^T\boldsymbol{\mu} = \tilde{\mathbf{f}}, \\ \mathbf{C}\tilde{\mathbf{u}} = \mathbf{0}, \end{cases} \tag{23}$$

where the superscript T denotes the transpose operation, $\mathbf{K}$ and $\mathbf{M}$ are the generalized stiffness and mass matrices, respectively, $\mathbf{M}_\Sigma$ is the matrix associated with the finite element discretization of the absorbing boundary condition (15) and is non-zero only for the degrees of freedom lying on Σ, $\tilde{\mathbf{u}}$ can be decomposed as follows

$$\tilde{\mathbf{u}} = \begin{bmatrix} \tilde{\mathbf{u}}_{-n_\theta} & \cdots & \tilde{\mathbf{u}}_n & \cdots & \tilde{\mathbf{u}}_{n_\theta} \end{bmatrix}^T, \tag{24}$$

where each block $\tilde{\mathbf{u}}_n$ is the vector of coefficients of the n-th Fourier mode of the fictitious solution, $\mathbf{C}$ is the constraint matrix and can be written in block form as

$$\mathbf{C} = \begin{bmatrix} \mathbf{C}_{-n_\theta} & \cdots & \mathbf{C}_n & \cdots & \mathbf{C}_{n_\theta} \end{bmatrix}, \tag{25}$$

$\boldsymbol{\mu}$ is the vector of Lagrange multipliers, and $\tilde{\mathbf{f}}$ is another vector of Fourier coefficients resulting from the discretization of the right hand-side of Eq. (19),

8   Ulrich Hetmaniuk and Charbel Farhat

and can also be expressed in block form as follows

$$\tilde{\mathbf{f}} = \begin{bmatrix} \tilde{\mathbf{f}}_{-n_\theta} & \cdots & \tilde{\mathbf{f}}_n & \cdots & \tilde{\mathbf{f}}_{n_\theta} \end{bmatrix}^T .$$
(26)

If pointwise Lagrange multipliers are used to enforce the constraint equations implied by (20), then each block $\mathbf{C}_n$ associated with the Fourier mode n depends only on the discretization of $\tilde{u}_n$. More specifically, each of the discrete equations embedded in the equation $\mathbf{C}\tilde{\mathbf{u}} = \mathbf{0}$ corresponds to the discretization of an equation of the form

$$\sum_{n=-n_\theta}^{n=n_\theta} \tilde{u}_n(r_k, z_k)e^{in\theta_k} = 0,$$
(27)

where (r_k, θ_k, z_k) denotes a point lying on Γ_F, and (r_k, z_k) the corresponding node of a_b^{ex}.

Because the Fourier basis $\{e^{in\theta}\}_{n=-n_\theta}^{n=n_\theta}$ is an orthogonal basis, $\mathbf{K}$ and $\mathbf{M}$ are block diagonal sparse matrices that can be written as

$$\mathbf{K} = \begin{bmatrix} \mathbf{K}_{-n_\theta} & 0 & 0 & 0 & 0 \\ 0 & \ddots & 0 & 0 & 0 \\ 0 & 0 & \mathbf{K}_n & 0 & 0 \\ 0 & 0 & 0 & \ddots & 0 \\ 0 & 0 & 0 & 0 & \mathbf{K}_{n_\theta} \end{bmatrix},$$
(28)

and

$$\mathbf{M} = \begin{bmatrix} \mathbf{M}_{-n_\theta} & 0 & 0 & 0 & 0 \\ 0 & \ddots & 0 & 0 & 0 \\ 0 & 0 & \mathbf{M}_n & 0 & 0 \\ 0 & 0 & 0 & \ddots & 0 \\ 0 & 0 & 0 & 0 & \mathbf{M}_{n_\theta} \end{bmatrix},$$
(29)

where each pair of blocks $\mathbf{K}_n$ and $\mathbf{M}_n$ is associated with the n-th Fourier mode.

Furthermore, if the M operator of the absorbing boundary condition (3) is chosen among a wide range of local operators — which is assumed in this paper — then $\mathbf{M}_\Sigma$ is also a block diagonal sparse matrix that we expand as follows

$$\mathbf{M}_\Sigma = \begin{bmatrix} \mathbf{M}_{\Sigma,-n_\theta} & 0 & 0 & 0 & 0 \\ 0 & \ddots & 0 & 0 & 0 \\ 0 & 0 & \mathbf{M}_{\Sigma,n} & 0 & 0 \\ 0 & 0 & 0 & \ddots & 0 \\ 0 & 0 & 0 & 0 & \mathbf{M}_{\Sigma,n_\theta} \end{bmatrix},$$
(30)

where, again, each block $\mathbf{M}_{\Sigma,n}$ is associated with the n-th Fourier mode.

The algebraic system of equations (23) is graphically interpreted in Figure 5 for a two-body scatterer composed of a main cylinder and a minor prism with a square cross section. Essentially, each node on Γ_F introduces a constraint of the form given in (27) in the solution of an otherwise axisymmetric scattering problem. As shown in the left portion of Figure 5, the nodes of Γ_F can be grouped into sets characterized by $\theta = constant$. The trace on a_b^{ex} of one of these sets is shown in the right portion of Figure 5.

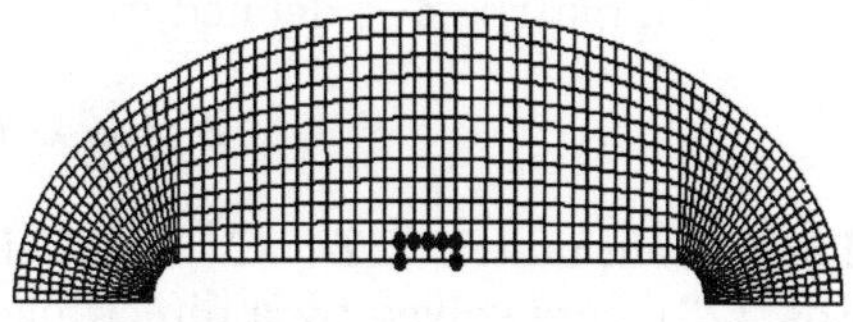

Fig. 5. Constrained nodes along $\theta = 9.6°$ (left: selected nodes on Γ_F — right: corresponding nodes in a_b^{ex})

The mathematical formulation (19,20) being a hybrid variational formulation, its finite element discretization is subject to the classical inf-sup condition [7]. Since in this work Γ_F and a_b^{ex} are assumed to have matching discrete interfaces and discrete Lagrange multipliers are introduced at these interfaces, the inf-sup condition can be expected to relate mainly the highest Fourier mode n_θ and the parameters of the finite element discretization (number of elements, degree of the polynomial shape functions, etc.) along the intersection of Γ_F and a_b^{ex}. For example, satisfying the inf-sup condition can be expected to guarantee that the constraint matrix $\mathbf{C}$ (25) has full column rank.

Analyzing the inf-sup condition governing the hybrid variational formulation (19,20) is a difficult issue that will be addressed in a sequel paper. Here, we *guide* the selection of the number of Fourier modes $2n_\theta + 1$ simply by an accuracy consideration for the highest Fourier mode to be represented in the sought-after solution. For example, the Nyquist sampling theorem states that samples spaced apart by h perfectly represent functions whose shortest wavelengths are $4h$. However, since our objective is not to represent accurately a single Fourier mode but to superpose these to reproduce the sought-after solution, we relax the $4h$ requirement in the Nyquist condition to $2h$, which gives

$$\frac{2\pi}{n_\theta} \approx 2h \iff n_\theta \approx \frac{\pi}{h}, \tag{31}$$

where h is the mesh size along the intersection of Γ_F and a_b^{ex}.

4   A Blended Fictitious/Real Domain Decomposition Method

In a previous work [4], the algebraic system of equations (23) was rewritten as

$$(\mathbf{C}\mathbf{Z}^{-1}\mathbf{C}^T)\boldsymbol{\mu} = \mathbf{C}\mathbf{Z}^{-1}\tilde{\mathbf{f}}, \tag{32}$$

$$\tilde{\mathbf{u}} = \mathbf{Z}^{-1}(\tilde{\mathbf{f}} - \mathbf{C}^T\boldsymbol{\mu}), \tag{33}$$

where the matrix $\mathbf{Z}$ is defined by

$$\mathbf{Z} = \mathbf{K} - k^2\mathbf{M} - \mathbf{M}_\Sigma. \tag{34}$$

The system matrix $\mathbf{C}\mathbf{Z}^{-1}\mathbf{C}^T$ was built explicitly and factored, and therefore Eqs. (23) were solved by a direct algorithm.

However, there are two incentives for considering an iterative domain decomposition method with Lagrange multipliers for solving the system of equations (23)

- if n_μ exceeds a few thousands — for example, if there are many features and the wave number is relatively high — storing the system matrix $\mathbf{C}\mathbf{Z}^{-1}\mathbf{C}^T$ can become unfeasible,
- the structure of Eqs. (23) is similar to that of the equations governing the FETI-H method [5], which is a numerically scalable domain decomposition method with Lagrange multipliers whose efficiency at solving exterior Helmholtz problems, particularly in two dimensions, has already been established [5,8].

For both reasons outlined above, we consider here solving the system of equations (23) by the FETI-H iterative method equipped with the numerically scalable technique proposed in [9] for addressing linear constraints of the form $\mathbf{C}\tilde{\mathbf{u}} = 0$. This leads us to designing a blended fictitious/real domain decomposition method for the solution of partially axisymmetric exterior Helmholtz problems.

4.1   The FETI-H Method

First, we overview the FETI-H domain decomposition method for the solution of a system of equations of the form

$$(\mathbf{K} - k^2\mathbf{M} - \mathbf{M}_\Sigma)\tilde{\mathbf{u}} = \tilde{\mathbf{f}}, \tag{35}$$

which arises from the finite element discretization of an exterior Helmholtz problem in a bounded domain Ω_b^{ex}. For the sake of clarity, we consider only the case where Ω_b^{ex} is decomposed into two subdomains $\Omega_b^{ex^1}$ and $\Omega_b^{ex^2}$. We refer the reader to [5] for the extension to an arbitrary number of subdomains as well as for further details.

It is shown in [5] that solving the above system of equations (35) is equivalent to solving the system of subdomain equations

$$\begin{cases} (\mathbf{Z}^1 + ik\mathbf{M}^{1,2})\tilde{\mathbf{u}}^1 + {\mathbf{B}^1}^T\boldsymbol{\lambda} = \tilde{\mathbf{f}}^1, \\ (\mathbf{Z}^2 - ik\mathbf{M}^{1,2})\tilde{\mathbf{u}}^2 + {\mathbf{B}^2}^T\boldsymbol{\lambda} = \tilde{\mathbf{f}}^2, \\ \qquad\quad \mathbf{B}^1\tilde{\mathbf{u}}^1 + \mathbf{B}^2\tilde{\mathbf{u}}^2 = 0, \end{cases} \tag{36}$$

where $\mathbf{Z}^s$, $\tilde{\mathbf{u}}^s$ and $\tilde{\mathbf{f}}^s$ denote respectively the problem matrix, solution vector, and right-hand side vector associated with the subdomain $\Omega_b^{ex^s}$, $\mathbf{B}^s$ is a signed Boolean matrix that extracts from a subdomain vector the degrees of freedom associated with the nodes lying on the interface $\Omega_b^{ex^1} \cap \Omega_b^{ex^2}$, $\boldsymbol{\lambda}$ is a vector of discrete Lagrange multipliers defined on this interface, and $\mathbf{M}^{1,2}$ is a mass-like matrix with non-zero entries only for the degrees of freedom lying on this interface.

When $\Omega_b^{ex^s}$ intersects the artificial boundary Σ, $\mathbf{Z}^s$ is non-singular because of the contribution of the absorbing boundary condition, and $\mathbf{Z}^s \pm ik\mathbf{M}^{1,2}$ remains non-singular. When $\Omega_b^{ex^s}$ does not intersect Σ, k^2 may coincide with a generalized eigenvalue of the pencil $(\mathbf{K}^s, \mathbf{M}^s)$ in which case $\mathbf{Z}^s$ becomes singular. In [5], it is proved that in that case, $\mathbf{M}^{1,2}$ prevents $\mathbf{Z}^s \pm ik\mathbf{M}^{1,2}$ from becoming singular.

Eliminating $\tilde{\mathbf{u}}^1$ and $\tilde{\mathbf{u}}^2$ from Eqs. (36) leads to an interface problem of the form

$$\mathbf{F}_I\boldsymbol{\lambda} = \mathbf{d}, \tag{37}$$

where

$$\begin{cases} \mathbf{F}_I = \mathbf{B}^1(\mathbf{Z}^1 + ik\mathbf{M}^{1,2})^{-1}{\mathbf{B}^1}^T + \mathbf{B}^2(\mathbf{Z}^2 - ik\mathbf{M}^{1,2})^{-1}{\mathbf{B}^2}^T, \\ \mathbf{d} = \mathbf{B}^1(\mathbf{Z}^1 + ik\mathbf{M}^{1,2})^{-1}\tilde{\mathbf{f}}^1 + \mathbf{B}^2(\mathbf{Z}^2 - ik\mathbf{M}^{1,2})^{-1}\tilde{\mathbf{f}}^2. \end{cases} \tag{38}$$

Note that $\mathbf{F}_I$ is a symmetric but not hermitian matrix. The FETI-H method can then be defined as the solution of problem (37) by a preconditioned GCR algorithm [13]. Its preconditioner is constructed as follows.

Let $\mathbf{r}^p$ denote the residual at the p-th GCR iteration

$$\mathbf{r}^p = \mathbf{d} - \mathbf{F}_I\boldsymbol{\lambda}^p. \tag{39}$$

The convergence of the GCR algorithm can be accelerated by modifying this algorithm so that at each iteration p, the residual $\mathbf{r}^p$ is orthogonal to a subspace represented by a matrix $\mathbf{Q}$

$$\mathbf{Q}^T\mathbf{r}^p = 0. \tag{40}$$

Indeed, condition (40) is a weighted residual weak form of $\mathbf{r}^p = \mathbf{0}$, and therefore its effect at each iteration p is to reduce the error until $\mathbf{r}^p$ converges to

0. If n_I denotes the size of the interface problem, constructing an interface matrix $\mathbf{Q}$ with n_I linearly independent columns and enforcing (40) guarantees convergence in one iteration. However, computational efficiency requires choosing $\mathbf{Q}$ "coarse" enough to keep the overhead associated with enforcing (40) affordable.

A straightforward approach for enforcing at each GCR iteration the constraint (40) is to split the iterate $\boldsymbol{\lambda}^p$ as follows

$$\boldsymbol{\lambda}^p = \tilde{\boldsymbol{\lambda}}^p + \mathbf{Q}\boldsymbol{\gamma}^p. \tag{41}$$

While the objective of $\tilde{\boldsymbol{\lambda}}^p$ is to enforce at convergence $\mathbf{B}^1\tilde{\mathbf{u}}^1 + \mathbf{B}^2\tilde{\mathbf{u}}^2 = \mathbf{0}$, that of $\boldsymbol{\gamma}^p$ is to enforce at each iteration $\mathbf{Q}^T\mathbf{r}^p = 0$. Substituting the splitting of $\boldsymbol{\lambda}^p$ into (39, 40) gives

$$(\mathbf{Q}^T\mathbf{F}_I\mathbf{Q})\boldsymbol{\gamma}^p = \mathbf{Q}^T(\mathbf{d} - \mathbf{F}_I\tilde{\boldsymbol{\lambda}}^p). \tag{42}$$

which shows that at each iteration p, $\boldsymbol{\gamma}^p$ can be obtained from the solution of an auxiliary "second-level" coarse problem.

From (41, 42), it follows that $\boldsymbol{\lambda}^p$ can be computed as

$$\boldsymbol{\lambda}^p = \mathbf{P}\tilde{\boldsymbol{\lambda}}^p + \boldsymbol{\lambda}^0, \tag{43}$$

where $\mathbf{P}$ is the projector given by

$$\mathbf{P} = \mathbf{I} - \mathbf{Q}(\mathbf{Q}^T\mathbf{F}_I\mathbf{Q})^{-1}\mathbf{Q}^T\mathbf{F}_I, \tag{44}$$

and $\boldsymbol{\lambda}^0$ is given by

$$\boldsymbol{\lambda}^0 = \mathbf{Q}(\mathbf{Q}^T\mathbf{F}_I\mathbf{Q})^{-1}\mathbf{Q}^T\mathbf{d}. \tag{45}$$

The interface problem (37) is then transformed into

$$(\mathbf{P}^T\mathbf{F}_I\mathbf{P})\tilde{\boldsymbol{\lambda}}^p = \mathbf{P}^T(\mathbf{d} - \mathbf{F}_I\boldsymbol{\lambda}^0). \tag{46}$$

Hence, accelerating the convergence of the GCR algorithm by the introduction of (40) can be interpreted as preconditioning the interface problem (37) with $\mathbf{P}^T\mathbf{P}$. The major additional computational cost entailed by this preconditioner is that associated with the construction and factorization of the matrix $\mathbf{Q}^T\mathbf{F}_I\mathbf{Q}$. In [5], it is proved that

$$\mathbf{P}^T\mathbf{F}_I\mathbf{P} = \mathbf{F}_I\mathbf{P}, \tag{47}$$

so that only one projection is required per GCR iteration.

In summary, the FETI-H method applied to the solution of the original problem (35) consists of converting this global system of equations into the interface problem (46) after constructing $\mathbf{Q}$ with planar waves, and solving this interface problem by the GCR algorithm. More specifically, each column $\mathbf{Q}_j$ is built as follows

$$\mathbf{Q}_j = e^{ik\boldsymbol{\theta}_j\cdot\boldsymbol{x}}, \tag{48}$$

where $\boldsymbol{x}$ denotes the vector of nodal coordinates on the subdomain interface boundaries, and the directions $\boldsymbol{\theta}_j$ are uniformly distributed on the unit circle of $\mathrm{I\!R}^2$ in two dimensions, and the unit sphere of $\mathrm{I\!R}^3$ in three dimensions.

4.2   A One-Shot Iterative Method

To apply FETI-H to the solution of the fictitious problem (23), we proceed as follows. First, we decompose a_b^{ex} into, for example, two non-overlapping subdomains $a_b^{ex^1}$ and $a_b^{ex^2}$. Then, we reformulate Eqs. (23) at the subdomain level to obtain the following mixed fictitious/real system of subdomain equations

$$\begin{cases} (\mathbf{Z}^1 + ik\mathbf{M}^{1,2})\tilde{\mathbf{u}}^1 + {\mathbf{B}^1}^T\boldsymbol{\lambda} + {\mathbf{C}^1}^T\boldsymbol{\mu} = \tilde{\mathbf{f}}^1, \\ (\mathbf{Z}^2 - ik\mathbf{M}^{1,2})\tilde{\mathbf{u}}^2 + {\mathbf{B}^2}^T\boldsymbol{\lambda} + {\mathbf{C}^2}^T\boldsymbol{\mu} = \tilde{\mathbf{f}}^2, \\ \mathbf{B}^1\tilde{\mathbf{u}}^1 + \mathbf{B}^2\tilde{\mathbf{u}}^2 = \mathbf{0}, \\ \mathbf{C}^1\tilde{\mathbf{u}}^1 + \mathbf{C}^2\tilde{\mathbf{u}}^2 = \mathbf{0}, \end{cases} \tag{49}$$

where $\mathbf{C}^s$ denotes the constraint matrix associated with the subdomain $a_b^{ex^s}$, and all other quantities have the same meaning as before. Then, following the approach described in [9], we introduce the extended vector of Lagrange multipliers $\boldsymbol{\Lambda}$ and the extended matrix of subdomain constraints $\mathbb{B}^s$

$$\boldsymbol{\Lambda} = \begin{bmatrix} \boldsymbol{\lambda} \\ \boldsymbol{\mu} \end{bmatrix}, \quad \mathbb{B}^s = \begin{bmatrix} \mathbf{B}^s \\ \mathbf{C}^s \end{bmatrix}. \tag{50}$$

Using the above notation, problem (49) can be rewritten as

$$\mathbb{F}_I\,\boldsymbol{\Lambda} = \mathbf{d}, \tag{51}$$

where

$$\begin{cases} \mathbb{F}_I = \mathbb{B}^1(\mathbf{Z}^1 + ik\mathbf{M}^{1,2})^{-1}{\mathbb{B}^1}^T + \mathbb{B}^2(\mathbf{Z}^2 - ik\mathbf{M}^{1,2})^{-1}{\mathbb{B}^2}^T, \\ \mathbf{d} = \mathbb{B}^1(\mathbf{Z}^1 + ik\mathbf{M}^{1,2})^{-1}\tilde{\mathbf{f}}^1 + \mathbb{B}^2(\mathbf{Z}^2 - ik\mathbf{M}^{1,2})^{-1}\tilde{\mathbf{f}}^2. \end{cases} \tag{52}$$

Let $\mathbf{r}^p$ denote here the p-th residual associated with the solution of (51) by the GCR algorithm

$$\mathbf{r}^p = \mathbf{d} - \mathbb{F}_I\boldsymbol{\Lambda}^p. \tag{53}$$

This residual $\mathbf{r}^p$ can be partitioned as

$$\mathbf{r}^p = \begin{bmatrix} \mathbf{r}^p_\lambda \\ \mathbf{r}^p_\mu \end{bmatrix},$$

where $\mathbf{r}^p_\lambda$ measures the jump between the p-th iterates of $\tilde{\mathbf{u}}^1$ and $\tilde{\mathbf{u}}^2$ at the subdomain interfaces, and $\mathbf{r}^p_\mu$ measures the violation of the constraints enforced by $\mathbf{C}$. Since these two residuals have two different meanings, and following the recommendations formulated in [9], we propose to accelerate the convergence of the GCR algorithm by enforcing at each iteration the two following different constraints

$$\begin{cases} \mathbf{Q}^T_\lambda \mathbf{r}^p_\lambda = 0, \\ \boldsymbol{I}\mathbf{r}^p_\mu = 0, \end{cases} \tag{54}$$

where $\mathbf{Q}_\lambda^T$ is the matrix of planar waves as in (48). It follows that $\mathbb{Q}$ is given by

$$\mathbb{Q} = \begin{bmatrix} \mathbf{Q}_\lambda^T & 0 \\ 0 & I \end{bmatrix}. \tag{55}$$

Next, we transform the interface problem (51) into

$$(\mathbb{P}^T \mathbb{F}_I \mathbb{P})\tilde{\boldsymbol{\Lambda}} = \mathbb{P}^T(\mathbf{d} - \mathbb{F}_I \boldsymbol{\Lambda}^0), \tag{56}$$

where $\mathbb{P}$ is the extended projector defined by

$$\mathbb{P} = I - \mathbb{Q}(\mathbb{Q}^T \mathbb{F}_I \mathbb{Q})^{-1} \mathbb{Q}^T \mathbb{F}_I, \tag{57}$$

$\boldsymbol{\Lambda}^0$ is given by

$$\boldsymbol{\Lambda}^0 = \mathbb{Q}(\mathbb{Q}^T \mathbb{F}_I \mathbb{Q})^{-1} \mathbb{Q}^T \mathbf{d}, \tag{58}$$

and $\boldsymbol{\Lambda}$ has been split as follows

$$\boldsymbol{\Lambda} = \mathbb{P}\tilde{\boldsymbol{\Lambda}} + \boldsymbol{\Lambda}^0. \tag{59}$$

The solution of the system of equations (56) by the GCR algorithm defines a blended fictitious/real domain decomposition method for the solution of partially axisymmetric sound-soft acoustic scattering problems. In this method, the Lagrange multipliers $\boldsymbol{\lambda}$ and $\boldsymbol{\mu}$ are updated in one shot — that is, within the same loop.

4.3   Some Implementational Aspects

In the fictitious/real domain decomposition method summarized by Eqs. (49), the matrices $\mathbf{Z}^s \pm ik\mathbf{M}^{1,2}$ remain block diagonal. Each block is a sparse matrix associated with the finite element discretization of a two-dimensional problem. Therefore, for any reasonable mesh resolution associated with a low, medium or even reasonably high wave number k, storing each block is not an issue on most modern computational platforms. Besides, given that $\int_0^{2\pi} e^{in\theta} e^{-in\theta} d\theta = \int_0^{2\pi} e^{i(-n)\theta} e^{-i(-n)\theta} d\theta = 2\pi$, only $n_\theta + 1$ of these $2n_\theta + 1$ block need be formed, assembled and factored in each subdomain. All these operations can be performed in parallel by mapping each available processor onto a Fourier mode.

Furthermore, $\mathbf{Q}_\lambda$ is also block diagonal. Hence, the symmetric coarse matrix

$$\begin{bmatrix} \mathbf{Q}_\lambda^T & 0 \\ 0 & I \end{bmatrix} \mathbb{F}_I \begin{bmatrix} \mathbf{Q}_\lambda & 0 \\ 0 & I \end{bmatrix}, \tag{60}$$

has the following sparsity pattern

$$\begin{bmatrix} \times & & \times \\ & \times & & \times \\ & & \times & \times \\ \times & \times & \times & \otimes \end{bmatrix}. \tag{61}$$

We note that only the diagonal blocks of the above matrix need be assembled. To solve a system of equations of the form

$$\begin{bmatrix} \mathbf{Q}_\lambda^T & 0 \\ 0 & \mathbf{I} \end{bmatrix} \mathbb{F}_I \begin{bmatrix} \mathbf{Q}_\lambda & 0 \\ 0 & \mathbf{I} \end{bmatrix} \begin{bmatrix} \gamma_\lambda \\ \gamma_\mu \end{bmatrix} = \begin{bmatrix} \mathbf{Q}_\lambda^T & 0 \\ 0 & \mathbf{I} \end{bmatrix} \mathbb{F}_I \mathbf{r}^p, \tag{62}$$

which arises at each projection step, we form once for all the Schur complement on $\otimes$.

5   Applications

In this section, we apply the computational methodology presented in this paper to the prediction of the scattering of waves by two different sound-soft mockup submarines. We illustrate the convergence properties of this new methodology, assess its accuracy, and highlight its computational advantages.

In all examples discussed herein, we choose as a non-reflecting boundary condition the second-order Bayliss-Gunzburger-Turkel-like (BGTL) condition first designed in [6] as an "on surface" radiation condition, then developed and validated in [10,11] as an exterior absorbing boundary condition. The differential operator M (3) characterizing this second-order BGTL condition is given by

$$Mu = div_\Sigma \left(\tfrac{1}{2ik}(I + \tfrac{i\mathcal{R}}{k})^{-1} \nabla_\Sigma u \right)$$
$$+ \left(-ik + \mathcal{H} + \tfrac{i}{2k}(1 + i\tfrac{2\mathcal{H}}{k})^{-1}(\mathcal{K} - \mathcal{H}^2) - \tfrac{\Delta_\Sigma \mathcal{H}}{4k^2} \right) u, \tag{63}$$

where $\mathcal{H}$, $\mathcal{K}$, and ∇_Σ denote respectively the mean curvature of the artificial boundary Σ, its Gauss curvature, and its surface gradient operator, and $\mathcal{R}$ is the differential of the Gauss map — that is, a self-adjoint operator that maps the tangent plan to Σ onto the tangent plan to the unit sphere.

The operator M described above allows the implementation of a radiation condition on any arbitrarily shaped but convex artificial boundary Σ. Here, we choose Σ as the surface of either a cylinder or an ellipsoid whose main axis is aligned with the main axis of the mockup submarine.

We perform all computations in double precision complex arithmetic on an Origin 2000 parallel processor, using a number of processors between 1 and 16. The code used is an object-oriented C++ program.

5.1   Convergence, Accuracy, Storage Requirements, and Parallelism

First, we consider the scatterer graphically depicted in Figure 6. The tube of this mockup submarine has a length $L_A = 10$ and a diameter $D_A = 1.0$. Its conical tower has a length $L_F = 1$, a height $H_F = 0.5$, and an angular aperture equal to 45 degrees. We focus on predicting the scattering of the time-harmonic incident wave in the direction

$$(d_x, \ d_y, \ d_z) = (-\sqrt{3}/3, \ -\sqrt{3}/3, \ -\sqrt{3}/3),$$

in the frequency regime corresponding to $kD_A = 10$.

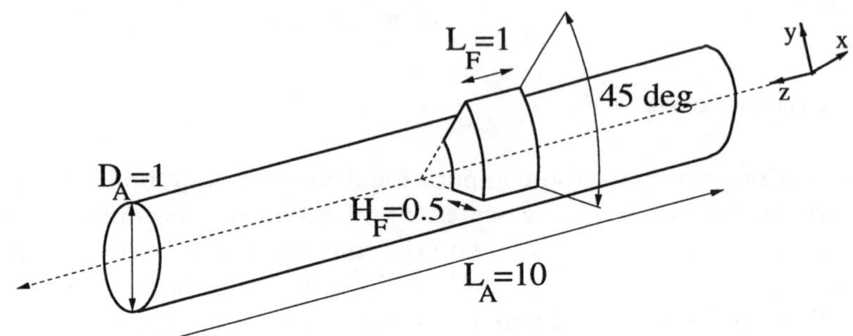

Fig. 6. Mockup submarine with a conical tower

For this scatterer, we choose as an artificial boundary Σ the surface of a cylinder with the same center and main axis as those of the tube of the scatterer, a radius $R_\Sigma = D_A/2 + H_F + m\lambda$, and a length $L_\Sigma = L_A + 2m\lambda$ (see Figure 7). Note that here, $\lambda = 2\pi/k$ denotes the wave length and not a Lagrange multiplier, and m is an integer. We construct three different computational domains a_b^{ex} corresponding to $m = 1$, $m = 2$, and $m = 4$. For each one of these computational domains, we generate three uniform finite element discretizations based on the standard $Q1$ element, and characterized by three different mesh resolutions (number of elements per wave length) $\lambda/h = 10$, $\lambda/h = 15$, and $\lambda/h = 20$. We report in Table 1 the computational sizes in number of grid points of all nine finite element discretizations of a_b^{ex}. We contrast these sizes with those of the corresponding instances of Ω_b^{ex} and their finite element discretizations that are required by the three-dimensional finite element analysis of this acoustic scattering problem when the partial axisymmetry of the scatterer is not exploited. The numbers reported in Table 1 suggest that even for a number of Fourier modes as high as 200, the proposed solution method should deliver significant computational savings over a standard three-dimensional finite element analysis.

For each considered mesh resolution, we apply the Nyquist criterion (31) to select n_θ, and report in Table 2 the number of Lagrange multipliers n_μ as well as the total memory required by the proposed computational method when equipped with a direct method for solving Eqs. (23) and applied to the solution of this acoustic scattering problem. We contrast this amount of memory with that needed in a standard three-dimensional finite element analysis just to store the generalized stiffness matrix $\mathbf{K} - k^2\mathbf{M} - \mathbf{M}_\Sigma$. The

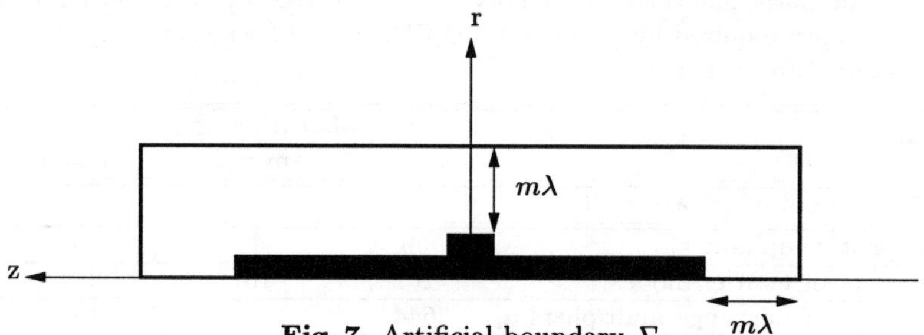

Fig. 7. Artificial boundary Σ

Table 1. Comparison of the computational sizes of a_b^{ex} and Ω_b^{ex} (number of grid points)

Mesh resolution	Computational domain	Location of Σ		
		$m = 1$	$m = 2$	$m = 4$
$\lambda/h = 10$	a_b^{ex}	3,615	6,165	12,465
	(Ω_b^{ex})	(559,923)	(1,348,065)	(4,303,089)
$\lambda/h = 15$	a_b^{ex}	7,972	13,687	27,817
	(Ω_b^{ex})	(1,887,016)	(4,538,155)	(14,469,673)
$\lambda/h = 20$	a_b^{ex}	14,029	24,169	49,249
	(Ω_b^{ex})	(4,456,129)	(10,743,537)	(34,284,385)

comparison of these memory requirements shows that even when the direct method is applied with 201 Fourier modes, its total memory requirements are two to almost three times less than the memory needed by a standard three-dimensional finite element analysis just for storing the problem's matrix in a sparse format.

For each different computational domain and mesh resolution, after applying the solution method proposed in this paper for predicting the acoustic scattered field u, we evaluate the far-field pattern (FFP)

$$\text{FFP}(\hat{x}) = u_\infty(\hat{x}), \tag{64}$$

where $\hat{x} \in S^1$, $S^1 = \{x \in \mathbb{R}^3 \ / \ \|x\|_2 = 1\}$ is the unit sphere, and u_∞ is the amplitude of the scattered field. We compute u_∞ by a numerical approximation of the following integral

$$u_\infty(\hat{x}) = \frac{1}{4\pi} \int_\Gamma (\frac{\partial u}{\partial \nu}(y) + ik \, \hat{x}.\nu \, u(y)) e^{-ik\hat{x}.y} d\sigma_y, \qquad \forall \hat{x} \in S^1. \tag{65}$$

For the case $m = 4$ and $\lambda/h = 20$, we label the computed solution and FFP as the "reference" solution and reference FFP, because that case corresponds to the largest computational domain a_b^{ex} and the finest mesh discretization considered herein. For each other case, we compute the relative error in the

Table 2. Comparison of the *total* memory required by the fictitious solution method and the memory required by a standard 3D FE method for storing only the generalized sparse stiffness matrix

| | Location of Σ | | |
	$m = 1$	$m = 2$	$m = 4$
Mesh resolution $\lambda/h = 10$			
Nyquist compliant n_θ	50	50	50
Number of Fourier modes	101	101	101
Number of Lagrange multipliers n_μ	644	644	644
Total memory requirements (Memory required for storing $\mathbf{K} - k^2\mathbf{M} - \mathbf{M}_\Sigma$ in a fully 3D FE analysis)	68 Mb. (125 Mb.)	132 Mb. (291 Mb.)	336 Mb. (930 Mb.)
Mesh resolution $\lambda/h = 15$			
Nyquist compliant n_θ	75	75	75
Number of Fourier modes	151	151	151
Number of Lagrange multipliers n_μ	1,446	1,446	1,446
Total memory requirements (Memory required for storing $\mathbf{K} - k^2\mathbf{M} - \mathbf{M}_\Sigma$ in a fully 3D FE analysis)	265 Mb. (423 Mb.)	515 Mb. (980 Mb.)	1,261 Mb. (3,125 Mb.)
Mesh resolution $\lambda/h = 20$			
Nyquist compliant n_θ	100	100	100
Number of Fourier modes	201	201	201
Number of Lagrange multipliers n_μ	2,568	2,568	2,568
Total memory requirements (Memory required for storing $\mathbf{K} - k^2\mathbf{M} - \mathbf{M}_\Sigma$ in a fully 3D 3D FE analysis)	669 Mb. (998 Mb.)	1,349 Mb. (2,320 Mb.)	3,253 Mb. (7,405 Mb.)

FFP

$$RelErr = \frac{\| \mathrm{FFP}^{ref} - \mathrm{FFP} \|_2}{\| \mathrm{FFP}^{ref} \|_2}, \tag{66}$$

where the superscript ref designates the reference solution, and report that error in Table 3. For $\lambda/h = 10$, the fact that increasing m barely decreases the relative error in the FFP is indicative that the mesh is almost underresolved. Indeed, it is known that in that case the computed solution is subject to the pollution effect [12], and the corresponding error increases with the size of the computational domain. Otherwise, the errors reported for $\lambda/h = 15$ and $\lambda/h = 20$ justify *a posteriori* the "reference" label for the solution computed with $m = 4$ and $\lambda/h = 20$. They also demonstrate that the less computationally intensive case $m = 2$ and $\lambda/h = 15$ delivers a converged solution.

Table 3. Relative errors in the far-field pattern: convergence for an increasing size of the computational domain and an increasing size of the mesh resolution

| | Location of Σ | | |
Mesh resolution $m = 1$	$m = 2$	$m = 4$
$\lambda/h = 10$	5.6%	4.6%	4.6%
$\lambda/h = 15$	3.9%	2.4%	1.3%
$\lambda/h = 20$	2.4%	2.3%	Reference

Table 4. Relative errors in the far-field pattern: convergence for $m = 2$, $\lambda/h = 15$, and an increasing value of n_θ

n_θ	20	30	40	50	60	70	75	90	120
$RelErr$	23.7%	17.7%	10.3%	5.2%	2.9%	2.4%	2.4%	2.4%	2.4%

Therefore, we fix next $m = 2$ and $\lambda/h = 15$, and solve repeatedly the target acoustic scattering problem using several values of n_θ ranging between $n_\theta = 20$ and $n_\theta = 120$. Again, for each case, we evaluate the FFP and report in Table 4 its relative error with respect to the reference value FFP^{ref} introduced above ($m = 4$, $\lambda/h = 20$). For a fixed computational domain and a fixed mesh resolution, the relative error in the FFP is shown to decay as $1/n_\theta^{2.5}$. The results reported in Table 4 also show that for $n_\theta = 75$ the solution is converged.

In summary, for the partially axisymmetric acoustic scattering problem described herein and a cylindrical artificial boundary Σ, the proposed solution methodology equipped with the second-order BGTL non-reflecting boundary condition (63) converges when Σ is positioned 2 wave lengths away from the surface of the scatterer in each direction, λ/h is chosen to be in the neighborhood of 15 elements per wave length, and about 150 Fourier modes are represented in the computed solution. For this reason, we adopt these discretization parameters in our sample assessment of the parallel performance of both fictitious and mixed fictitious/real versions of the proposed solution methodology. We remind the reader that in both cases, we parallelize the computations by mapping each available processor onto a set of Fourier modes. We carry out all the computational steps on a mode-by-mode basis, except for the solution of the system of equations (32) arising in the direct approach or the system of equations (62) in the blended approach, which we perform using once for all a parallel dense factorization algorithm. In the case of the blended fictitious/real domain decomposition method, we decompose the computational domain a_b^{ex} into two subdomains, one of which is the smallest set containing all the constraints defined by the matrix $\mathbf{C}$. We construct $\mathbf{Q}_\lambda$ using 16 plane waves, and achieve convergence with a relative residual less than 10^{-6} after 10 iterations.

Table 5. Parallel performance on an Origin 2000 of the proposed fictitious domain decomposition method ($m = 2$, $\lambda/h = 15$, and $n_\theta = 75$)

Number of processors	1	2	4	8	16
Total CPU time	13,701 s.	7,272 s.	3,861 s.	1,903 s.	964 s.
Speedup	1	1.9	3.6	7.2	14.2
Efficiency (Speedup per processor)	100 %	94 %	88 %	90 %	88 %

Table 6. Parallel performance on an Origin 2000 of the proposed blended fictitious/real domain decomposition method ($m = 2$, $\lambda/h = 15$, and $n_\theta = 75$)

Number of processors	1	2	4	8	16
Total CPU time	1,451 s.	777.6 s.	430.8 s.	231.2 s.	136 s.
Speedup	1	1.87	3.36	6.27	10.6
Efficiency (Speedup per processor)	100 %	93 %	84 %	78 %	66 %

Table 7. CPU time comparison of the fictitious and fictitious/real solution methods ($m = 2$, 16 processors, 2 subdomains)

	$\lambda/h = 10$	$\lambda/h = 15$	$\lambda/h = 20$
CPU time (fictitious method)	137.4 s.	963.6 s.	4,859.7 s.
CPU time (fictitious/real method)	27.5 s.	135.9 s.	489.2 s.
Nb. of iterations	7	10	10
Nb. of directions	18	16	18

The performance results reported in Table 5 and Table 6 demonstrate an excellent parallel scalability of the fictitious domain decomposition method, and a reasonable one for the blended fictitious/real one.

Finally, we compare in Table 7 the performance results on a 16-processor Origin 2000 of the fictitious and blended fictitious/real solvers for three different mesh resolutions and an artificial boundary positioned 2 wavelengths away from the scatterer. As hoped for, the blended method is found to be 5 to 10 times faster than the basic fictitious domain decomposition method.

5.2   Significance, Validation, and Computational Performance

Next, we consider the scattering of an incident wave at

$$(d_x,\ d_y,\ d_z) = (-\sqrt{2}/2,\ 0,\ -\sqrt{2}/2),$$

by the mockup submarine shown in Fig. 8, in the frequency regime corresponding to $k = 5$. This scatterer is composed of a tube and a prismatic tower with a rectangular cross section. The length of the tube is $L_A = 10$, and its diameter is $D_A = 1.0$. The height of the tower is $H_F = 0.5$, its length is $L_F = 1$, and its width is $W_F = 0.25$. We delimit the computational domain around this mockup submarine by an ellipsoid of revolution as graphically depicted in Figure 9 with $m_1 = m_2 = 0.8$.

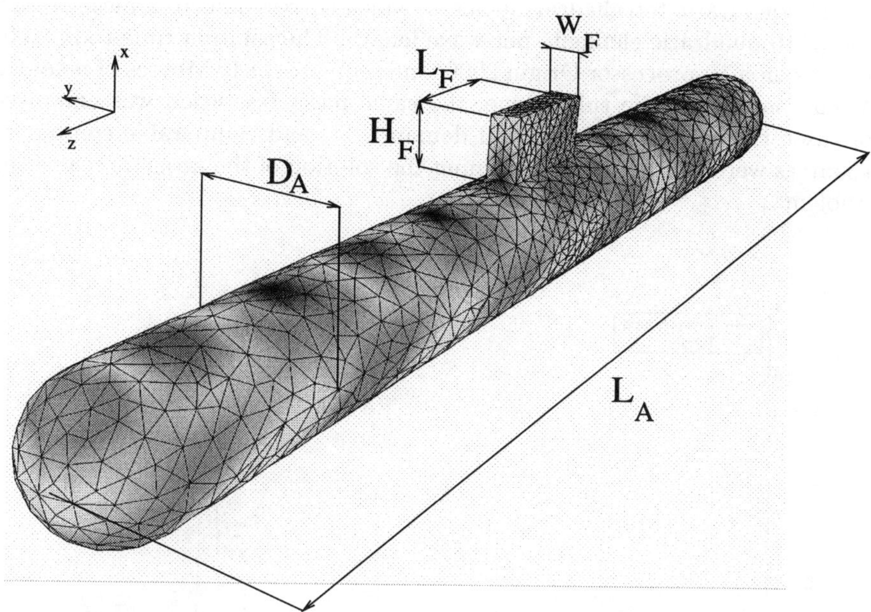

Fig. 8. Mockup submarine — Prismatic tower with a rectangular cross section

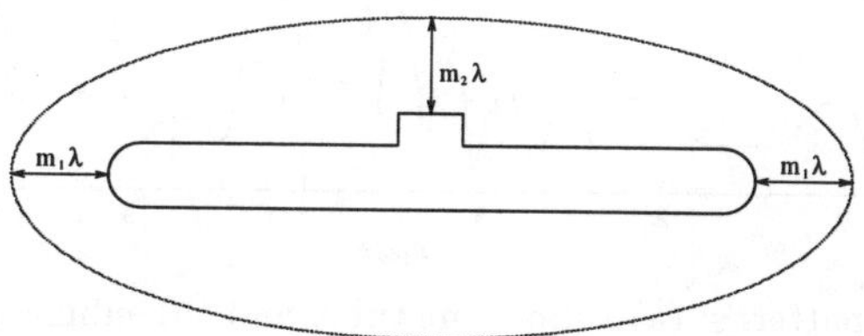

Fig. 9. Computational domain delimited by an ellipsoidal artificial boundary (view in a cutting plane)

First, we report in Figure 10 the traces on the intersection of the unit sphere S^1 and the plane YoZ of the far-field patterns associated with the scattering of the incident wave specified above by (a) the tube alone, and (b) the scattering system composed of the tube and the tower. The reader can observe that the differences between these far-field patterns are significant ($\| \mathrm{FFP}^{tube+tower} - \mathrm{FFP}^{tube} \|_2 / \| \mathrm{FFP}^{tube+tower} \|_2 = 25\%$), which highlights the effect of the tower on the scattered field.

The FFPs reported in Figure 10 have been obtained by analyzing the corresponding acoustic scattering problems by a three-dimensional finite element method, using the FETI-H [5] iterative solver. For that purpose, the three-dimensional computational domain Ω_b^{ex} was discretized by 792,694 nodes and 566,416 quadratic tetrahedral elements, which corresponds to a mesh resolution of 10 quadratic elements per wave length. The solution time took 2,004 seconds on a 10-processor Origin 2000, and required 3,849 Mbytes of memory. When considering the tube alone, the same mesh resolution was employed for discretizing the computational domain A_b^{ex}, and comparable computing resources were consumed for obtaining the solution of the acoustic scattering problem.

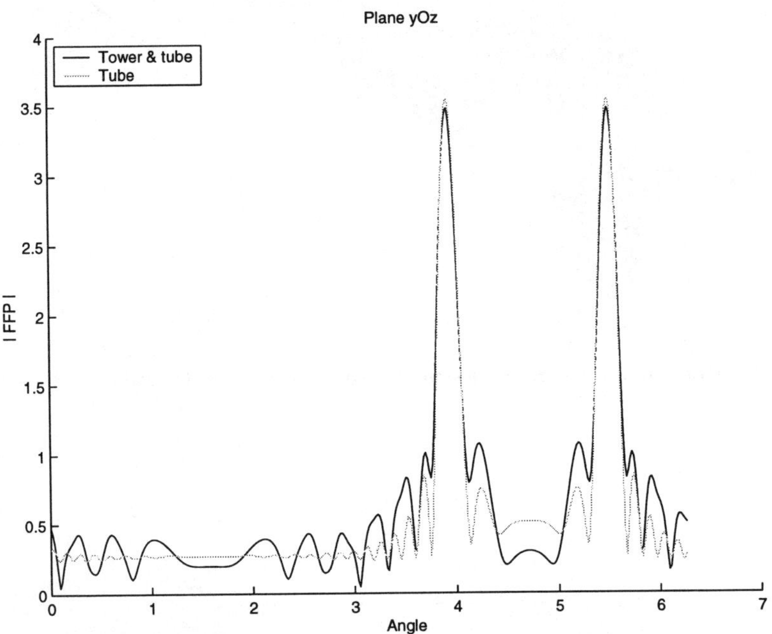

Fig. 10. Far-field patterns: (a) submarine tube, and (b) submarine tube and tower

We have also applied the fictitious/real domain decomposition method described in this paper to the prediction of the scattering of the incident wave specified above by the entire mockup submarine (tube and tower). For that purpose, we have discretized the two-dimensional computational domain a_b^{ex} by 4,661 nodes and 1,464 quadratic eight-noded elements, which also corresponds to a resolution of 10 quadratic elements per wave length, and have employed 332 Lagrange multipliers for enforcing the Dirichlet boundary condition on the surface of the tower. We have also decomposed a_b^{ex} in two subdomains. For this discretization, the Nyquist criterion (31) suggests $n_\theta = 50$. We report in Table 8 the performance results obtained on a 10-processor Origin 2000 for $n_\theta = 25$, $n_\theta = 50$, and $n_\theta = 75$, and contrast them with those of the three-dimensional finite element analysis performed with the FETI-H solver. In order to verify the accuracy delivered by the fictitious/real domain decomposition methodology, we also display in Figure 11 the three traces on the intersection between the unit sphere S^1 and the plane YoZ of the three FFPs corresponding to the three different values of n_θ, and compare them to the trace on that circle of the FFP obtained by postprocessing the result of the three-dimensional finite element analysis of this problem. We also report in Table 8 the relative discrepancies $\| \text{FFP}^{n_\theta} - \text{FFP}^{3D} \|_2 / \| \text{FFP}^{3D} \|_2$, where the superscripts $3D$ and n_θ designate the three-dimensional finite element analysis and the blended fictitious/real domain decomposition method with $2n_\theta + 1$ Fourier modes, respectively.

Table 8. Parallel performance results on a 10-processor Origin 2000 system

	$n_\theta = 25$	$n_\theta = 50$	$n_\theta = 75$
$\dfrac{\|\text{FFP}^{n_\theta} - \text{FFP}^{3D}\|_2}{\|\text{FFP}^{3D}\|_2}$	5.04 %	3.02 %	2.99 %
CPU time 3D FE method	2,004 s.	2,004 s.	2,004 s.
CPU time Fictitious/real method	17.3 s.	33.9 s.	56.4 s.
Improvement factor	115	59	35
Nb. of iterations	13	15	20
Memory 3D FE method	3,849 Mb.	3,849 Mb.	3,849 Mb.
Memory Fictitious/real method	86.7 Mb.	168.9 Mb.	253.0 Mb.
Improvement factor	44	22	15

Both quantitative and qualitative comparisons reported in Table 8 and Figure 11, respectively, show that for $n_\theta = 50$, the FFP obtained by the blended domain decomposition method is in perfect agreement with that obtained by the three-dimensional finite element analysis. Furthermore, the performance results reported in Table 8 show that, for this acoustic scattering problem, the blended fictitious/real domain decomposition method is from one order up to two orders of magnitude faster than the standard three-dimensional finite element method and one order of magnitude leaner in memory requirements. These results highlight the significant potential of the proposed methodology for the solution of partially axisymmetric acoustic scattering problems.

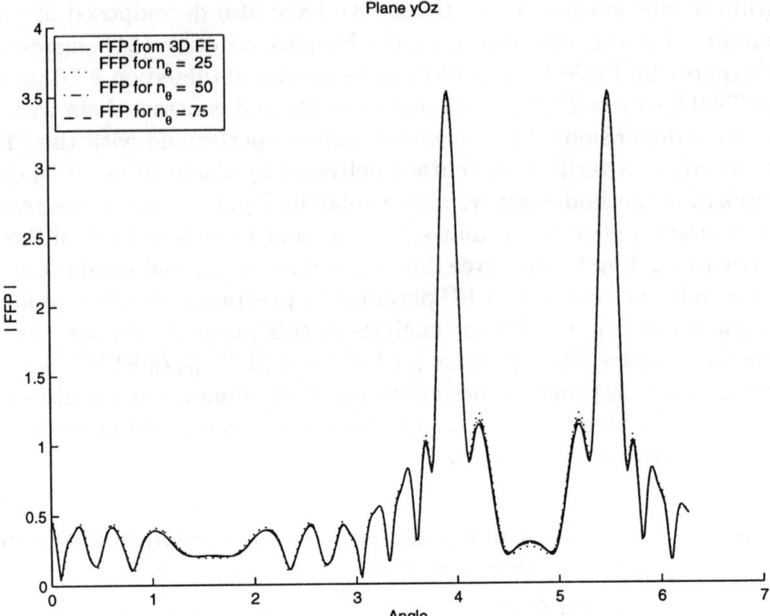

Fig. 11. Comparison of the far-field patterns predicted by the three-dimensional and fictitious domain analyses

6   Conclusions

Many scatterers, particularly in aerospace and military applications, are neither entirely axisymmetric, nor completely arbitrarily shaped. Rather, they consist of the assembly of one or two major axisymmetric components and a few features. For such scatterers, an axisymmetric acoustic scattering analysis is not applicable. On the other hand, a straightforward three-dimensional analysis is inefficient because it does not exploit the geometrical properties of the axisymmetric components. Ignoring the features in order to simplify the problem to a purely axisymmetric one can lead to significant errors in the prediction of the acoustic signature of the scatterer, as illustrated in this paper for a two-body, main tube and tower, submarine structure. For these reasons, we have presented in this paper a fictitious domain decomposition method aimed at solving efficiently partially axisymmetric acoustic scattering problems. In this method, which is currently limited to sound-soft obstacles, the exterior Helmholtz problem is extended into an axisymmetric exterior problem, and parts of the Dirichlet boundary conditions are enforced by Lagrange multipliers. The axisymmetry of the enlarged domain is then exploited by expanding the sought-after solution into a Fourier series. The Fourier modes of the solution are obtained by solving either by a direct or an iterative approach a series of two-dimensional problems that are coupled by the Lagrange multipliers. In both cases, the result is a fast Helmholtz solver which, when applied to the prediction of the scattering of time-harmonic acoustic waves by a partially axisymmetric sound-soft obstacle, is more than one order of magnitude faster than a three-dimensional finite element analysis, and considerably less memory greedy. When the fictitious method is equipped with the FETI-H iterative solver, it results in a blended fictitious/real domain decomposition method that increases again the speed of the basic fictitious method by a factor ranging between 5 and 10, depending on the problem size and configuration. Hence, the extension of the proposed method to sound-hard and lossy boundary conditions offers a significant potential for speeding up the solution of high-frequency partially axisymmetric acoustic scattering problems of practical interest.

Acknowledgment

The authors acknowledge the support by the Office of Naval Research under Grant N-00014-95-1-0663. The first author also acknowledges the support by the Corps des Ponts et Chaussées, France.

References

1. Farhat C., Hetmaniuk U., Rixen D. (1999) An efficient substructuring method for analyzing structures with major axisymmetric components. AIAA Paper 99-1283, 40th AIAA/ASME/ASCE/AHS/ASC Structures, Structural Dynamics, and Materials Conference, St Louis, MO, April 12–15
2. Bernardi C., Dauge M., Maday Y. (1999) Spectral methods for axisymmetric domains. Series in Applied Mathematics, P.G. Ciarlet et P.-L. Lions eds., Northholland et Gauthier-Villars
3. Dihn Q. V., Glowinski R., He J., Kwock V., Pan T. W., Périaux J. (1992) Lagrange multiplier approach to fictitious domain methods: application to fluid dynamics and electromagnetics. In: Fifth International Symposium on Domain Decomposition Methods for Partial Differential Equations, Philadelphia, PA, SIAM, 151–194.
4. Farhat C., Hetmaniuk U. (in press) A fictitious domain decomposition method for the solution of partially axisymmetric acoustic scattering problems – Part I: Dirichlet boundary conditions. Internat. J. Numer. Meths Engrg.
5. Farhat C., Macedo A., Lesoinne M. (2000) A two-level domain decomposition method for the iterative solution of high frequency exterior Helmholtz problems. Numer. Math. **85**, 283–308
6. Antoine X., Barucq H., Bendali A. (1999) Bayliss–Turkel like radiation conditions on surfaces of arbitrary shape. J. Math. Anal. Appl. **229**, 184–211
7. Brezzi F., Fortin M. (1991) Mixed and hybrid finite element methods. Springer Series in Computational Mathematics, 15, Springer-Verlag
8. Farhat C., Macedo A., Lesoinne M., Roux F. X., Magoulès F., de La Bourdonnaie A. (2000) Two-level domain decomposition methods with Lagrange multipliers for the fast iterative solution of acoustic scattering problems. Comput. Meths. Appl. Mech. Engrg. **184**, 213–240

9. Farhat C., Lacour C., Rixen D. (1998) Incorporation of linear multipoint constraints in substructure based iterative solvers - Part I: A numerically scalable algorithm. Internat. J. Numer. Meths. Engrg. **43**, 997–1016
10. Djellouli R., Farhat C., Macedo A., Tezaur R. (2000) Finite element solution of two-dimensional acoustic scattering problems using arbitrarily shaped convex artificial boundaries. J. of Comput. Acou. **8**, 81–100
11. Tezaur R., Macedo A., Farhat C., Djellouli R. (submitted) Three-dimensional finite element calculations in acoustic scattering using arbitrarily shaped convex artificial boundaries. Internat. J. Numer. Meths. Engrg.
12. Babuska I., Sauter S. (1997) Is the pollution effect of the FEM avoidable for the Helmholtz equation considering high wave numbers ? SIAM J. Numer. Anal. **34**, 2392–2423
13. Saad Y. (1995) Iterative methods for sparse linear systems. PWS Publishing Company, Boston

Dual-Primal FETI Methods with Face Constraints

Axel Klawonn[1], Olof B. Widlund[2], and Maksymilian Dryja[3]

[1] Fraunhofer Institute for Algorithms and Scientific Computing (SCAI), Schloss Birlinghoven, D–53754 Sankt Augustin, Germany
[2] Courant Institute of Mathematical Sciences, New York University, 251 Mercer Street, New York, NY 10012, USA
[3] Department of Mathematics, Warsaw University, Banacha 2, 02-097 Warsaw, Poland

Abstract. In this paper, an iterative substructuring method with Lagrange multipliers is considered for elliptic problems in three dimensions. The algorithm belongs to the family of dual–primal FETI methods using vertex and face average constraints. It is shown that the condition number of the dual–primal FETI method can be bounded polylogarithmically as a function of the dimension of the individual subregion problems and that the bounds are otherwise independent of the number of subdomains and the mesh size. Our bound also depends on a parameter TOL, which measures the variation of the coefficient of the elliptic problem. These results are obtained within a framework which was already used successfully to analyze other dual–primal FETI methods.

1 Introduction

Dual–primal FETI (FETI–DP) methods are iterative substructuring algorithms. They are used to define preconditioned conjugate gradient methods to solve the huge algebraic linear systems which arise from finite element discretizations.

The FETI–DP methods were introduced by Farhat, Lesoinne, Le Tallec, Pierson, and Rixen [8]. Their work was followed by a significant contribution to the theory of two dimensional second and fourth order problems by Mandel and Tezaur [14], by a paper by Farhat, Lesoinne, and Pierson [9] which specifically addresses an algorithm for three–dimensional problems, and by Pierson's doctoral dissertation [15]. The algorithm presented in [9], [15], uses constraints on the averages over faces, similarly to the algorithm considered in this paper. In our recent work, Klawonn, Widlund, and Dryja [11], we have extended the family of dual-primal FETI methods for elliptic problems in three dimensions considering different algorithms using constraints on the averages over edges and faces. In the present work, motivated by the algorithm for three dimensional problems used in [8], we analyze a variant which only uses face constraints. For problems with constant coefficients, our method only depends polylogarithmically on the number of unknowns on each subdomain. Under the assumption of an acceptable face path, cf. section 3, our estimate is also independent on the jumps of the coefficients of the elliptic model problem. Comparing the theoretical estimates obtained for the dual-primal FETI algorithm with just face constraints with the algorithms based on edge constraints which are proposed and analyzed in [11], reveals subtle differences which seem to indicate that the edge based algorithms are more powerful in the case of general distributions of the jumps in the coefficients.

The term dual–primal refers to the idea of enforcing some continuity constraints, across the interface between the subregions, throughout the iteration, as in a primal method, while all other constraints are enforced by using dual variables, i.e., Lagrange multipliers, as in a dual method.

Recently, there have been also extensions of the dual-primal FETI method to mortar finite element methods, cf. Dryja and Widlund [5], and to Stokes' equations, cf. Li [12].

The remainder of this paper is organized as follows. In section 2, we introduce our elliptic problems and the basic geometry of the decomposition. In section 3, we formulate a dual–primal FETI method using vertex and face average constraints. Finally, we present our convergence analysis.

2   Elliptic Model Problem, Finite Elements, and Geometry

Let $\Omega \subset \mathbf{R}^3$, be a bounded, polyhedral region, let $\partial\Omega_D \subset \partial\Omega$ be a closed set of positive measure, and let $\partial\Omega_N := \partial\Omega \setminus \partial\Omega_D$ be its complement. We impose homogeneous Dirichlet and general Neumann boundary conditions, respectively, on these two subsets and introduce the Sobolev space $H_0^1(\Omega, \partial\Omega_D) := \{v \in H^1(\Omega) : v = 0 \text{ on } \partial\Omega_D\}$.

We decompose Ω into non-overlapping subdomains $\Omega_i, i = 1, \ldots, N$, also known as substructures, and each of which is the union of shape-regular elements with the finite element nodes on the boundaries of neighboring subdomains matching across the interface $\Gamma := \left(\bigcup_{i=1}^{N} \partial\Omega_i\right) \setminus \partial\Omega$. The interface Γ is decomposed into subdomain faces, regarded as open sets, which are shared by two subregions, edges which are shared by more than two subregions and the vertices which form the endpoints of edges. If Γ intersects $\partial\Omega_N$ along an edge common to the boundaries of only two subdomains, we will regard it as part of the face common to this pair of subdomains. We denote faces of Ω_i by $\mathcal{F}^{ij}$, edges by $\mathcal{E}^{ik}$, and vertices by $\mathcal{V}^{i\ell}$.

For simplicity, we will only consider a piecewise linear, conforming finite element approximation of the following scalar, second order model problem:

Find $u \in H_0^1(\Omega, \partial\Omega_D)$, such that

$$a(u,v) = f(v) \quad \forall v \in H_0^1(\Omega, \partial\Omega_D), \tag{1}$$

where

$$a(u,v) = \sum_{i=1}^{N} \rho_i \int_{\Omega_i} \nabla u \cdot \nabla v \, dx, \quad f(v) = \sum_{i=1}^{N} \Big(\int_{\Omega_i} f v \, dx + \int_{\partial \Omega_i \cap \partial \Omega_N} g_N v \, ds \Big).$$

$$(2)$$

where g_N is the Neumann boundary data defined on $\partial \Omega_N$; it provides a contribution to the load vector of the finite element problem. We assume that the coefficient ρ_i is a positive constant on each subregion Ω_i.

In our theoretical analysis, we assume that each subregion Ω_i is the union of a number of shape regular tetrahedral coarse elements and that the number of tetrahedra is uniformly bounded for each subdomain. Thus, the subregions are not very thin and we can also easily show that the diameters of any pair of neighboring subdomains are comparable.

We also make a number of technical assumptions on the intersection of the boundary of the substructures and $\partial \Omega_D$; see [11]. The sets of nodes in Ω_i, on $\partial \Omega_i$, and on Γ are denoted by $\Omega_{i,h}, \partial \Omega_{i,h}$, and Γ_h, respectively.

We denote the standard finite element space of continuous, piecewise linear functions on Ω_i by $W^h(\Omega_i)$. For simplicity, we assume that the triangulation of each subdomain is quasi uniform. The diameter of Ω_i is H_i, or generically, H. We denote the corresponding finite element trace spaces by $W_i := W^h(\partial \Omega_i), i = 1, \ldots, N$, and by $W := \prod_{i=1}^{N} W_i$ the associated product space. We will often consider elements of W which are discontinuous across the interface.

The finite element approximation of the elliptic problem is continuous across Γ and we denote the corresponding subspace of W by $\widehat{W}$. We note that while the stiffness matrix K and its Schur complement S, obtained from K by elimination of the interior subdomain variables, which correspond to the product space W generally are singular those of $\widehat{W}$ are not. The stiffness matrix K is a direct sum of local stiffness matrices $K^{(i)}$ which correspond to the subdomains Ω_i and to the appropriate terms in the first formula of (2). Eliminating the interior variables of $K^{(i)}$ by block Gaussian elimination results in local Schur complement matrices $S^{(i)}$. We note that S is again a direct sum of the local Schur complement matrices $S^{(i)}$.

For the dual–primal FETI methods, we will use additional, intermediate subspaces $\widetilde{W}$ of W for which a relatively small number of continuity constraints are enforced across the interface throughout the iteration. One of the benefits of working in $\widetilde{W}$, rather than in W, is that certain related Schur complements $\widetilde{S}$ and S_Δ are positive definite.

As in previous work on Neumann–Neumann and FETI algorithms, a crucial role is played by *the weighted counting functions* $\mu_i \in \widehat{W}$, which are associated with the individual subdomain boundaries $\partial \Omega_i$; cf., e.g., [3,7]. In the present context they will be used in the definition of certain diagonal scaling matrices. These functions are defined, for $\gamma \in [1/2, \infty)$, and for $x \in \Gamma_h \cup \partial \Omega_h$,

30   A. Klawonn, O.B. Widlund, and M. Dryja

by a sum of contributions from Ω_i, and its relevant next neighbors

$$
\mu_i(x) = \begin{cases} \displaystyle\sum_{j \in \mathcal{N}_x} \rho_j^\gamma(x) & x \in \partial\Omega_{i,h} \cap \partial\Omega_{j,h}, \\ \rho_i^\gamma(x) & x \in \partial\Omega_{i,h} \cap (\partial\Omega_h \setminus \Gamma_h), \\ 0 & x \in (\Gamma_h \cup \partial\Omega_h) \setminus \partial\Omega_{i,h}. \end{cases} \tag{3}
$$

Here, $\mathcal{N}_x$ is the set of indices of the subregions which have x on its boundary. We note that any node of Γ_h belongs either to two faces, more than two edges, or to the vertices of several substructures.

The pseudo inverses $\mu_i^\dagger$ are defined, for $x \in \Gamma_h \cup \partial\Omega_h$, by

$$
\mu_i^\dagger(x) = \begin{cases} \mu_i^{-1}(x) & \text{if } \mu_i(x) \neq 0, \\ 0 & \text{if } \mu_i(x) = 0. \end{cases}
$$

3 New Dual–Primal FETI Methods

In previous studies of dual–primal FETI methods for problems in two dimensions, cf. Farhat, Lesoinne, Le Tallec, Pierson, and Rixen [8] and Mandel and Tezaur [14], the constraints on the degrees of freedom associated with the vertices of the substructures are enforced, i.e., the corresponding degrees of freedom belong to the primal set of variables, while all the constraints associated with the edge nodes are enforced only at the convergence of the iterative method. In each step of the iteration a fully assembled linear subsystem is solved. In a simple two–dimensional case, this subsystem corresponds to all the interior and cross point variables; these variables can be eliminated at a modest expense since we can first eliminate all the interior variables, in parallel across the subdomains, resulting in a Schur complement for the cross point variables which can be shown to be sparse. This Schur complement has a dimension which equals the number of subdomain vertices which do not belong to $\partial\Omega_D$.

In their recent paper, Mandel and Tezaur [14] established a condition number bound of the form $C(1 + \log(H/h))^2$ for the resulting FETI method equipped with a Dirichlet preconditioner which is very similar to those used for the older FETI methods and which is built from local solvers on the subregions with zero Dirichlet conditions at the vertices of the subregions. They also established a corresponding result for a fourth-order elliptic problem in the plane.

The same algorithm is also defined for three dimensions but it does not perform well. This is undoubtedly related to the poor performance of vertex-based iterative substructuring methods; see [4, Section 6.1] and [11]. Recently, Farhat, Lesoinne, and Pierson added edge and face constraints to this basic algorithm, see [9], and improved the performance.

Following the approach in Klawonn, Widlund, and Dryja [11], it is convenient to work in subspaces $\widetilde{W} \subset W$ for which sufficiently many constraints are enforced so that the resulting leading diagonal block matrix of the saddle point problem, though no longer block diagonal, is strictly positive definite. We will explain how this can be accomplished and also introduce two subspaces, $\widehat{W}_\Pi \subset \widehat{W}$ and $\widetilde{W}_\Delta$, corresponding to a primal and a dual part of the space $\widetilde{W}$. These subspaces will play an important role in the description and analysis of our iterative method. The direct sum of these spaces equals $\widetilde{W}$, i.e.,

$$\widetilde{W} = \widehat{W}_\Pi \oplus \widetilde{W}_\Delta. \tag{4}$$

The second subspace, $\widetilde{W}_\Delta$, is the direct sum of local subspaces $\widetilde{W}_{\Delta,i}$ of $\widetilde{W}$ where each subdomain Ω_i contributes a subspace $\widetilde{W}_{\Delta,i}$; only its $i-th$ component in the sense of the product space $\widetilde{W}$ is nontrivial.

In the description of our algorithm and in its analysis, we will need certain standard finite element cutoff functions $\theta_{\mathcal{E}^{ik}}$, $\theta_{\mathcal{F}^{ij}}$, and $\theta_{\mathcal{V}^{i\ell}}$. The first two are the discrete harmonic functions which equal 1 on $\mathcal{E}_h^{ik}$ and $\mathcal{F}_h^{ij}$, respectively, and which vanish elsewhere on Γ_h; $\theta_{\mathcal{V}^{i\ell}}$ denotes the piecewise discrete harmonic extension of the standard nodal basis function associated with the vertex $\mathcal{V}^{i\ell}$.

We are now ready to describe our algorithm in terms of pairs of subspaces. In our recent work [11], we analyzed a family of FETI-DP methods which are denoted by Algorithms A-D. The present paper can be viewed as an extension of [11], thus we denote the algorithm discussed here by Algorithm E.

Algorithm E: The primal subspace, $\widehat{W}_\Pi$, is spanned by the vertex nodal finite element basis functions $\theta_{\mathcal{V}^{i\ell}}$ and the cutoff functions $\theta_{\mathcal{F}^{ij}}$ associated with all the faces of the interface. The local subspace $\widetilde{W}_{\Delta,i}$ is defined as the subspace of W_i where the values at the subdomain vertices vanish together with the averages $\overline{u}_{\mathcal{F}^{ij}}$, i.e., by

$$\widetilde{W}_{\Delta,i} := \{u \in W_i : u(\mathcal{V}^{i\ell}) = 0, \overline{u}_{\mathcal{F}^{ij}} = 0 \ \forall \mathcal{V}^{i\ell}, \mathcal{F}^{ij} \subset \partial\Omega_i\}.$$

Here,

$$\overline{u}_{\mathcal{F}^{ij}} = \frac{\int_{\mathcal{F}^{ij}} u \, dx}{\int_{\mathcal{F}^{ij}} 1 \, dx}. \tag{5}$$

Hence, $\widetilde{W} = \widetilde{W}_E$ is the subspace of W of functions that are continuous at the subdomain vertices and have the same values of $\overline{u}_{\mathcal{F}^{ij}}$ independently of which component of $u \in \widetilde{W}_E$ is used in the evaluation of these averages.

For all pairs of substructures Ω_i, Ω_k, which have an edge $\mathcal{E}^{ik}$ in common, we need an *acceptable face path*. An acceptable face path for such a pair is a path from Ω_i to Ω_k, possibly via several other substructures Ω_j, which do not necessarily touch the edge in question, and such that the associated coefficients ρ_j, ρ_i, and ρ_k satisfy

$$TOL * \rho_j \geq \min(\rho_i, \rho_k).$$

Let us note that this concept is less general than that of an acceptable edge path introduced in Klawonn, Widlund, and Dryja [11] but more general than the concept of quasi–monotonicity introduced in Dryja, Sarkis, and Widlund [3]. The latter concept could also have been used in our analysis.

It is useful to distinguish between the continuity constraints at the vertices and the other constraints. The latter are sometimes called optional constraints since they are not needed to guarantee solvability of the subproblems if there are enough vertex constraints. The vertex constraints are enforced in the subassembly process, for the primal problem, outlined above. The optional constraints could be similarly incorporated after a change of variables. Another possibility, advocated by Farhat, Lesoinne, and Pierson [9], is to introduce an additional set of Lagrange multipliers which are computed exactly in each iteration to enforce the required optional constraints of the primal subspace. For a more detailed description of this approach, we refer to [9, section 4.2], especially formulae (24)-(28).

We can now formulate our FETI–DP algorithm. The primal part of the algorithm is based on the exact elimination of all unknowns of the primal subspace as well as the interior variables. The remaining system is written in terms of a Schur complement $\widetilde{S}$. Thus, for our algorithms, we arrive at this reduced problem after eliminating the primal variables associated with the interior nodes, the vertex nodes designated as primal, as well as the Lagrange multipliers related to the optional constraints. Analogously, we get from the load vectors associated with each subdomain a reduced right hand side $\tilde{f}_\Delta$. The Schur complement $\widetilde{S}$ satisfies the following minimum property, cf. [11]: $\forall w_\Delta \in \widetilde{W}_\Delta$,

$$\langle \widetilde{S} w_\Delta, w_\Delta \rangle = \min \langle S w, w \rangle, \tag{6}$$

where we take the minimum over all $w \in \widetilde{W}$ of the form $w = w_\Pi + w_\Delta, w_\Pi \in \widehat{W}_\Pi$. Here, $\langle \cdot, \cdot \rangle$ denotes the ℓ_2-inner product. We note that any Schur complement of a positive definite, symmetric matrix is always associated with such a variational problem.

We can now reformulate the original finite element problem, reduced to the degrees of freedom of the second subspace $\widetilde{W}_\Delta$, as a minimization problem with constraints given by the requirement of continuity across all of Γ_h:

Find $u_\Delta \in \widetilde{W}_\Delta$, such that

$$\left. \begin{array}{c} J(u_\Delta) := \frac{1}{2} \langle \widetilde{S} u_\Delta, u_\Delta \rangle - \langle \tilde{f}_\Delta, u_\Delta \rangle \to \min \\ B_\Delta u_\Delta = 0 \end{array} \right\}. \tag{7}$$

The matrix B_Δ is constructed from $\{0, 1, -1\}$ such that the values of the solution u_Δ, associated with more than one subdomain, coincide when $B_\Delta u_\Delta = 0$. These constraints are very simple and just express that the nodal values coincide across the interface; in comparison with the one-level FETI method, see, e.g., [10], we can drop some of the constraints, in particular those associated with the vertex nodes of the primal space. However, we will otherwise use all possible constraints and thus work with a fully redundant set of Lagrange multipliers as in [10, section 5] and [11].

By introducing a set of Lagrange multipliers $\lambda \in V := \mathit{range}\,(B_\Delta)$, to enforce the constraints $B_\Delta u_\Delta = 0$, we obtain a saddle point formulation of (7), which is similar to that of the one-level FETI method; see, e.g., Klawonn and Widlund [10]. We use that $\widetilde{S}$ is invertible and eliminate the subvector u_Δ, and obtain the following system for the dual variable:

$$F\lambda = d, \qquad (8)$$

where

$$F := B_\Delta \widetilde{S}^{-1} B_\Delta^t$$

and the right hand side

$$d := B_\Delta \widetilde{S}^{-1} \tilde{f}_\Delta.$$

Algorithmically, the matrix $\widetilde{S}$ is only needed in terms of $\widetilde{S}^{-1}$ times a vector and such an operation can be computed relatively inexpensively. While it is natural to describe a Schur complement in terms of a second set of variables and resulting from the elimination of a first set, the action of its inverse on a vector can often advantageously be obtained by solving the entire linear system from which it originates after augmenting the given right hand side with zeros. Full advantage can then be taken of algorithms that symmetrically reorder the larger matrix so as to preserve sparsity. In the case at hand, it is thus advantageous to group all the interior and dual variables of each subdomain together and to factor the resulting blocks in parallel across the subdomains using a good ordering algorithm. The contributions to the remaining Schur complement, of the primal variables, can also be computed locally prior to subassembly and factorization of this final, global part of the linear system of equations; we note that this is a quite small system.

To define the FETI–DP Dirichlet preconditioner, we need to introduce an additional set of Schur complement matrices, $S_\Delta^{(i)}, i = 1, \ldots, N$, which is obtained by restricting $S^{(i)}$ to the space $\widetilde{W}_{\Delta,i}$. The associated block–diagonal matrix is denoted by

$$S_\Delta := \mathit{diag}_{i=1}^{N}(S_\Delta^{(i)}).$$

We can compute the action of S_Δ on a vector from the second subspace $\widetilde{W}_\Delta$ by solving local Dirichlet problems with solutions in $\widetilde{W}_{\Delta,i}, i = 1, \ldots, N$, and then multiplying them by the stiffness matrix of the respective subdomain. These solutions are constrained to vanish at the subdomain vertices and to have zero face averages.

We also introduce diagonal scaling matrices $D_\Delta^{(i)}$ that operate on the Lagrange multiplier space. Each element on the main diagonal corresponds to a Lagrange multiplier which enforces continuity between the nodal values of some $w_i \in \widetilde{W}_i$ and $w_j \in \widetilde{W}_j$ at some point $x \in \Gamma_h$. This diagonal element is defined as $\rho_j^\gamma(x)\mu_j^\dagger(x)$. Finally, we define a scaled jump operator by

$$B_{D,\Delta} := [D_\Delta^{(1)} B_\Delta^{(1)}, \ldots, D_\Delta^{(N)} B_\Delta^{(N)}].$$

As in Klawonn, Widlund, and Dryja [11, section 4], we solve the dual system (8) using the preconditioned conjugate gradient algorithm with the preconditioner

$$M^{-1} := B_{D,\Delta} S_\Delta B_{D,\Delta}^t. \tag{9}$$

The dual–primal FETI method is now the standard preconditioned conjugate gradient algorithm for solving the preconditioned system

$$M^{-1} F \lambda = M^{-1} d.$$

This definition of M clearly depends on the choice of the subspaces $\widehat{W}_\Pi$ and $\widetilde{W}_\Delta$.

4   Some Auxiliary Lemmas

The purpose of this section is to provide, in most cases without proofs, the few auxiliary results that are required for a complete proof of Lemma 7, which provide the core of the proofs of our main result. Some of these results are borrowed from [4,7,6]. Here, we formulate them using trace spaces on the subdomain boundaries, i.e., $H^{1/2}(\partial\Omega_i)$ instead of the spaces $H^1(\Omega_i)$ and discrete harmonic extensions; given the well–known equivalence of the norms, nothing essentially new needs to be proven. In our proofs, we will work with the S–norm defined by $|u|_S^2 = \sum_{i=1}^N |u_i|_{S^{(i)}}^2$ and $|u_i|_{S^{(i)}}^2 = \langle S^{(i)} u_i, u_i \rangle$. A proof of the equivalence of the $S^{(i)}-$ and the $H^{1/2}(\partial\Omega_i)-$semi–norms of elements of W_i can be found in [1] for the case of piecewise linear elements and two dimensions and the tools necessary to extend this result to more general finite elements are provided in [16]; in our case, we of course have to multiply $|u_i|_{H^{1/2}(\partial\Omega_i)}^2$ by the factor ρ_i.

We also recall that we can define the $H_{00}^{1/2}(\tilde{\Gamma})-$norm, $\tilde{\Gamma} \subset \partial\Omega_i$, of an element of W_i which is supported in $\tilde{\Gamma}$, as the $H^{1/2}(\partial\Omega_i)-$norm of the function extended by zero onto $\partial\Omega_i \setminus \tilde{\Gamma}$.

The first lemma can, essentially, be found in Dryja, Smith, and Widlund [4, Lemma 4.4].

Lemma 1 *Let* $\theta_{\mathcal{F}^{ij}}$ *be the finite element function that is equal to 1 at the nodal points on the face* $\mathcal{F}^{ij}$*, which is common to two subregions* Ω_i *and* Ω_j*, and that vanishes on* $(\partial\Omega_{i,h} \cup \partial\Omega_{j,h}) \setminus \mathcal{F}_h^{ij}$*. Then,*

$$|\theta_{\mathcal{F}^{ij}}|_{H^{1/2}(\partial\Omega_i)}^2 \le C(1 + \log(H_i/h_i))H_i.$$

The same bounds also hold for the other subregion Ω_j*.*

The following result can, essentially, be found in Dryja, Smith, and Widlund [4, Lemma 4.5] or in Dryja [2, Lemma 3].

Lemma 2 *Let $\theta_{\mathcal{F}^{ij}}$ be the function introduced in Lemma 1 and let I^h denote the interpolation operator onto the finite element space $W^h(\Omega_i)$. Then, $\forall u \in W_i$,*

$$\|I^h(\theta_{\mathcal{F}^{ij}} u)\|^2_{H^{1/2}_{00}(\mathcal{F}^{ij})} \leq C(1 + \log(H_i/h_i))^2 \left(|u|^2_{H^{1/2}(\mathcal{F}^{ij})} + \frac{1}{H_i}\|u\|^2_{L_2(\mathcal{F}^{ij})}\right).$$

We will also need two additional results which are used to estimate the contributions to our bounds from the edges of Ω_i. For the next lemma, see Dryja, Smith, and Widlund [4, Lemma 4.7].

Lemma 3 *Let $\theta_{\mathcal{E}^{ik}}$ be the cutoff function associated with the edge $\mathcal{E}^{ik}$. Then, $\forall u \in W_i$,*

$$|I^h(\theta_{\mathcal{E}^{ik}} u)|^2_{H^{1/2}(\partial\Omega_i)} \leq C\|u\|^2_{L_2(\mathcal{E}^{ik})}.$$

This result follows by an elementary estimate of the energy norm of the zero extension of the boundary values and by noting that the harmonic extension has a smaller energy.

We will also need a Sobolev-type inequality for finite element functions, see Dryja and Widlund [6, Lemma 3.3] or Dryja [2, Lemma 1].

Lemma 4 *Let $\mathcal{E}^{ik}$ be any edge of Ω_i which forms part of the boundary of a face $\mathcal{F}^{ij} \subset \partial\Omega_i$. Then, $\forall u \in W_i$,*

$$\|u\|^2_{L_2(\mathcal{E}^{ik})} \leq C(1 + \log(H_i/h_i))\left(|u|^2_{H^{1/2}(\mathcal{F}^{ij})} + \frac{1}{H_i}\|u\|^2_{L_2(\mathcal{F}^{ij})}\right).$$

We also state a nonstandard version of Friedrichs' inequality that is given in a somewhat different form in [7, Lemma 6].

Lemma 5 *Let $\mathcal{E}^{ik}$ be an edge of $\mathcal{F}^{ij}$. Then, $\forall u \in W_i$ that vanish on $\mathcal{E}^{ik}$,*

$$\|u\|^2_{L_2(\mathcal{F}^{ij})} \leq CH_i(1 + \log(H_i/h_i))|u|^2_{H^{1/2}(\mathcal{F}^{ij})}.$$

5 Convergence Analysis

Our analysis follows the approach of our recent article on dual-primal FETI methods [11].

As in [11,14], the two different Schur complements, $\widetilde{S}$ and S_Δ, introduced in section 3, play an important role in the analysis of the dual–primal iterative algorithm. Both operate on the second subspace $\widetilde{W}_\Delta$ and we also recall that $\widetilde{S}$ represents a global problem while S_Δ does not.

Let $V := range\,(B_\Delta)$ be the space of Lagrange multipliers. As in [10, Section 5], we introduce a projection

$$P_\Delta := B^t_{D,\Delta} B_\Delta.$$

A simple computation shows, see [10, Lemma 4.2], that P_Δ preserves the jump of any function $u_\Delta \in \widetilde{W}_\Delta$, i.e., $B_\Delta P_\Delta u_\Delta = B_\Delta u_\Delta$ and we also have $P_\Delta u = 0 \ \forall u \in \widehat{W}$.

Analogously to [10, Lemma 5.2], we have

Lemma 6 *For any $\mu \in V$, there exists a $w_\Delta \in range\,(P_\Delta)$, such that $\mu = B_\Delta w_\Delta$.*

Proof. We note that for any $\mu \in V = range\,(B_\Delta)$, there exists a w'_Δ, such that $\mu = B_\Delta w'_\Delta$. Choosing $w_\Delta := P_\Delta w'_\Delta$, we have $B_\Delta w_\Delta = B_\Delta w'_\Delta = \mu$.

Let $x \in \Gamma_h$ and let $w_\Delta \in \widetilde{W}_\Delta$. We borrow the following formula from [10]:

$$P_\Delta w_\Delta(x) \;=\; \sum_{j \in \mathcal{N}_{\Delta,x}} \rho_j^\gamma \mu_j^\dagger (w_{\Delta,i}(x) - w_{\Delta,j}(x)), \quad x \in \partial\Omega_{i,h} \cap \Gamma_h. \tag{10}$$

Here, $\mathcal{N}_{\Delta,x}$ is the set of indices of the subregions which have the node x on its boundary. We note that the coefficients in this expression are constant on the set of the nodal points of each face and each edge of $\partial\Omega_i$, and that this formula is independent of the particular choice of B_Δ.

We now analyze Algorithm E and begin by proving the following core estimate.

Lemma 7 (Algorithm E) *For all $w_\Delta \in \widetilde{W}_{\Delta,E}$, we have,*

$$|P_\Delta w_\Delta|^2_{S_\Delta} \leq C \, \max((1 + \log(H/h))^2, TOL * (1 + \log(H/h)))|w_\Delta|^2_{\widetilde{S}},$$

where $C > 0$ is independent of $h, H, TOL, \rho_i,$ and γ.

Proof. We consider an arbitrary $w_\Delta \in \widetilde{W}_{\Delta,E}$. In order to compute its $\widetilde{S}$−norm, cf. (6), we determine the element $w = w_\Pi + w_\Delta \in \widetilde{W}_E, w_\Pi \in \widehat{W}_{\Pi,E}$, with the correct minimal property. Then, by the definition of $\widetilde{S}$, $|w_\Delta|_{\widetilde{S}} = |w|_S$. We next note that we can subtract any continuous function from w_Δ without changing the values of $P_\Delta w_\Delta$; thus, $P_\Delta w = P_\Delta w_\Delta$. It is also easy to see, by carrying out a simple computation and by using formula (10), that $P_\Delta w_\Delta \in \widetilde{W}_{\Delta,B}$. We also note that the S_Δ−norm of any element of $\widetilde{W}_\Delta$ equals its S−norm.

We model our proof on [10, Lemmas 4.7, 5.4] but note that the arguments need to be modified to some extent. We also note that we only have contributions from faces and edges since all elements in $\widetilde{W}_E$ are continuous at the vertices. Here, in contrast to the proof in [10], we do not need to assume that there are not any subdomains with boundaries which only intersects $\partial\Omega_D$ only in isolated points.

We introduce the notation $(v_i)_{i=1,\ldots,N} := P_\Delta w$. Then, we have to estimate

$$|P_\Delta w|^2_S = \sum_{i=1}^{N} |v_i|^2_{S^{(i)}}.$$

We can therefore focus on the estimate of the contribution from a single subdomain Ω_i. We first assume that its boundary and the boundaries of its relevant neighbors do not intersect $\partial\Omega_D$.

We cut the function v_i using the functions $\theta_{\mathcal{F}^{ij}}$ and $\theta_{\mathcal{E}^{ik}}$ and write it as a sum of terms which vanish at all the interface nodes outside individual faces and edges; cf., e.g., [4,7,6]. We then have, since the v_i vanish at the subdomain vertices,

$$v_i = \sum_{\mathcal{F}^{ij}\subset\partial\Omega_i} I^h(\theta_{\mathcal{F}^{ij}}v_i) + \sum_{\mathcal{E}^{ik}\subset\partial\Omega_i} I^h(\theta_{\mathcal{E}^{ik}}v_i).$$

We find that the face $\mathcal{F}^{ij}$ contributes

$$I^h(\theta_{\mathcal{F}^{ij}}\rho_j^\gamma \mu_j^\dagger(w_i - w_j))$$

and we have to estimate its $H_{00}^{1/2}(\mathcal{F}^{ij})$−norm; this formula follows from (10).

With $\gamma \geq 1/2$, we can easily prove that

$$\rho_i(\rho_j^\gamma \mu_j^\dagger)^2 \leq \min(\rho_i, \rho_j). \tag{11}$$

We note that $\rho_j^\gamma \mu_j^\dagger$ is constant on $\mathcal{F}_h^{ij}$ and that w has common face averages, i.e., $\overline{w}_{i,\mathcal{F}^{ij}} = \overline{w}_{j,\mathcal{F}^{ij}}$. Using inequality (11), these observations, and Lemma 2, we obtain,

$$\begin{aligned}
&\rho_i\|I^h(\theta_{\mathcal{F}^{ij}}\rho_j^\gamma \mu_j^\dagger(w_i - w_j))\|^2_{H_{00}^{1/2}(\mathcal{F}^{ij})}\\
&= \rho_i\|I^h(\theta_{\mathcal{F}^{ij}}\rho_j^\gamma \mu_j^\dagger((w_i - \overline{w}_{i,\mathcal{F}^{ij}}) - (w_j - \overline{w}_{j,\mathcal{F}^{ij}})))\|^2_{H_{00}^{1/2}(\mathcal{F}^{ij})}\\
&\leq C\,(1 + \log(H_i/h_i))^2 \min(\rho_i, \rho_j)\left(|w_i - w_j|^2_{H^{1/2}(\mathcal{F}^{ij})} + \right.\\
&\qquad\qquad \left. + \tfrac{1}{H_i}\|(w_i - \overline{w}_{i,\mathcal{F}^{ij}}) - (w_j - \overline{w}_{j,\mathcal{F}^{ij}})\|^2_{L_2(\mathcal{F}^{ij})}\right).
\end{aligned} \tag{12}$$

We can estimate this expression by

$$C\,(1 + \log(H_i/h_i))^2 \left(\rho_i|w_i|^2_{H^{1/2}(\mathcal{F}^{ij})} + \rho_j|w_j|^2_{H^{1/2}(\mathcal{F}^{ij})}\right),$$

as desired, by applying a Poincaré inequality. We note that, by assumption, H_j and H_i are comparable and so are h_j and h_i, since the triangulations of Ω_i and Ω_j are quasi uniform.

By using Lemma 3, we can estimate the contributions of the edges of Ω_i to the energy of v_i in terms of L_2−norms over the edges. These L_2−terms are then estimated by using Lemma 4. If four subdomains, e.g., $\Omega_i, \Omega_j, \Omega_k$, and Ω_ℓ, have an edge $\mathcal{E}^{ik}$ in common, then, according to (10), there are three contributions to the estimate of the contribution of Ω_i to $|P_\Delta w|^2_S$, namely

$$\begin{aligned}
&\rho_i\,\|I^h(\rho_j^\gamma \mu_j^\dagger \theta_{\mathcal{E}^{ik}}(w_i - w_j))\|^2_{L_2(\mathcal{E}^{ik})} + \rho_i\,\|I^h(\rho_k^\gamma \mu_k^\dagger \theta_{\mathcal{E}^{ik}}(w_i - w_k))\|^2_{L_2(\mathcal{E}^{ik})} +\\
&\qquad\qquad + \rho_i\,\|I^h(\rho_\ell^\gamma \mu_\ell^\dagger \theta_{\mathcal{E}^{ik}}(w_i - w_\ell))\|^2_{L_2(\mathcal{E}^{ik})}.
\end{aligned} \tag{13}$$

38   A. Klawonn, O.B. Widlund, and M. Dryja

We first consider the second term in detail assuming that Ω_i shares a face with each of Ω_j and Ω_ℓ, but only an edge with Ω_k. We assume that we have an acceptable face path through the subdomain Ω_j via the faces $\mathcal{F}^{ij}$ and $\mathcal{F}^{jk}$, i.e., $TOL * \rho_j \geq \min(\rho_i, \rho_k)$. In general the acceptable face path could be more complicated but such a case could be analyzed similarly. We obtain

$$
\begin{aligned}
\rho_i \, &\|\rho_k^\gamma \mu_k^\dagger I^h(\theta_{\mathcal{E}^{ik}}(w_i - w_k))\|_{L_2(\mathcal{E}^{ik})}^2 \\
= \rho_i \, &\|\rho_k^\gamma \mu_k^\dagger \left(I^h(\theta_{\mathcal{E}^{ik}}(w_i - \overline{w}_{i,\mathcal{F}^{ij}})) - I^h(\theta_{\mathcal{E}^{ik}}(w_j - \overline{w}_{j,\mathcal{F}^{ij}})) + \right. \\
&\left. + I^h(\theta_{\mathcal{E}^{ik}}(w_j - \overline{w}_{j,\mathcal{F}^{jk}})) - I^h(\theta_{\mathcal{E}^{ik}}(w_k - \overline{w}_{k,\mathcal{F}^{jk}})) \right) \|_{L_2(\mathcal{E}^{ik})}^2 \\
\leq C \min(\rho_i, \rho_k) &\left(\|I^h(\theta_{\mathcal{E}^{ik}}(w_i - \overline{w}_{i,\mathcal{F}^{ij}}))\|_{L_2(\mathcal{E}^{ik})}^2 + \right. \\
&\quad + \|I^h(\theta_{\mathcal{E}^{ik}}(w_j - \overline{w}_{j,\mathcal{F}^{ij}}))\|_{L_2(\mathcal{E}^{ik})}^2 + \\
&\quad + \|I^h(\theta_{\mathcal{E}^{ik}}(w_j - \overline{w}_{j,\mathcal{F}^{jk}}))\|_{L_2(\mathcal{E}^{ik})}^2 + \\
&\left. \quad + \|I^h(\theta_{\mathcal{E}^{ik}}(w_k - \overline{w}_{k,\mathcal{F}^{jk}}))\|_{L_2(\mathcal{E}^{ik})}^2 \right) .
\end{aligned}
\tag{14}
$$

It is sufficient to estimate the first term, the remaining three terms can be treated completely analogously. Using Lemma 4 and a Poincaré inequality, we have

$$
\|I^h(\theta_{\mathcal{E}^{ik}}(w_i - \overline{w}_{i,\mathcal{F}^{ij}}))\|_{L_2(\mathcal{E}^{ik})}^2 \leq C \left(1 + \log(H_i/h_i)\right) |w_i|_{H^{1/2}(\mathcal{F}^{ij})}^2 .
$$

Using these estimates, yields

$$
\begin{aligned}
\rho_i \, &\|\rho_k^\gamma \mu_k^\dagger I^h(\theta_{\mathcal{E}^{ik}}(w_i - w_k))\|_{L_2(\mathcal{E}^{ik})}^2 \\
\leq C \left(1 + \log(H/h)\right) &\left(\rho_i |w_i|_{H^{1/2}(\mathcal{F}^{ij})}^2 + \rho_k |w_k|_{H^{1/2}(\mathcal{F}^{jk})}^2 + \right. \\
&\left. + \min(\rho_i, \rho_k)(|w_j|_{H^{1/2}(\mathcal{F}^{ij})}^2 + |w_j|_{H^{1/2}(\mathcal{F}^{jk})}^2) \right) .
\end{aligned}
$$

From $TOL * \rho_j \geq \min(\rho_i, \rho_k)$, we obtain

$$
\begin{aligned}
\rho_i \, &\|\rho_k^\gamma \mu_k^\dagger I^h(\theta_{\mathcal{E}^{ik}}(w_i - w_k))\|_{L_2(\mathcal{E}^{ik})}^2 \\
\leq C \left(1 + \log(H/h)\right) &\left(\rho_i |w_i|_{H^{1/2}(\mathcal{F}^{ij})}^2 + \rho_k |w_k|_{H^{1/2}(\mathcal{F}^{jk})}^2 \right) + \\
+ C * TOL * \left(1 + \log(H/h)\right) &\left(\rho_j |w_j|_{H^{1/2}(\mathcal{F}^{ij})}^2 + \rho_j |w_j|_{H^{1/2}(\mathcal{F}^{jk})}^2 \right) .
\end{aligned}
$$

Since Ω_i and Ω_j, as well as Ω_i and Ω_ℓ, have a face in common, the argument given above can be simplified for the first and third edge contributions, see (13); they can be reduced to estimates of face terms.

We finally have to consider boundary subregions which have a nonempty intersection with $\partial \Omega_D$ and show that we can obtain bounds of the same quality. We then need different arguments to eliminate the $L_2(\mathcal{F}^{ij})$−terms. In case this intersection is a face or an edge, we can use exactly the same arguments as in [10, p. 71] which includes using Lemma 5. If the boundary of a substructure intersects $\partial \Omega_D$ in just one or a few single points, the shifting can be done exactly as above for the face and edge terms of an interior subregion. Since these bounds are quite technical, we conclude our proof by referring to our earlier papers [10,11] for more details.

We now prove our condition number estimate for Algorithm E, which only depends polylogarithmically on the dimension of the subproblems.

Theorem 1 (Algorithm E) *The condition number satisfies*

$$\kappa(M_E^{-1}F_E) \leq C \max((1+\log(H/h))^2, TOL * (1+\log(H/h))).$$

Here, C is independent of h, H, γ, TOL, and the values of the ρ_i.

Proof. We have to estimate the smallest eigenvalue $\lambda_{min}(M_E^{-1}F_E)$ from below and the largest eigenvalue $\lambda_{max}(M_E^{-1}F_E)$ from above. We will show that

$$\langle M_E\lambda, \lambda\rangle \leq \langle F_E\lambda, \lambda\rangle \leq \overline{C}\langle M_E\lambda, \lambda\rangle \quad \forall\lambda \in V, \tag{15}$$

with $\overline{C} := C \max((1+\log(H/h))^2, TOL * (1+\log(H/h)))$.

Lower bound: This bound is derived using purely algebraic arguments. As in the analysis of the one–level FETI methods, we can use the following formula, see Mandel and Tezaur [13] or Klawonn and Widlund [10, p. 73],

$$\langle F_E\lambda, \lambda\rangle = \sup_{0\neq v_\Delta \in \widetilde{W}_\Delta} \frac{\langle \lambda, B_\Delta v_\Delta\rangle^2}{|v_\Delta|_S^2}.$$

Let $\mu \in V$ be arbitrary. It then follows from Lemma 6 that there exists a $w_\Delta \in range\,(P_\Delta)$ with $\mu = B_\Delta w_\Delta$. Since $w_\Delta = P_\Delta w_\Delta$ and $|u_\Delta|_{\widetilde{S}} \leq |u_\Delta|_{S_\Delta} \quad \forall u_\Delta \in \widetilde{W}_\Delta$, we obtain

$$\langle F_E\lambda, \lambda\rangle \geq \frac{\langle \lambda, B_\Delta w_\Delta\rangle^2}{|w_\Delta|_{\widetilde{S}}^2} \geq \frac{\langle \lambda, B_\Delta w_\Delta\rangle^2}{|w_\Delta|_{S_\Delta}^2} = \frac{\langle \lambda, \mu\rangle^2}{|B_{D,\Delta}^t\mu|_{S_\Delta}^2} = \frac{\langle \lambda, \mu\rangle^2}{\langle M_E^{-1}\mu, \mu\rangle}.$$

The left inequality of (15) follows by choosing $\mu := M_E\lambda$.

Upper bound: Using Lemma 7, we obtain $\forall\lambda \in V$,

$$\langle F_E\lambda, \lambda\rangle = \sup_{0\neq w_\Delta \in \widetilde{W}_\Delta} \frac{\langle \lambda, B_\Delta w_\Delta\rangle^2}{|w_\Delta|_{\widetilde{S}}^2} \leq \overline{C} \sup_{w_\Delta \neq 0} \frac{\langle \lambda, B_\Delta w_\Delta\rangle^2}{|P_\Delta w_\Delta|_{S_\Delta}^2}$$

$$= \overline{C} \sup_{w_\Delta \neq 0} \frac{\langle \lambda, B_\Delta w_\Delta\rangle^2}{\langle M_E^{-1}B_\Delta w_\Delta, B_\Delta w_\Delta\rangle} = \overline{C} \sup_{\mu \in V} \frac{\langle \lambda, \mu\rangle^2}{\langle M_E^{-1}\mu, \mu\rangle}$$

$$= \overline{C}\langle M_E\lambda, \lambda\rangle.$$

Acknowledgments: The work of the authors was supported in part by the National Science Foundation under Grants NSF-CCR-9732208 and in part by the US Department of Energy under Contract DE-FG02-92ER25127 and that of the third author also in part by the Polish Science Foundation under Grant 2P03A 021 16.

References

1. Bjørstad P. E., Widlund O. B. (1986) Iterative methods for the solution of elliptic problems on regions partitioned into substructures. SIAM J. Numer. Anal., **23**:1093–1120.
2. Dryja M. (1988) A method of domain decomposition for 3-D finite element problems. In Glowinski R., Golub G. H., Meurant G. A., Périaux J., editors, First International Symposium on Domain Decomposition Methods for Partial Differential Equations, pages 43–61, Philadelphia, PA. SIAM.
3. Dryja M., Sarkis M. V., Widlund O. B. (1996) Multilevel Schwarz methods for elliptic problems with discontinuous coefficients in three dimensions. Numer. Math., **72**:313–348.
4. Dryja M., Smith B. F., Widlund O. B. (1994) Schwarz analysis of iterative substructuring algorithms for elliptic problems in three dimensions. SIAM J. Numer. Anal., **31**:1662–1694.
5. Dryja M., Widlund, O. B. (2001) A FETI–DP method for a mortar discretization of elliptic problems. In Pavarino L., Toselli A., editors, Proceedings of a workshop on domain decomposition methods held in Zürich, Switzerland June 7–8, 2001. Lecture Notes in Computational Science and Engineering, Springer-Verlag.
6. Dryja M., Widlund O. B. (1994) Domain decomposition algorithms with small overlap. SIAM J. Sci. Comput., **15**:604–620.
7. Dryja M., Widlund O. B. (1995) Schwarz methods of Neumann-Neumann type for three-dimensional elliptic finite element problems. Comm. Pure Appl. Math., **48**:121–155.
8. Farhat Ch., Lesoinne M., Le Tallec P., Pierson K., and Rixen D.J. (2001) FETI-DP: A dual-primal unified FETI method – part I: A faster alternative to the two-level FETI method. Int. J. Numer. Meth. Engng., **50**:1523–1544.
9. Farhat Ch., Lesoinne M., Pierson K. (2000) A scalable dual-primal domain decomposition method. Numer. Lin. Alg. Appl., **7**:687–714.
10. Klawonn A., Widlund O. B. (2001) FETI and Neumann–Neumann Iterative Substructuring Methods: Connections and New Results. Comm. Pure Appl. Math., **54**:57–90.
11. Klawonn A., Widlund O. B., Dryja M. (2001) Dual-Primal FETI Methods for Three-Dimensional Elliptic Problems with Heterogeneous Coefficients. Technical Report TR 815, Courant Institute of Mathematical Sciences.
12. Li J. (2001) A dual-primal FETI method for incompressible Stokes equations. Technical Report TR-816, Courant Institute of Mathematical Sciences.
13. Mandel J., Tezaur, R. (1996) Convergence of a Substructuring Method with Lagrange Multipliers. Numer. Math., **73**:473–487, 1996.
14. Mandel J., Tezaur R. (2001) On the convergence of a dual-primal substructuring method. Numer. Math., **88**:543–558.
15. Pierson K. H. (2000) A family of domain decomposition methods for the massively parallel solution of computational mechanics problems. PhD thesis, University of Colorado at Boulder, Aerospace Engineering.
16. Widlund O. B. (1986) An extension theorem for finite element spaces with three applications. In Hackbusch W., Witsch K., editors, *Numerical Techniques in Continuum Mechanics*, pages 110–122, Braunschweig/Wiesbaden, 1987. Notes on Numerical Fluid Mechanics, v. 16, Friedr. Vieweg und Sohn. Proceedings of the Second GAMM-Seminar, Kiel, January, 1986.

A FETI - DP Method for a Mortar Discretization of Elliptic Problems

Maksymilian Dryja[1] and Olof B. Widlund[2]

[1] Warsaw University, Warsaw, Banacha 2, 02-097 Warsaw, Poland
[2] Courant Institute of Mathematical Sciences, New York University, 251 Mercer Street, New York, NY 10012, USA

Abstract. In this paper, an iterative substructuring method with Lagrange multipliers is proposed for discrete problems arising from approximation of elliptic problem in two dimensions on non-matching meshes. The problem is formulated using a mortar technique. The algorithm belongs to the family of dual-primal FETI (Finite Element Tearing and Interconnecting) methods which has been analyzed recently for discretization on matching meshes. In this method the unknowns at the vertices of substructures are eliminated together with those of the interior nodal points of these substructures. It is proved that the preconditioner proposed is almost optimal; it is also well suited for parallel computations.

1 Introduction

We will consider a dual-primal FETI (FETI-DP) method, see [3],[6] and [5], for solving discrete problems arising from the approximation of the Dirichlet problem defined on a union of substructures Ω_i. Each substructure is the union of a number of elements of a coarse, shape-regular triangulation and the number of these triangles, which form such a substructure, is assumed to be uniformly bounded. The discretization is obtained by a mortar method on nonmatching meshes across the interface Γ; see [1,2]. As in all other iterative substructuring methods, the unknowns corresponding to the interior nodal points are eliminated; in this dual-primal FETI method those of the vertices of Ω_i are eliminated as well. The remaining Schur complement system is solved by a FETI method; see section 3 for details.

A full analysis of the convergence of several FETI-DP methods has been worked out for finite element approximations on matching meshes; see [6] for the two-dimensional case and [5] for three dimensions. Our goal is to extend these results to mortar discretizations.

In this paper, we analyze two cases when there is a Neumann-Dirichlet $(N - D)$ ordering of Ω_i, and when we do not have such an ordering. A Neumann substructure Ω_i is one where all sides are chosen as mortars; for a Dirichlet substructure all sides are nonmortars. In the case when there is a Neumann-Dirichlet ordering, we establish a bound for the condition number of FETI-DP method which is proportional to $(1 + \log(H/h))^2$, while we need four such factors in the general case.

The remainder of this paper is organized as follows. In section 2 differential and discrete problems are formulated while in section 3 the dual-primal formulation is introduced. Sections 4 and 5 are devoted to the analysis of the proposed preconditioner.

2   Differential and Discrete Problems

We will consider the following elliptic problem: find $u^* \in H_0^1(\Omega)$ such that

$$a(u^*, v) = f(v), \qquad v \in H_0^1(\Omega), \tag{1}$$

where

$$a(u, v) = \int_\Omega \nabla u \nabla v dx, \qquad f(v) = \int_\Omega f v \, dx,$$

and Ω is a polygonal 2-D region which is a union of polygons Ω_i, $i = 1, ... N$. These subregions form a coarse partitioning of Ω with subdomains with diameters of the order of H. In each Ω_i, we introduce a quasi-uniform, but otherwise arbitrary, triangulation of the subregion with a mesh parameter h_i; generally the resulting triangulations do not match across the edges of Ω_i.

Let

$$W(\Omega) = W_1(\Omega_1) \times ... \times W_N(\Omega_N),$$

where $W_i(\Omega_i)$ are the finite element spaces of piecewise linear continuous functions on the triangulation of Ω_i and which vanish on $\partial\Omega$ and let the interface be defined by $\Gamma = (\cup\partial\Omega_i)\backslash\partial\Omega$. We choose mortar and nonmortar edges of Γ, and denote them by $\gamma_{m(i)}$ and $\delta_{m(i)}$. In the analysis of the proposed preconditioner, we need a uniform bound on the ratios $h_{\gamma_{m(j)}}/h_{\delta_{m(i)}}$ where $h_{\gamma_{m(j)}}$ and $h_{\delta_{m(i)}}$ are the mesh parameters of $\gamma_{m(j)} \subset \partial\Omega_j$ and $\delta_{m(i)} \subset \partial\Omega_i$, $\gamma_{m(j)} = \delta_{m(i)}$, respectively. The problem (1) is approximated in X, a subspace of W, of functions which satisfy the mortar condition, see [1,2],

$$b(u, \psi) \equiv \sum_{i=1}^N \sum_{\delta_{m(i)} \subset \partial\Omega_i} \int_{\delta_{m(i)}} (u_i - u_j)\psi ds = 0, \qquad \psi \in M(\Gamma), \tag{2}$$

where $M(\Gamma) = \Pi_i \Pi_{\delta_{m(i)} \subset \partial\Omega_i} M(\delta_{m(i)})$ and $M(\delta_{m(i)})$ is the standard mortar space defined on $\delta_{m(i)}$, i.e., piecewise linear continuous functions which are constant on the elements which intersect $\partial\delta_{m(i)}$. Additionally, we assume that functions of X are continuous at the vertices of Ω_i, i.e., they take the same values, see [2]. Here, in (2) $u_i \in W_i$ and $u_j \in W_j$ are the restrictions to $\delta_{m(i)} = \gamma_{m(j)}$, respectively.

3   A Dual-Primal Formulation of the Problem

We will use some of the notations of [5]. Let

$$K := diag_{j=1}^{N}(K^{(j)}),$$ (3)

where $K^{(j)}$ is the local stiffness matrix with respect to the standard basis functions of $W_j(\Omega_j)$. We eliminate the unknown variables corresponding to the interior nodal points and the vertices of Ω_i . A Schur complement $\tilde{S}$ results which is of the form:

$$\tilde{S} := K_{rr} - \begin{pmatrix} K_{ri} & K_{rc} \end{pmatrix} \begin{pmatrix} K_{ii} & K_{ic} \\ K_{ci} & K_{cc} \end{pmatrix}^{-1} \begin{pmatrix} K_{ir} \\ K_{cr} \end{pmatrix}.$$ (4)

Here,

$$\tilde{K} := \begin{pmatrix} K_{ii} & K_{ic} & K_{ir} \\ K_{ci} & K_{cc} & K_{cr} \\ K_{ri} & K_{rc} & K_{rr} \end{pmatrix},$$

where the rows correspond to the interior, vertex, and remaining (edge) nodal points, respectively. It is obtained from K by reordering the unknowns and taking into account that functions of X take the same values at the vertices of Ω_i.

Let

$$W(\Gamma) = W_1(\partial\Omega_1) \times ... \times W_N(\partial\Omega_N),$$

and let $W_r(\Gamma)$ denote the space of functions defined at edge nodal points and which vanish at the vertices of Ω_i, and let W_c be the subspace of W of functions that are continuous at the vertices.

The dual-primal formulation of the mortar discretization of (1) is: find $u_r^* \in W_r$ such that

$$J(u_r^*) = \min_{\substack{v_r \in W_r \\ Bv_r = 0}} J(v_r), \quad J(v) := 1/2\langle \tilde{S}v, v\rangle - \langle f_r, v\rangle,$$ (5)

where $< , >$ means a scalar product in l_2 and B is defined by the mortar condition (2) as follows: on $\delta_{m(i)} \subset \partial\Omega_i$, $\delta_{m(i)} = \gamma_{m(j)}$, the matrix form of (2) is

$$B_{\delta_{m(i)}}\underline{u}_{i|\delta_{m(i)}} - B_{\gamma_{m(j)}}\underline{u}_{j|\gamma_{m(j)}} = 0,$$ (6)

where

$$B_{\delta_{m(i)}} = \{(\psi_l, \varphi_p)_{L^2(\delta_{m(i)})}\}, \qquad l, p = 1, ..., n_{m(i)},$$

and $\varphi_p \in W_i(\partial\Omega_i)_{|\delta_{m(i)}}, \psi_l \in M(\delta_{m(i)})$;

$$B_{\gamma_{m(j)}} = \{(\psi_l, \varphi_k)_{L^2(\delta_{m(i)})}\}, \qquad l = 1, ..., n_{m(i)}, k = 1, ..., n_{m(j)}$$

and $\varphi_k \in W_j(\partial\Omega_j)_{|\gamma_{m(j)}}; n_{m(i)}$ and $n_{m(j)}$ are the number of interior nodal points of $\delta_{m(i)}$ and $\gamma_{m(j)}$, respectively. Condition (6) can be rewritten as

$$\underline{u}_{i|\delta_{m(i)}} - B^{-1}_{\delta_{m(i)}} B_{\gamma_{m(j)}} \underline{u}_{j|\gamma_{m(i)}} = 0 \tag{7}$$

since the matrix $B_{\delta_{m(i)}} = B^T_{\delta_{m(i)}} > 0$. We note that $B_{\gamma_{m(j)}}$ is generally a rectangular matrix.

The matrix B is block-diagonal,

$$B = blockdiag\{D_{\delta_{m(i)}}\} \tag{8}$$

for $i = 1,...,N$, and $\delta_{m(i)} \subset \partial\Omega_i$ where

$$D_{\delta_{m(i)}} \begin{pmatrix} \underline{u}_{i|\delta_{m(i)}} \\ \underline{u}_{j|\gamma_{m(j)}} \end{pmatrix} \equiv (I \ (-B^{-1}_{\delta_{m(i)}} B_{\gamma_{m(j)}})) \begin{pmatrix} \underline{u}_{i|\delta_{m(i)}} \\ \underline{u}_{j|\gamma_{m(j)}} \end{pmatrix}. \tag{9}$$

Introducing a space of Lagrange multipliers $V := Im(B)$ to enforce the constraints $Bv_r = 0$, we obtain a saddle point formulation of (5),

$$\begin{pmatrix} \tilde{S} & B^T \\ B & 0 \end{pmatrix} \begin{pmatrix} u^*_r \\ \lambda^* \end{pmatrix} = \begin{pmatrix} \tilde{f}_r \\ 0 \end{pmatrix}, \tag{10}$$

where $u^*_r \in W_r$ and $\lambda^* \in V$. We obtain the problem

$$F\lambda^* = d, \tag{11}$$

where

$$F = B\tilde{S}^{-1}B^T, \qquad d = B\tilde{S}^{-1}\tilde{f}_r.$$

We now define a preconditioner for F. Let

$$S^{(j)} = K^{(j)}_{bb} - K^{(j)}_{bi}(K^{(j)}_{ii})^{-1}K^{(j)}_{ib}, \tag{12}$$

be the standard Schur complement of $K^{(j)}$ where $K^{(j)}_{ii}$ and $K^{(j)}_{bb}$ are the submatrices of $K^{(j)}$ corresponding to the interior and boundary nodal unknowns of $\bar{\Omega}_j$, respectively. Let

$$S^{(j)}_{rr} = K^{(j)}_{rr} - K^{(j)}_{ri}(K^{(j)}_{ii})^{-1}K^{(j)}_{ir} \tag{13}$$

denote the Schur complement of $K^{(j)}$, without the rows and columns corresponding to the vertices. This is the restriction of $S^{(j)}$ to the space of functions which vanish at the vertices. Let

$$S := diag^N_{i=1}(S^{(i)}), \qquad S_{rr} := diag^N_{i=1}(S^{(i)}_{rr}).$$

We take a preconditioner M of F of the form

$$M = (BS_{rr}B^T)^{-1}, \qquad M^{-1} = BS_{rr}B^T. \tag{14}$$

Remark We could also take, see [4],

$$\widehat{M}^{-1} = (BB^T)^{-1}BS_{rr}B^T(BB^T)^{-1}$$

as the preconditioner of F but its full analysis is not known for the mortar discretization.

4   Convergence Analysis: the Upper Bound

In this and the next section, we prove estimates, from above as well as below, of $< F\lambda, \lambda >$ in terms of $< M\lambda, \lambda >$, $\lambda \in V$. We follow the approach of [6,5] and begin with the upper bound. We first prove the following auxiliary result.

Lemma 1 For $w_r \in W_r$

$$|B^T B w_r|^2_{S_{rr}} \leq C(1 + \log \frac{H}{h})^2 |w_r|^2_{\tilde{S}}, \tag{15}$$

where C is independent of $H = max_i H_i$, $h = min_i h_i$ and where H_i is the diameter of Ω_i.

Proof Let $w = \{w_i\}_{i=1}^N$ be the discrete harmonic extension of w_r to the interior points and to the vertices of Ω_i in the sense of $< \tilde{S}\cdot, \cdot >$. We have

$$|w_r|^2_{\tilde{S}} = |w|^2_S, \quad w \in W_c. \tag{16}$$

Using this fact, we will estimate $|B^T B w_r|^2_{S_{rr}}$ in terms of $|w|^2_S$. We select $u = I^H w$, the function which is linear on the edges and which takes the value of w at the vertices. Setting $w_r = (w - u) + u$ on Γ and noting that $Bu = 0$, we have

$$|B^T B w_r|^2_{S_{rr}} = |B^T B(w - u)|^2_{S_{rr}} = \sum_{i=1}^N |B^T B(w - u)|^2_{S^{(i)}}. \tag{17}$$

We note that $B^T B(w - u) = 0$ at the vertices. Using that, we obtain

$$|B^T B(w - u)|^2_{S^{(i)}} \leq C\{ \sum_{\delta_{m(i)} \subset \partial \Omega_i} |B^T B(w - u)|^2_{S_{\delta_{m(i)}}} +$$

$$+ \sum_{\gamma_{m(i)} \subset \partial \Omega_i} |B^T B(w - u)|^2_{S_{\gamma_{m(i)}}} \}, \tag{18}$$

where $S_{\delta_{m(i)}}$ and $S_{\gamma_{m(i)}}$ are the matrix representations of the $H^{1/2}_{00}$–norm on $\delta_{m(i)}$ and $\gamma_{m(i)}$, respectively. From the structure of B follows, see (9),

$$|B^T Bz|^2_{S_{\delta_{m(i)}}} \leq 2\{|z_i|^2_{S_{\delta_{m(i)}}} + |B_{ij} z_j|^2_{S_{\delta_{m(i)}}} \}, \tag{19}$$

where here and below $z \equiv w - u$, $z = \{z_i\}_{i=1}^N \subset W_c$, $B_{ij} \equiv B^{-1}_{\delta_{m(i)}} B_{\gamma_{m(j)}}$, and $\delta_{m(i)} = \gamma_{m(j)}$, $\gamma_{m(j)} \subset \partial \Omega_j$; also

$$|B^T Bz|^2_{S_{\gamma_{m(i)}}} \leq 2\{|B^T_{ki} z_k|^2_{S_{\gamma_{m(i)}}} + |B^T_{ki} B_{ki} z_i|^2_{S_{\gamma_{m(i)}}} \}, \tag{20}$$

where $B_{ki} \equiv B^{-1}_{\delta_{m(k)}} B_{\gamma_{m(i)}}$, $\gamma_{m(i)} = \delta_{m(k)}$, $\delta_{m(k)} \subset \partial \Omega_k$. We now estimate each term of (19) and (20).

It is known that,

$$|z_i|^2_{S_{\delta_{m(i)}}} \leq C(1 + \log \frac{H}{h})^2 |w_i|^2_{H^{1/2}(\partial\Omega_i)} \leq C(1 + \log \frac{H}{h})^2 |w_i|^2_{S^{(i)}}; \qquad (21)$$

see, e.g., [4]. To estimate the second term of (19), we use the stability of the mortar projection, see [1,2]. Let $\pi_{\delta_{m(i)}}(z_j, 0)$ correspond to $B_{ij}(z_{j|\gamma_{m(j)}})$ for z_j restricted to $\gamma_{m(i)}$. Using that, we have

$$|B_{ij}z_j|^2_{S_{\delta_{m(i)}}} \leq C\|\pi_{\delta_{m(i)}}(z_j, 0)\|^2_{H^{1/2}_{00}(\delta_{m(i)})} \leq$$

$$\leq C\|z_j\|^2_{H^{1/2}_{00}(\gamma_{m(j)})} \leq C(1 + \log \frac{H}{h})^2 |w_j|^2_{S^{(j)}}. \qquad (22)$$

The first term of (20) is estimated as follows:

$$|B^T_{ki}z_k|^2_{S_{\gamma_{m(i)}}} = \max_t \frac{|< z_k, B_{ki}t >|^2}{|S^{-1/2}_{\gamma_{m(i)}}t|^2}$$

$$\leq \max_t \frac{|z_k|^2_{S_{\delta_{m(k)}}} |B_{ki}t|^2_{S^{-1}_{\delta_{m(k)}}}}{|S^{-1/2}_{\gamma_{m(i)}}t|^2}. \qquad (23)$$

We will show that

$$|B_{ki}t|^2_{S^{-1}_{\delta_{m(k)}}} \leq C|t|^2_{S^{-1}_{\gamma_{m(i)}}}. \qquad (24)$$

It is known that the norm generated by $h^2_{\delta_{m(k)}} S^{-1}_{\delta_{m(k)}}$ is equivalent to the norm of $H^{-1/2}(\delta_{m(k)})$, the dual to $H^{1/2}_{00}(\delta_{m(k)})$, where $h_{\delta_{m(k)}}$ is the mesh size on $\delta_{m(k)}$; see, e.g., [7]. Using that, we have

$$h^2_{\delta_{m(k)}} |B_{ki}t|^2_{S^{-1}_{\delta_{m(k)}}} \leq C\|\pi_{\delta_{m(k)}}(t, 0)\|^2_{H^{-1/2}(\delta_{m(k)})}. \qquad (25)$$

We now show that

$$\|\pi_{\delta_{m(k)}}(t, 0)\|^2_{H^{-1/2}(\delta_{m(k)})} \leq C(1 + h_{\delta_{m(k)}}/h_{\gamma_{m(i)}})\|t\|^2_{H^{-1/2}(\gamma_{m(i)})}.$$

To see that, we note that

$$\|\pi_{\delta_{m(k)}}(t, 0)\|_{H^{-1/2}(\delta_{m(k)})} \leq \|t\|_{H^{-1/2}(\delta_{m(k)})} + \|\pi_{\delta_{m(k)}}(t, 0) - t\|_{H^{-1/2}(\delta_{m(k)})}$$

and that

$$\|\pi_{\delta_{m(k)}}(t, 0) - t\|_{H^{-1/2}(\delta_{m(k)})} = \sup_g \frac{|(\pi_{\delta_{m(k)}}(t, 0) - t, g - Q_{\delta_{m(k)}}(g, 0))_{L^2(\delta_{m(k)})}|}{\|g\|_{H^{1/2}_{00}(\delta_{m(k)})}},$$

where $Q_{\delta_{m(k)}}$ is the L_2 - orthogonal projection onto the mortar space $M(\delta_{m(k)})$. Using a known estimate for $g - Q_{\delta_{m(k)}}g$, the L_2 - stability of $\pi_{\delta_{m(k)}}$, see, e.g., [1,2], and an inverse inequality, we get

$$\|\pi_{\delta_{m(k)}}(t, 0) - t\|_{H^{-1/2}(\delta_{m(k)})} \leq C(h^{1/2}_{\delta_{m(k)}}/h^{1/2}_{\gamma_{m(i)}})\|t\|_{H^{-1/2}(\delta_{m(k)})}.$$

Using this in (25) and that $h_{\gamma_{m(i)}}/h_{\delta_{m(k)}}$ is uniformly bounded, and the equivalence of $h^2_{\gamma_{m(i)}}S^{-1}_{\gamma_{m(i)}}$ to the norm of $H^{-1/2}(\gamma_{m(i)})$, we get (24). Substituting now (24) into (23), we have

$$|B^T_{ki}z_k|^2_{S_{\gamma_{m(i)}}} \le C|z_k|^2_{S_{\delta_{m(k)}}} \le$$

$$\le C(1 + \log \frac{H}{h})^2 |w_k|^2_{S^{(k)}}. \tag{26}$$

There remains to estimate the second term of (20). We have

$$|B^T_{ki}B_{ki}z_i|^2_{S_{\gamma_{m(i)}}} = \max_t \frac{|<B_{ki}z_i, B_{ki}t>|^2}{|S^{-1/2}_{\gamma_{m(i)}}t|^2}$$

$$\le \max_t \frac{|B_{ki}z_i|^2_{S_{\delta_{m(k)}}} |B_{ki}t|^2_{S^{-1}_{\delta_{m(k)}}}}{|S^{-1/2}_{\gamma_{m(i)}}t|^2}. \tag{27}$$

Using the proof of (22), we show that

$$|B_{ki}z_i|^2_{S_{\delta_{m(k)}}} \le C(1 + \log \frac{H}{h})^2 |w_i|^2_{S^{(i)}}. \tag{28}$$

Using this and (24) in (27), we obtain

$$|B^T_{ki}B_{ki}z_i|^2_{S_{\gamma_{m(i)}}} \le C(1 + \log \frac{H}{h})^2 |w_i|^2_{S^{(i)}}. \tag{29}$$

We now substitute (21) and (22) into (19), and (26) and (29) into (20), and the resulting estimates into (18). This gives

$$|B^T B(w - u)|^2_{S^{(i)}} \le C(1 + \log \frac{H}{h})^2 \{|w_i|^2_{S^{(i)}} + \sum_j |w_j|^2_{S^{(j)}}\}, \tag{30}$$

where the sum is taken over those $\partial\Omega_j$ which have an edge in common with $\partial\Omega_i$. Using this in (17) and then in (16), we obtain (15) and the proof is complete.

Theorem 1 For $\lambda \in V = Im(B)$ holds

$$<F\lambda, \lambda> \le C(1 + \log \frac{H}{h})^2 <M\lambda, \lambda>, \tag{31}$$

where C is independent of h and H.
Proof We have, cf. e.g., [4],

$$<F\lambda, \lambda> = \max_{w_r \in W_r} \frac{|<\lambda, Bw_r>|^2}{|w_r|^2_{\tilde{S}}};$$

see (11). By Lemma 1, see also (14),

$$< F\lambda, \lambda > \le C(1 + \log \frac{H}{h})^2 \max_{w_r} \frac{|< \lambda, Bw_r >|^2}{|B^T Bw_r|^2_{S_{rr}}} =$$

$$= C(1 + \log \frac{H}{h})^2 \max_{w_r} \frac{|< \lambda, Bw_r >|^2}{< BS_{rr}B^T Bw_r, Bw_r >} =$$

$$= C(1 + \log \frac{H}{h})^2 \max_{w_r} \frac{|< M^{1/2}\lambda, M^{-1/2}Bw_r >|^2}{< M^{-1/2}Bw_r, M^{-1/2}Bw_r >} =$$

$$= C(1 + \log \frac{H}{h})^2 < M\lambda, \lambda >$$

which proves (31).

5   Convergence Analysis: the Lower Bound

In this section, we estimate $< F\lambda, \lambda >$ from below in terms of $< M\lambda, \lambda >$.
We first discuss the case when there is a Neumann-Dirichlet $(N - D)$ ordering
of the substructures Ω_i. Here a Neumann substructure is one with all mortar
edges while a Dirichlet substructure has only nonmortar edges. In this special
case we establish a constant lower bound. We then discuss the case without a
$N - D$ ordering and show a weaker bound containing two logarithmic factors.

5.1   The Neumann-Dirichlet Case

We note that for the $N - D$ ordering the matrix B in (8), cf. (9), can be
rewritten in the form

$$B = (I \quad - B_N), \tag{32}$$

where I is an identity matrix of order equal to the dimension of $V = Im(B)$;
it corresponds to the Dirichlet substructures Ω_i. The block diagonal matrix
B_N corresponds to the Neumann substructures and the only nonzero blocks
equal $B_{\delta_{m(i)}}^{-1} B_{\gamma_{m(j)}}$ which correspond to the edges $\delta_{m(i)} = \gamma_{m(j)}$; see (9).

The matrix $S_{rr} = diag_{i=1}^{N}(S_{rr}^{(i)})$ is reordered as

$$S_{rr} = \begin{pmatrix} S_{rr}^D & 0 \\ 0 & S_{rr}^N \end{pmatrix}, \tag{33}$$

where the first block corresponds to the Dirichlet substructures and $S_{rr}^D = diag_{i \in I_D}(S_{rr}^{(i)})$, while the second corresponds to the Neumann subdomains, and $S_{rr}^N = diag_{i \in I_N}(S_{rr}^{(i)})$. Here I_D and I_N are the sets of Dirichlet and

Neumann substructures, respectively.

We note that using the form of B, cf. (32), we have

$$M^{-1} = BS_{rr}B^T = S_{rr}^D + B_N S_{rr}^N B_N^T. \tag{34}$$

We note also that for $w \in W_r$,

$$< (S_{rr})^{-1}w, w > \leq < \tilde{S}^{-1}w, w > \tag{35}$$

since by (16),

$$< \tilde{S}w, w > \leq < S_{rr}w, w > . \tag{36}$$

Theorem 2 For $\lambda \in V = Im(B)$ holds

$$< M\lambda, \lambda > \leq < F\lambda, \lambda > \tag{37}$$

Proof The proof reduces to showing that

$$\lambda_{min}(M^{-1/2}FM^{-1/2}) = \min_{\lambda} \frac{< F\lambda, \lambda >}{< M\lambda, \lambda >} \geq 1. \tag{38}$$

We note that, according to (33) and (34),

$$< S_{rr}^D\lambda, \lambda > \leq < M^{-1}\lambda, \lambda >$$

and therefore

$$< M\lambda, \lambda > \leq < (S_{rr}^D)^{-1}\lambda, \lambda > .$$

Using this and (35) in (38), we have

$$\lambda_{min}(M^{-1/2}FM^{-1/2}) \geq \min_{\lambda} \frac{< (S_{rr})^{-1}B^T\lambda, B^T\lambda >}{< M\lambda, \lambda >} \geq$$

$$\geq \min_{\lambda} \frac{< (S_{rr})^{-1}B^T\lambda, B^T\lambda >}{< (S_{rr}^D)^{-1}\lambda, \lambda >} \geq 1$$

since

$$< (S_{rr})^{-1}B^T\lambda, B^T\lambda > = < (S_{rr}^D)^{-1}\lambda, \lambda > + < (S_{rr}^N)^{-1}B_N^T\lambda, B_N^T\lambda > .$$

5.2   The General Case

In this subsection, we discuss the case when there is no $N - D$ ordering of the substructures. In this case

$$B = (I \quad -B_m),$$

50   Maksymilian Dryja and Olof B. Widlund

where I is a block diagonal matrix corresponding to the nonmortar sides while B_m is a block matrix which corresponds to the mortar sides. Each block row of B_m has one nonzero element equal to $B_{\delta_{m(i)}}^{-1} B_{\gamma_{m(j)}}$, cf. (9).

We note that in the case considered S_{rr} cannot be represented as a block diagonal matrix of the form (33) but is of the form

$$S_{rr} = \begin{pmatrix} S_{rr}^{nn} & S_{rr}^{nm} \\ S_{rr}^{mn} & S_{rr}^{mm} \end{pmatrix}, \tag{39}$$

where the first row corresponds to the nonmortar sides and the second to the mortar sides. We note that the matrices S_{rr}^{nn} and S_{rr}^{mm} are block diagonal and they are of the form:

$$S_{rr}^{nn} = diag_i(\tilde{S}_{rr,i}^{nn}) \ , \quad \tilde{S}_{rr,i}^{nn} = diag_{\delta_{m(i)}}(S_{rr,\delta_{m(i)}}^{nn}), \tag{40}$$

$$S_{rr}^{mm} = diag_i(\tilde{S}_{rr,i}^{mm}) \ , \quad \tilde{S}_{rr,i}^{mm} = diag_{\gamma_{m(i)}}(S_{rr,\gamma_{m(i)}}^{mm}), \tag{41}$$

where $diag_{\delta_{m(i)}}$ and $diag_{\gamma_{m(i)}}$ are taken for $\delta_{m(i)} \subset \partial\Omega_i$ and $\gamma_{m(i)} \subset \partial\Omega_i$, respectively.

We introduce an auxiliary matrix

$$\tilde{S}_{rr} = diag(S_{rr}) = \begin{pmatrix} S_{rr}^{nn} & 0 \\ 0 & S_{rr}^{mm} \end{pmatrix}. \tag{42}$$

Lemma 2 For $w \in W_r$ holds

$$< S_{rr}w, w > \leq 2 < \tilde{S}_{rr}w, w > \tag{43}$$

and

$$< \tilde{S}_{rr}w, w > \leq C(1 + \log \frac{H}{h})^2 < S_{rr}w, w >, \tag{44}$$

where C is independent of H and h.
Proof Using the observation that

$$\pm \begin{pmatrix} 0 & S_{rr}^{nm} \\ S_{rr}^{mn} & 0 \end{pmatrix} \leq \begin{pmatrix} S_{rr}^{nn} & 0 \\ 0 & S_{rr}^{mm} \end{pmatrix},$$

which follows from the fact that $S_{rr} = (S_{rr})^T > 0$, we obtain (43).

The inequality (44) follows from the fact, see, e.g., [4], that

$$< \tilde{S}_{rr}w, w > = \sum_i \sum_{E_k \subset \partial\Omega_i} ||w_i||^2_{H_{00}^{1/2}(E_k)} \leq$$

$$\leq C \sum_i (1 + \log \frac{H_i}{h_i})^2 |w_i|^2_{H^{1/2}(\partial\Omega_i)} \leq$$

$$\leq C(1 + \log \frac{H}{h})^2 < S_{rr}w, w >,$$

where E_k is an edge.

Theorem 3 For $\lambda \in V = Im(B)$ holds

$$< M\lambda, \lambda > \le C(1 + \log \frac{H}{h})^2 < F\lambda, \lambda >, \tag{45}$$

where C is independent of H and h.

Proof The proof of (45) reduces to showing that

$$\lambda_{min}(M^{-1/2}FM^{-1/2}) = \min_{\lambda} \frac{< F\lambda, \lambda >}{< M\lambda, \lambda >} = \tag{46}$$

$$\min_{\lambda} \frac{< \tilde{S}^{-1}B^T\lambda, B^T\lambda >}{< (BS_{rr}B^T)^{-1}\lambda, \lambda >} \ge C(1 + \log \frac{H}{h})^{-2}.$$

By (35),

$$\lambda_{min}(M^{-1/2}FM^{-1/2}) \ge \min_{\lambda} \frac{< S_{rr}^{-1}B^T\lambda, B^T\lambda >}{< (BS_{rr}B^T)^{-1}\lambda, \lambda >}. \tag{47}$$

In addition, using (42), (44) and the form of B, we obtain

$$< S_{rr}^{nn}\lambda, \lambda > \le < S_{rr}^{nn}\lambda, \lambda > + < S_{rr}^{mm}B_m^T\lambda, B_m^T\lambda > =$$

$$= < \tilde{S}_{rr}B^T\lambda, B^T\lambda > \le C(1 + \log \frac{H}{h})^2 < S_{rr}B^T\lambda, B^T\lambda > =$$

$$= C(1 + \log \frac{H}{h})^2 < BS_{rr}B^T\lambda, \lambda > .$$

Hence,

$$< (BS_{rr}B^T)^{-1}\lambda, \lambda > \le C(1 + \log \frac{H}{h})^2 < (S_{rr}^{nn})^{-1}\lambda, \lambda > . \tag{48}$$

By (43), we have

$$< (\tilde{S}_{rr})^{-1}\lambda, \lambda > \le 2 < S_{rr}^{-1}\lambda, \lambda > . \tag{49}$$

Using (48) and (49) in (47), we obtain

$$\lambda_{min}(M^{-1/2}FM^{-1/2}) \ge \min_{\lambda} \frac{\frac{1}{2} < (\tilde{S}_{rr})^{-1}\lambda, \lambda >}{C(1 + \log \frac{H}{h})^2 < (S_{rr}^{nn})^{-1}\lambda, \lambda >}.$$

From this (46) follows. The proof is complete.

Acknowledgments: The work of the authors was supported in part by the National Science Foundation under Grant NSF - CCR - 9732208 and that of the first author also in part by the Polish Science Foundation under Grant 2 P03A 02116.

References

1. Ben   Belgacem, F. (1999) The mortar finite element method with Lagrange multipliers. Numer. Math. **84**, 173–197
2. Bernardi, C., Maday, Y., Patera, A. T. (1989) A new nonconforming approach to domain decomposition. The mortar element method. College de France Seminar, H. Brezis and J. L. Lions, eds., Pitman.
3. Farhat, C., Lesoinne, M., Pierson, K. (2000) A scalable dual - primal domain decomposition. Numer. Lin. Alg. App. Vol. **7(7-8)**, 687–714
4. Klawonn, A., Widlund, O. (2001) FETI and Neumann - Neumann iterative substructuring methods: Connections and new results. Comm. Pure. Appl. Math. **54**, 57–90
5. Klawonn, A., Widlund, O., Dryja, M. (2001) Dual-primal FETI methods for three-dimensional elliptic problems with heterogeneous coefficients. Technical report **2001-815**, Computer Science Department, Courant Institute of Mathematical Sciences
6. Mandel, J., Tezaur, R. (2001) On the convergence of a dual-primal substructuring method. Numer. Math. **88:3**, 543–558
7. Peisker, P. (1988) On the numerical solution of the first biharmonic equation. MAN Mathematical Modelling and Numerical Analysis, **22**, #4, 655–676

Balancing Neumann-Neumann Methods
for Mixed Approximations of Linear Elasticity

Paulo Goldfeld[1], Luca F. Pavarino[2], and Olof B. Widlund[1]

[1] Courant Institute of Mathematical Sciences, 251 Mercer Street, New York, NY 10012

[2] Università di Milano, Dipartimento Di Matematica, Via Saldini 50, 20133 Milano, Italy

Abstract. Balancing Neumann-Neumann methods are introduced and analyzed for the algebraic systems of linear equations for mixed finite element approximations of linear elasticity for incompressible and almost incompressible materials as well as composite materials with different Lamé parameters in different parts of the domain. These methods solve iteratively the saddle point Schur complement, resulting from the implicit elimination of the interior degrees of freedom, using a hybrid preconditioner based on a coarse mixed elasticity problem and local mixed elasticity problems with natural and essential boundary conditions. The resulting algorithm is very efficient, parallel, and robust with respect to material heterogeneities.

1 Introduction

In our previous paper [28], we introduced and analyzed a balancing Neumann-Neumann domain decomposition method for incompressible Stokes equations. Here, we extend the algorithm to the equations of linear elasticity in mixed form for incompressible and almost incompressible materials. We also cover the case of composite materials where the Lamé parameters (and therefore the Poisson ratio) have different arbitrary positive values in different subdomains. Our algorithm is very efficient, parallel, and robust with respect to jumps in the Lamé parameters.

After decomposing the original domain of the problem into nonoverlapping subdomains, the interior unknowns, which are the interior displacement components and, on each subdomain, all except the constant pressure component, are implicitly eliminated. The resulting saddle point Schur complement is solved with a Krylov space method with a balancing Neumann-Neumann preconditioner based on the solution of a coarse mixed elasticity problem with a few degrees of freedom per subdomain and on the solution of local mixed elasticity problems with natural and essential boundary conditions on the subdomain boundaries. This preconditioner is of hybrid form in which the coarse problem is treated multiplicatively while the local problems are treated additively. The condition number of the preconditioned operator is independent of the number of subdomains, the jumps in the Lamé parameters, and is bounded from above by the product of the square of the logarithm of the local number of unknowns in each subdomain, the inverse of the square of the inf-sup constant of the discrete problem, and the inverse of the inf-sup constant of the coarse subproblem.

Previous theoretical and numerical work for Neumann-Neumann methods has been carried out for second order elliptic problems; see Mandel [22], Mandel and Brezina [23], Cowsar, Mandel and Wheeler [10], Dryja and Widlund [11], and Pavarino [26]. More recently, this family of methods has been extended to plate and shell problems, see Le Tallec, Mandel, and Vidrascu [20], to convection-diffusion problems, see Alart, Barboteu, Le Tallec, and Vidrascu [3], and Achdou, Le Tallec, Nataf, and Vidrascu [1], and to vector field problems, see Toselli [35]. We also note that the connection between Neumann-Neumann and FETI methods has been considered recently by Klawonn and Widlund [18].

Different domain decomposition methods for incompressible Stokes equations have been proposed previously. Iterative substructuring methods have been studied by Bramble and Pasciak [7], Pasciak [25], Quarteroni [29], Marini and Quarteroni [24], Fischer and Rønquist [14], Casarin [9], Rønquist [31], Le Tallec and Patra [19], Pavarino and Widlund [27], and Ainsworth and Sherwin [2]. Overlapping Schwarz methods have been studied by Gervasio [15], Fischer [12], Fischer, Miller, and Tufo [13], Klawonn and Pavarino [16], and Rønquist [32].

For a general introduction to domain decomposition methods we refer to Smith, Bjørstad, and Gropp [33] and Quarteroni and Valli [30].

The paper is organized as follows. In Section 2, we describe the mixed linear elasticity system, its discretization with both finite and spectral elements and the resulting discrete system. In Section 3, the substructuring process, also known as static condensation, is described in both matrix and variational form. The balancing Neumann-Neumann preconditioner is introduced in Section 4 and the main result on its convergence rate is given in Section 5. In Section 6, the algorithm is extended to the case of composite materials with discontinuous Lamé coefficients in different subdomains. Section 7 concludes the paper with numerical experiments for problems in the plane.

2   Mixed Methods for Linear Elasticity: Continuous and Discrete Problems

Let $\Omega \subset R^3$ be a polyhedral domain and let Γ_0 be a nonempty subset of its boundary. Let $\mathbf{V}$ be the Sobolev space $\mathbf{V} = \{\mathbf{v} \in H^1(\Omega)^3 : \mathbf{v}|_{\Gamma_0} = 0\}$.

The linear elasticity problem, with constant Lamé parameters, consists in finding the displacement $\mathbf{u} \in \mathbf{V}$ of the domain Ω, fixed along Γ_0, subject to a surface force of density $\mathbf{g}$, along $\Gamma_1 = \partial\Omega \setminus \Gamma_0$, and a body force $\mathbf{f}$:

$$2\mu \int_\Omega \varepsilon(\mathbf{u}) : \varepsilon(\mathbf{v})\, dx + \lambda \int_\Omega \mathrm{div}\mathbf{u}\ \mathrm{div}\mathbf{v}\, dx\ =\ <\mathbf{F}, \mathbf{v}> \quad \forall \mathbf{v} \in \mathbf{V}. \tag{1}$$

Here λ and μ are the Lamé constants, $\varepsilon_{ij}(\mathbf{u}) = \frac{1}{2}(\frac{\partial u_i}{\partial x_j} + \frac{\partial u_j}{\partial x_i})$ the linearized strain tensor, and the bilinear forms are defined as

$$\varepsilon(\mathbf{u}) : \varepsilon(\mathbf{v}) = \sum_{i=1}^{3}\sum_{j=1}^{3} \varepsilon_{ij}(\mathbf{u})\varepsilon_{ij}(\mathbf{v}), \quad < \mathbf{F}, \mathbf{v} > = \int_{\Omega} \sum_{i=1}^{3} f_i v_i \, dx + \int_{\Gamma_1} \sum_{i=1}^{3} g_i v_i \, ds.$$

In Sections 6 and 7, we will consider the case of variable Lamé parameters and show that our algorithms are quite robust. The Lamé parameters can alternatively be expressed in terms of the Poisson ratio ν and Young's modulus E:

$$\lambda = \frac{E\nu}{(1+\nu)(1-2\nu)}, \quad \mu = \frac{E}{2(1+\nu)}.$$

When the material is almost incompressible, the Poisson ratio ν approaches the value $1/2$, i.e., λ/μ approaches infinity. In such cases, finite or spectral element discretizations of this pure displacement formulation suffer increasingly from locking phenomena and the resulting stiffness matrices become increasingly ill-conditioned. A possible remedy is based on introducing the new variable $p = -\lambda \mathrm{div}\mathbf{u} \in L^2(\Omega) = U$ that we will call pressure and replacing the pure displacement problem with a mixed formulation: find $(\mathbf{u}, p) \in \mathbf{V} \times U$ such that

$$\begin{cases} 2\mu \int_{\Omega} \varepsilon(\mathbf{u}) : \varepsilon(\mathbf{v}) \, dx - \int_{\Omega} \mathrm{div}\mathbf{v} \, p \, dx = < \mathbf{F}, \mathbf{v} > \quad \forall \mathbf{v} \in \mathbf{V} \\[2ex] -\int_{\Omega} \mathrm{div}\mathbf{u} \, q \, dx - 1/\lambda \int_{\Omega} pq \, dx = \quad 0 \quad \forall q \in U; \end{cases} \tag{2}$$

see Brezzi and Fortin [8]. In the case of homogeneous Dirichlet boundary conditions for $\mathbf{u}$, we choose $U = L_0^2(\Omega) = \{q \in L^2(\Omega) : \int_{\Omega} q dx = 0\}$, since it can be shown that the pressure will have zero mean value.

We can also consider more general saddle point problems with a penalty term: find $(\mathbf{u}, p) \in \mathbf{V} \times U$ such that

$$\begin{cases} a(\mathbf{u}, \mathbf{v}) + \quad b(\mathbf{v}, p) \quad = < \mathbf{F}, \mathbf{v} > \quad \forall \mathbf{v} \in \mathbf{V} \\[2ex] b(\mathbf{u}, q) - 1/\lambda \, c(p, q) = \quad 0 \quad \forall q \in U; \end{cases} \tag{3}$$

see Brezzi and Fortin [8]. In our specific case, we have

$$a(\mathbf{u}, \mathbf{v}) = 2\mu \int_{\Omega} \varepsilon(\mathbf{u}) : \varepsilon(\mathbf{v}) \, dx, \quad b(\mathbf{v}, q) = -\int_{\Omega} \mathrm{div}\mathbf{v} \, q \, dx, \quad c(p, q) = \int_{\Omega} pq \, dx.$$

By letting $\lambda/\mu \to \infty$, we obtain the limiting problem for incompressible linear elasticity or the classical Stokes system for an incompressible fluid. Often the Stokes system is alternatively written using the bilinear form

$a(\mathbf{u}, \mathbf{v}) = \mu \int_{\Omega} \nabla \mathbf{u} : \nabla \mathbf{v} \, dx$. A penalty term as in the compressible case can also originate from stabilization techniques or penalty formulations for Stokes problems.

We will also need to consider problems with natural boundary conditions on all of $\partial\Omega$,

$$2\mu \sum_{j=1}^{d} \frac{1}{2}\left(\frac{\partial u_i}{\partial x_j} + \frac{\partial u_j}{\partial x_i}\right)n_j - pn_i = g_i \quad \text{on } \partial\Omega, \; i = 1, \ldots, d \qquad (4)$$

derived by using Green's formula. In this case, as for the Laplace operator, the bilinear form $a(\cdot, \cdot)$ has a nontrivial nullspace $\ker(a)$ consisting of the rigid body motions (a three-dimensional nullspace in two dimensions and a six-dimensional nullspace in three). Therefore there is a compatibility condition between $\mathbf{f}$ and $\mathbf{g}$, namely,

$$\int_{\Omega} \mathbf{f} \cdot \mathbf{v} dx + \int_{\partial\Omega} \mathbf{g} \cdot \mathbf{v} \, ds = 0 \quad \forall \mathbf{v} \in \ker(a).$$

We note that if the boundary conditions are mixed (part essential and part natural), then there is a unique solution without any compatibility conditions.

Using Korn's inequality on the subspace orthogonal to the rigid body motions, we have the following equivalence between the bilinear forms of the Stokes and elasticity equations (see, e.g., Klawonn and Widlund [17] for a proof):

Lemma 1. *There exists a constant $c > 0$ such that*

$$c\|\nabla\mathbf{u}\|_{L^2(\Omega)} \leq \|\varepsilon(\mathbf{u})\|_{L^2(\Omega)} \leq \|\nabla\mathbf{u}\|_{L^2(\Omega)}, \qquad \forall \mathbf{u} \in (H^1(\Omega))^d, \; \mathbf{u} \perp \ker(a).$$

Here $\|\varepsilon(\mathbf{u})\|_{L^2(\Omega)}^2 = \int_{\Omega} \varepsilon(\mathbf{u}) : \varepsilon(\mathbf{u}) dx$.

We will consider conforming discretizations of Stokes and mixed elasticity equations using finite as well as spectral finite elements, all with discontinuous pressures.

2.1   Finite Element Methods with Discontinuous Pressures

We assume that the domain Ω can be decomposed into N nonoverlapping subdomains Ω_i of characteristic size H forming a hexahedral (quadrilateral) finite element mesh τ_H, which is assumed to be shape regular but not necessarily quasi uniform. This coarse triangulation is further refined into a fine quadrilateral finite element triangulation τ_h of characteristic size h. Among the many choices of mixed finite elements available for Stokes and mixed elasticity equations, we consider the following:

a) $Q_2(h) - Q_0(h)$ mixed finite elements: the displacement space $\mathbf{V}$ is discretized by continuous, piecewise bi-quadratic displacements:

$$\mathbf{V}^h = \{\mathbf{v} \in \mathbf{V} : v_k|_T \in Q_2(T) \; \forall T \in \tau_h, \; k = 1, 2, \ldots, d\},$$

while the pressure space is discretized by discontinuous piecewise constant functions on τ_h

$$U^h = \{q \in U : q|_T \in Q_0(T) \ \forall T \in \tau_h\}\,.$$

These elements satisfy the uniform inf-sup condition

$$\sup_{\mathbf{v} \in \mathbf{V}^h} \frac{(\mathrm{div}\,\mathbf{v}, q)}{\|\mathbf{v}\|_{H^1}} \geq \beta_h \|q\|_{L^2} \quad \forall q \in U^h, \tag{5}$$

with $\beta_h \geq c > 0$ independent of h, but they lead to nonoptimal error estimates; see Brezzi and Fortin [8, chap. VI.4, p. 221].

b) $Q_2(h) - P_1(h)$ mixed finite elements: the displacement space is as before, while the pressure space consists of piecewise linear discontinuous pressures:

$$U^h = \{q \in U : q|_T \in P_1(T) \ \forall T \in \tau_h\}\,.$$

These elements satisfy a uniform inf-sup condition (5) as well; there are also optimal $O(h^2)$ error estimates for both displacements and pressures; see Brezzi and Fortin [8, chap. VI, p. 216].

We note that while finite element methods based on hexahedra and quadrilaterals enjoy popularity, our theory applies equally well to stable mixed methods based on triangles or tetrahedra.

2.2   Spectral Element Methods: $Q_n - Q_{n-2}$

Let Ω_{ref} be the reference square or cube $(-1,1)^d, d = 2,3$, and let $Q_n(\Omega_{\mathrm{ref}})$ be the set of polynomials on Ω_{ref} of degree n in each variable. We assume that the domain Ω can be decomposed into N nonoverlapping finite elements Ω_i, each of which is an image $\Omega_i = \phi_i(\Omega_{\mathrm{ref}})$, with ϕ_i an affine mapping. $\mathbf{V}$ is discretized, component by component, by continuous, piecewise tensor product polynomials of degree n:

$$\mathbf{V}^n = \{\mathbf{v} \in \mathbf{V} : v_k|_{\Omega_i} \circ \phi_i \in Q_n(\Omega_{\mathrm{ref}}),\ i = 1, 2, \ldots, N,\ k = 1, 2, \ldots, d\}\,.$$

The pressure space is discretized by piecewise tensor product polynomials of degree $n - 2$, which are discontinuous across the boundaries of the elements Ω_i:

$$U^n = \{q \in U : q|_{\Omega_i} \circ \phi_i \in Q_{n-2}(\Omega_{\mathrm{ref}}),\ i = 1, 2, \ldots, N\}\,.$$

We use Gauss-Lobatto-Legendre (GLL(n)) quadrature in the implementation, which also allows for the construction of a very convenient nodal tensor-product basis for $\mathbf{V}^n$. Denote by $\{\xi_i\}_{i=0}^n$ the set of GLL(n) points of $[-1,1]$, and by σ_i the quadrature weight associated with ξ_i. Let $l_i(x)$ be the Lagrange interpolating polynomial of degree n that vanishes at all the GLL(n) nodes except at ξ_i, where it equals 1. Each element of $Q_n(\Omega_{\mathrm{ref}})$ is expanded in the

GLL(n) basis, and each L^2 inner product of two scalar components u and v is replaced, in the three-dimensional case, by

$$(u,v)_{n,\Omega} = \sum_{s=1}^{N} \sum_{i,j,k=0}^{n} (u \circ \phi_s)(\xi_i,\xi_j,\xi_k)(v \circ \phi_s)(\xi_i,\xi_j,\xi_k)|J_s|\sigma_i\sigma_j\sigma_k \, ,$$

where $|J_s|$ is the determinant of the Jacobian of ϕ_s. The mass matrix based on these basis elements and GLL(n) quadrature is diagonal. Similarly, a very convenient basis for U^n consists of the tensor-product Lagrangian nodal basis functions associated with the internal GLL(n) nodes, i.e., the endpoints -1 and $+1$ are excluded. We will call these the *pressure GLL(n) nodes*.

The $Q_n - Q_{n-2}$ method satisfies a nonuniform inf-sup condition

$$\sup_{\mathbf{v} \in \mathbf{V}^n} \frac{(\mathrm{div}\mathbf{v},q)}{\|\mathbf{v}\|_{H^1}} \geq \beta_n \|q\|_{L^2} \quad \forall q \in U^n, \tag{6}$$

where $\beta_n = Cn^{-(d-1)/2}$, $d = 2,3$, and the constant C is independent of n and q; see Maday, Meiron, Patera, and Rønquist [21] and Stenberg and Suri [34]. However, numerical experiments, reported in [21], have also shown that for practical values of n, e.g., $n \leq 16$, the inf-sup constant β_n of the $Q_n - Q_{n-2}$ method decays much slower than what would first be expected from the theoretical bound.

An alternative, with a uniform bound on the inf-sup constant, is provided by the $Q_n - P_{n-1}$ method; see Bernardi and Maday [6]. However, this pressure space is less convenient than Q_{n-2} as far as implementation is concerned.

2.3   The Discrete System

Let $\widetilde{\mathbf{V}}$ and $\widetilde{U}$ be the discrete displacement and pressure spaces. In the finite element case, we write $\widetilde{\mathbf{V}} \times \widetilde{U} = \mathbf{V}^h \times U^h$, while in the spectral element case we have $\widetilde{\mathbf{V}} \times \widetilde{U} = \mathbf{V}^n \times U^n$. The discrete system obtained from (3) using finite or spectral elements is: find $\mathbf{u} \in \widetilde{\mathbf{V}}$ and $p \in \widetilde{U}$ such that

$$\begin{cases} a(\mathbf{u},\mathbf{v}) + b(\mathbf{v},p) = \mathbf{F}(\mathbf{v}) \; \forall \mathbf{v} \in \widetilde{\mathbf{V}} \\[2mm] b(\mathbf{u},q) - 1/\lambda \, c(p,q) = 0 \quad \forall q \in \widetilde{U}, \end{cases} \tag{7}$$

where we denote with the same letters the bilinear forms obtained using the appropriate quadrature rule described above. In matrix form, we have

$$K \begin{bmatrix} \mathbf{u} \\ p \end{bmatrix} = \begin{bmatrix} A & B^T \\ B & -1/\lambda \, C \end{bmatrix} \begin{bmatrix} \mathbf{u} \\ p \end{bmatrix} = \begin{bmatrix} \mathbf{b} \\ 0 \end{bmatrix} .$$

On the *benign* subspace

$$(\widetilde{\mathbf{V}} \times \widetilde{U})_B = \{(\mathbf{u},p) \in \widetilde{\mathbf{V}} \times \widetilde{U} : b(\mathbf{u},q) - 1/\lambda \, c(p,q) = 0 \; \forall q \in \widetilde{U}\}, \tag{8}$$

problem (7) is equivalent to the positive definite problem: find $(\mathbf{u}, p) \in (\widetilde{\mathbf{V}} \times \widetilde{U})_B$ such that

$$a(\mathbf{u}, \mathbf{v}) + 1/\lambda \, c(p, q) = \mathbf{F}(\mathbf{v}) \quad \forall (\mathbf{v}, q) \in (\widetilde{\mathbf{V}} \times \widetilde{U})_B .$$

3 Substructuring for Saddle Point Problems

The domain Ω is decomposed into open, nonoverlapping hexahedral (quadrilateral) subdomains Ω_i and the interface Γ, i.e.,

$$\Omega = \cup_{i=1}^{N} \Omega_i \cup \Gamma.$$

Here $\Gamma = \left(\cup_{i=1}^{N} \partial \Omega_i \right) \setminus \partial \Omega$. Each Ω_i typically consists of one, or a few, spectral elements of degree n or of many finite elements. We denote by Γ_h and $\partial \Omega_h$ the set of nodes belonging to the interface Γ and $\partial \Omega$, respectively. The starting point of our algorithm is the implicit elimination of the interior degrees of freedom, i.e., the displacement component that is supported in the open subdomains and what we will call the interior pressure component which has zero average over the individual subdomains. This process, also known as static condensation, is carried out by solving decoupled local saddle point problems on each subdomain Ω_i with Dirichlet boundary conditions for the displacements given on $\partial \Omega_i$. We then obtain a saddle point Schur complement problem for the interface displacements and a constant pressure in each subdomain. This reduced problem will be solved by a preconditioned Krylov space iteration, normally the preconditioned conjugate gradient method.

For simplicity, we will use the same letters to denote both functions and their associated vector representations; the same convention will also be used for linear operators and their associated matrix forms.

In order to eliminate the interior degrees of freedom, we reorder the vector of unknowns as

$$\begin{bmatrix} \mathbf{u}_I \\ p_I \\ \mathbf{u}_\Gamma \\ p_0 \end{bmatrix} \quad \begin{array}{l} \text{interior displacements} \\ \text{interior pressures with zero average} \\ \text{interface displacements} \\ \text{constant pressures in each } \Omega_i. \end{array}$$

Then, after using the same permutation, the discrete system matrix can be written as

$$\begin{bmatrix} K_{II} & K_{\Gamma I}^T \\ & \\ K_{\Gamma I} & K_{\Gamma\Gamma} \end{bmatrix} = \left[\begin{array}{cc|cc} A_{II} & B_{II}^T & A_{\Gamma I}^T & 0 \\ B_{II} & -1/\lambda \, C_{II} & B_{I\Gamma} & 0 \\ \hline A_{\Gamma I} & B_{I\Gamma}^T & A_{\Gamma\Gamma} & B_0^T \\ 0 & 0 & B_0 & 1/\lambda \, C_0 \end{array} \right],$$

where the zero blocks are due to interior displacements having zero flux across the subdomain boundaries and the interior pressure having zero average.

60   Paulo Goldfeld et al.

Eliminating the interior unknowns $\mathbf{u}_I$ and p_I by static condensation, we obtain the saddle point Schur complement system

$$S_\lambda \begin{bmatrix} \mathbf{u}_\Gamma \\ p_0 \end{bmatrix} = \begin{bmatrix} \tilde{\mathbf{b}} \\ 0 \end{bmatrix}, \tag{9}$$

where

$$S_\lambda = K_{\Gamma\Gamma} - K_{\Gamma I} K_{II}^{-1} K_{\Gamma I}^T =$$

$$= \begin{bmatrix} A_{\Gamma\Gamma} & B_0^T \\ B_0 & -1/\lambda\, C_0 \end{bmatrix} - \begin{bmatrix} A_{\Gamma I} & B_{I\Gamma}^T \\ 0 & 0 \end{bmatrix} \begin{bmatrix} A_{II} & B_{II}^T \\ B_{II} & -1/\lambda\, C_{II} \end{bmatrix}^{-1} \begin{bmatrix} A_{\Gamma I}^T & 0 \\ B_{I\Gamma} & 0 \end{bmatrix} =$$

$$= \begin{bmatrix} S_{\Gamma,\lambda} & B_0^T \\ B_0 & -1/\lambda\, C_0 \end{bmatrix},$$

and

$$\begin{bmatrix} \tilde{\mathbf{b}} \\ 0 \end{bmatrix} = \begin{bmatrix} \mathbf{b}_\Gamma \\ 0 \end{bmatrix} - \begin{bmatrix} A_{\Gamma I} & B_{I\Gamma}^T \\ 0 & 0 \end{bmatrix} \begin{bmatrix} A_{II} & B_{II}^T \\ B_{II} & -1/\lambda\, C_{II} \end{bmatrix}^{-1} \begin{bmatrix} \mathbf{b}_I \\ 0 \end{bmatrix}.$$

By using a second permutation that reorders the interior displacements and pressures subdomain by subdomain, we note that K_{II}^{-1} represents the solution of N decoupled saddle point problems, one for each subdomain and all uniquely solvable, with Dirichlet data given on $\partial\Omega_i$

$$K_{II}^{-1} = \begin{bmatrix} {K_{II}^{(1)}}^{-1} & & 0 \\ & \ddots & \\ 0 & & {K_{II}^{(N)}}^{-1} \end{bmatrix}.$$

This is the matrix associated with the discrete extension operator $\mathcal{SH}_\lambda$ described below.

The Schur complement S_λ does not need to be explicitly assembled since only its action $S_\lambda v$ on a vector v is needed in a Krylov iteration. This operation essentially only requires the action of K_{II}^{-1} on a vector, i.e., the solution of N decoupled saddle point problems. In other words, $S_\lambda v$ is computed by subassembling the actions of the subdomain Schur complements $S_\lambda^{(i)}$ defined for Ω_i, by

$$S_\lambda^{(i)} = K_{\Gamma\Gamma}^{(i)} - K_{\Gamma I}^{(i)} (K_{II}^{(i)})^{-1} {K_{\Gamma I}^{(i)}}^T =$$

$$= \begin{bmatrix} A_{\Gamma\Gamma}^{(i)} & {B_0^{(i)}}^T \\ B_0^{(i)} & -1/\lambda\, C_0^{(i)} \end{bmatrix} - \begin{bmatrix} A_{\Gamma I}^{(i)} & {B_{I\Gamma}^{(i)}}^T \\ 0 & 0 \end{bmatrix} \begin{bmatrix} A_{II}^{(i)} & {B_{II}^{(i)}}^T \\ B_{II}^{(i)} & -1/\lambda\, C_{II}^{(i)} \end{bmatrix}^{-1} \begin{bmatrix} {A_{\Gamma I}^{(i)}}^T & 0 \\ B_{I\Gamma}^{(i)} & 0 \end{bmatrix} =$$

$$= \begin{bmatrix} S_{\Gamma,\lambda}^{(i)} & {B_0^{(i)}}^T \\ B_0^{(i)} & -1/\lambda\, C_0^{(i)} \end{bmatrix}.$$

Once $\begin{bmatrix} \mathbf{u}_\Gamma \\ p_0 \end{bmatrix}$ is known, $\begin{bmatrix} \mathbf{u}_I \\ p_I \end{bmatrix}$ can be found by back-substitution,

$$\begin{bmatrix} \mathbf{u}_I \\ p_I \end{bmatrix} = \begin{bmatrix} A_{II} & B_{II}^T \\ B_{II} & -1/\lambda\, C_{II} \end{bmatrix}^{-1} \left(\begin{bmatrix} \mathbf{b}_I \\ 0 \end{bmatrix} - \begin{bmatrix} A_{\Gamma I}^T & 0 \\ B_{I\Gamma} & 0 \end{bmatrix} \begin{bmatrix} \mathbf{u}_\Gamma \\ p_0 \end{bmatrix} \right).$$

The substructuring procedure described in the previous section is associated with the space decomposition

$$\widetilde{\mathbf{V}} \times \widetilde{U} = \oplus_{i=1}^N \mathbf{V}_i \times U_i \oplus \mathbf{V}_\Gamma \times U_0,$$

where the interior spaces are defined as

$$\mathbf{V}_i = \widetilde{\mathbf{V}} \cap H_0^1(\Omega_i), \qquad\qquad U_i = \widetilde{U} \cap L_0^2(\Omega_i),$$

and the spaces of interface displacements and coarse pressures, constant in each subdomain, are defined as

$$\mathbf{V}_\Gamma = \mathcal{SH}_\lambda(\widetilde{\mathbf{V}}_{|\Gamma}) = \{\mathbf{v} \in \widetilde{\mathbf{V}} : \mathbf{v}_{|\Omega_i} = \mathcal{SH}_\lambda(\mathbf{v}_{|\partial\Omega_i}),\ i = 1,\cdots,N\},$$

$$U_0 = \{q \in \widetilde{U} : q_{|\Omega_i} = \text{constant},\ i = 1,\cdots,N\}.$$

Here $\mathcal{SH}_\lambda : \widetilde{\mathbf{V}}_{|\Gamma} \longrightarrow \widetilde{\mathbf{V}}$, is the displacement component of the discrete saddle point harmonic extension operator that maps an interface displacement $\mathbf{u}_\Gamma \in \widetilde{\mathbf{V}}_{|\Gamma}$ onto the solution $\begin{bmatrix} \tilde{\mathbf{u}} \\ \tilde{p} \end{bmatrix}$ of the following homogeneous saddle point problem, defined on each subdomain separately: find $\tilde{\mathbf{u}} \in \widetilde{\mathbf{V}}$ and $\tilde{p} \in \widetilde{U}$ such that on each Ω_i,

$$\begin{cases} a(\tilde{\mathbf{u}}, \mathbf{v}) + b(\mathbf{v}, \tilde{p}) = 0 \ \ \forall \mathbf{v} \in \mathbf{V}_i \\[2mm] b(\tilde{\mathbf{u}}, q) - 1/\lambda\, c(\tilde{p}, q) = 0 \ \ \forall q \in U_i \\[2mm] \tilde{\mathbf{u}} = \mathbf{u}_\Gamma \ \text{on} \ \ \partial\Omega_i\,. \end{cases} \tag{10}$$

If we define the interface inner product by

$$s_\lambda(\mathbf{u}_\Gamma, \mathbf{v}_\Gamma) = a(\mathcal{SH}_\lambda(\mathbf{u}_\Gamma), \mathcal{SH}_\lambda(\mathbf{v}_\Gamma)) = \mathbf{u}_\Gamma^T S_{\Gamma,\lambda} \mathbf{v}_\Gamma,$$

and by $b_0(\mathbf{u}_\Gamma, p_0)$ and $c_0(p_0, q_0)$ the restrictions of the other bilinear forms to the saddle point harmonic extensions and the coarse piecewise constant pressures, then the variational formulation of the saddle point Schur complement problem (9) can be given by: find $\mathbf{u}_\Gamma \in \mathbf{V}_\Gamma$ and $p_0 \in U_0$ such that,

$$\begin{cases} s_\lambda(\mathbf{u}_\Gamma, \mathbf{v}_\Gamma) + b_0(\mathbf{v}_\Gamma, p_0) = \widetilde{\mathbf{F}}(\mathbf{v}_\Gamma) \ \forall \mathbf{v}_\Gamma \in \mathbf{V}_\Gamma \\[2mm] b_0(\mathbf{u}_\Gamma, q_0) - 1/\lambda\, c_0(p_0, q_0) = 0 \ \ \ \ \forall q_0 \in U_0. \end{cases} \tag{11}$$

On the benign subspace $(\mathbf{V}_\Gamma \times U_0)_B$ defined by

$$(\mathbf{V}_\Gamma \times U_0)_B = \{(\mathbf{u}_\Gamma, p_0) \in \mathbf{V}_\Gamma \times U_0 : B_0\mathbf{u}_\Gamma - 1/\lambda\, C_0 p_0 = 0\}$$
$$= \{(\mathbf{u}_\Gamma, p_0) \in \mathbf{V}_\Gamma \times U_0 : b_0(\mathbf{u}_\Gamma, q_0) - 1/\lambda\, c_0(p_0, q_0) = 0\},$$

cf. (8), problem (11) is equivalent to the positive definite problem: find $(\mathbf{u}_\Gamma, p_0) \in (\mathbf{V}_\Gamma \times U_0)_B$ such that

$$s_\lambda(\mathbf{u}_\Gamma, \mathbf{v}_\Gamma) + 1/\lambda c_0(p_0, q_0) = \widetilde{\mathbf{F}}(\mathbf{v}_\Gamma) \quad \forall(\mathbf{v}_\Gamma, q_0) \in (\mathbf{V}_\Gamma \times U_0)_B. \tag{12}$$

4   A Neumann-Neumann Preconditioner

We will solve the saddle point Schur complement problem

$$S_\lambda \begin{bmatrix} \mathbf{u}_\Gamma \\ p_0 \end{bmatrix} = \begin{bmatrix} S_{\Gamma,\lambda} & B_0^T \\ B_0 & -1/\lambda\, C_0 \end{bmatrix} \begin{bmatrix} \mathbf{u}_\Gamma \\ p_0 \end{bmatrix} = \begin{bmatrix} \widetilde{\mathbf{b}} \\ 0 \end{bmatrix} \tag{13}$$

by a preconditioned Krylov space method such as GMRES or PCG. The latter can be applied to this indefinite problem because we will start and keep the iterates in the subspace of benign functions.

The matrix form of the preconditioner is

$$Q_\lambda = Q_H + (I - Q_H S_\lambda) \sum_{i=1}^{N} Q_i(I - S_\lambda Q_H), \tag{14}$$

where the coarse operator Q_H and local operators Q_i are defined below. The preconditioned operator is then

$$T_\lambda = Q_\lambda S_\lambda = T_0 + (I - T_0) \sum_{i=1}^{N} T_i(I - T_0), \tag{15}$$

where $T_0 = Q_H S_\lambda$ and $T_i = Q_i S_\lambda$. The operators Q_H, Q_i, T_0, and T_i also depend on λ (and μ) but we leave them without subscripts in order to keep the notation simple. Q_λ can also be written as a three-step preconditioner as in [28]. In order to keep the notation simple, we will use the same symbol (e.g., $\mathbf{v}_\Gamma$) for both the interface vector and the function of $\mathbf{V}_\Gamma$ obtained by extension inside each subdomain using the discrete saddle point harmonic extension operator $\mathcal{SH}_\lambda$. In addition, we will avoid writing explicitly finite or spectral element interpolants; therefore, when writing a product of functions (for example $\delta_i \mathbf{v}_\Gamma$) we mean the finite or spectral element function with nodal values equal to the product of those of the two functions.

The balancing Neumann-Neumann preconditioner T_λ is associated with further decomposing the interface space $\mathbf{V}_\Gamma \times U_0$ as

$$\mathbf{V}_\Gamma \times U_0 = \mathbf{V}_0 \times U_0 + \sum_{i=1}^{N} \mathbf{V}_{\Gamma_i} \times U_{0,i}.$$

Here, the coarse displacement space $\mathbf{V}_0$ is defined in terms of special functions $\delta_i^\dagger$ introduced below and is given by either one of the three following choices:

$\mathbf{V}_0^0 = \{\mathbf{v} \in \mathbf{V}_\Gamma : \mathbf{v} \in span\{\delta_i^\dagger\}$ *times the functions spanning* $\ker(a)\}$,
$\mathbf{V}_0^1 = \mathbf{V}_0^0 \cup span\{$*normal direction quadratic edge/face bubble functions*$\}$,
$\mathbf{V}_0^2 = \mathbf{V}_0^0 \cup span\{$*bi- or tri-linear coarse piecewise* Q_1 *functions*$\}$,

while the local spaces are defined by:

$$\mathbf{V}_{\Gamma,i} = \{\mathbf{v} \in \mathbf{V}_\Gamma : \quad \mathbf{v}(\mathbf{x}) = 0 \;\; \forall x \in \Gamma_h \setminus \partial\Omega_{i,h}\}, \qquad U_{0,i} = span\{1\}.$$

We could also consider richer coarse spaces obtained, e.g., by adding to $\mathbf{V}_0^0$ functions of $\mathbf{V}_\Gamma$ that are piecewise tri- or bi-quadratic polynomials on Γ, as we did in our study [28] of the Stokes case.

We now describe the coarse and local problems in more detail.

Coarse problem: Given a residual vector r, the coarse term $Q_H r$ is the solution of a coarse, global saddle point problem with a few displacement degrees of freedom and one constant pressure per subdomain Ω_i:

$$Q_H = R_H^T S_{0,\lambda}^{-1} R_H,$$

where

$$R_H = \begin{bmatrix} L_0^T & 0 \\ 0 & I \end{bmatrix},$$

and

$$S_{0,\lambda} = R_H S_\lambda R_H^T = \begin{bmatrix} L_0^T S_{\Gamma,\lambda} L_0 & L_0^T B_0^T \\ B_0 L_0 & -1/\lambda\, C_0 \end{bmatrix}. \tag{16}$$

The columns of the matrix L_0 span the coarse space $\mathbf{V}_0$ and in order to define them, we need to define the Neumann-Neumann counting functions $\delta_i \in \mathbf{V}_\Gamma$ associated with each subdomain Ω_i and their pseudo inverses $\delta_i^\dagger$:

- δ_i is zero at the interface nodes outside $\partial\Omega_{i,h}$ while its value at any node on $\partial\Omega_{i,h}$ equals the number of subdomains shared by that node;

- the pseudo inverse $\delta_i^\dagger$ is the function $1/\delta_i(x)$ for all nodes where $\delta_i(x) \neq 0$, and it vanishes at all other points of $\Gamma_h \cup \partial\Omega_h$.

Then, the columns of L_0 are defined by one of the following three choices:

0) the inverse counting functions $\delta_i^\dagger$ multiplied by the functions of $\ker(a)$ (the associated coarse space is $\mathbf{V}_0^0$);

1) as in 0) with the addition of the quadratic coarse edge/face bubble functions for the normal direction (the associated coarse space is $\mathbf{V}_0^1$);

2) as in 0) with the addition of the continuous piecewise bi- or tri-linear functions on the coarse mesh τ_H (the associated coarse space is $\mathbf{V}_0^2$).

Choice 0) corresponds to the standard choice for second order scalar elliptic problems and it provides a quite minimal coarse displacement space. It turns out to be far from uniformly inf-sup stable and therefore it leads to a non-scalable algorithm in the incompressible case. However, in the compressible case where λ/μ is bounded, it still leads to a scalable algorithm; see our main theorem and the numerical results. Choices 1) and 2) are enrichments of 0) that turn out to be inf-sup stable uniformly in N and λ/μ.

In order to avoid linearly dependent $\delta_i^\dagger$ functions, and hence a singular coarse space problem, we might have to drop all of the components of these functions for one subdomain, depending on the coarse triangulation.

In variational terms, the coarse problem is defined as follows: Given $\begin{bmatrix} \mathbf{u}_\Gamma \\ p_0 \end{bmatrix} \in \mathbf{V}_\Gamma \times U_0$, define $\begin{bmatrix} \mathbf{w}_\Gamma \\ q_0 \end{bmatrix} = T_0 \begin{bmatrix} \mathbf{u}_\Gamma \\ p_0 \end{bmatrix} \in \mathbf{V}_0 \times U_0$ as the solution of the coarse saddle point problem

$$\begin{cases} s_\lambda(\mathbf{w}_\Gamma, \mathbf{v}) + \quad b_0(\mathbf{v}, q_0) \quad = s_\lambda(\mathbf{u}_\Gamma, \mathbf{v}) + \quad b_0(\mathbf{v}, p_0) \quad \forall \mathbf{v} \in \mathbf{V}_0 \\ b_0(\mathbf{w}_\Gamma, q) - 1/\lambda \, c_0(q_0, q) = b_0(\mathbf{u}_\Gamma, q) - 1/\lambda \, c_0(p_0, q) \quad \forall q \in U_0. \end{cases}$$ (17)

i.e.,

$$\begin{cases} s_\lambda(\mathbf{w}_\Gamma - \mathbf{u}_\Gamma, \mathbf{v}) + \quad b_0(\mathbf{v}, q_0 - p_0) \quad = 0 \quad \forall \mathbf{v} \in \mathbf{V}_0 \\ b_0(\mathbf{w}_\Gamma - \mathbf{u}_\Gamma, q) - 1/\lambda \, c_0(q_0 - p_0, q) = 0 \quad \forall q \in U_0. \end{cases}$$ (18)

It can be established that the coarse space correction is independent of p_0 and we can therefore drop the terms containing this quantity:

$$(I - T_0) \begin{bmatrix} \mathbf{u}_\Gamma \\ p_0 \end{bmatrix} = (I - T_0) \begin{bmatrix} \mathbf{u}_\Gamma \\ 0 \end{bmatrix} = \begin{bmatrix} (I - T_0^u)\,\mathbf{u}_\Gamma \\ -T_0^p \mathbf{u}_\Gamma \end{bmatrix}.$$

Local problems: The local operators Q_i will only be applied to residuals of benign displacement fields and thus the second residual component will vanish. It is also possible to show that the pressure components obtained in this step of the preconditioner plays no further role when we next apply the operator $(I - T_0)$. Each local operator Q_i is based on the solution of a local saddle point problem on Ω_i with natural boundary condition, given below. This local problem is singular for any subdomain Ω_i the boundary of which does not intersect the Dirichlet boundary Γ_0; the rigid body motions are in the null space. Such a subregion is called a *floating* subdomain. To avoid possible complications with singular problems, we modify the local saddle point problems on the floating subdomains by adding ε times the displacement mass matrix to the local stiffness matrix $K^{(i)}$. We could also make these solutions unique by requiring that each displacement component is orthogonal to the nullspace of $a(\cdot, \cdot)$ over Ω_i; the right hand sides will always be compatible.

Given a residual vector r with a first component r_Γ and a zero second component, $Q_i r$ is the weighted solution of a local saddle point problem on subdomain Ω_i with a natural boundary condition on $\partial\Omega_i \setminus \Gamma_0$:

$$Q_i r = \begin{bmatrix} R_i^T D_i^{-1} & 0 \\ 0 & 0 \end{bmatrix} \begin{bmatrix} S_{\Gamma,\varepsilon,\lambda}^{(i)} & B_0^{(i)^T} \\ B_0^{(i)} & -1/\lambda\, C_0^{(i)} \end{bmatrix}^{-1} \begin{bmatrix} D_i^{-1} R_i & 0 \\ 0 & 0 \end{bmatrix} \begin{bmatrix} r_\Gamma \\ 0 \end{bmatrix}. \tag{19}$$

Here R_i are $0, 1$ restriction matrices mapping r_Γ into r_{Γ_i} and D_i are diagonal matrices representing multiplication by the counting functions δ_i. Moreover,

$$S_\varepsilon^{(i)} = \begin{bmatrix} S_{\Gamma,\varepsilon,\lambda}^{(i)} & B_0^{(i)^T} \\ B_0^{(i)} & -1/\lambda\, C_0^{(i)} \end{bmatrix}$$

is the local saddle point Schur complement, associated with the subdomain Ω_i, of the regularized local stiffness matrix

$$K_\varepsilon^{(i)} = \begin{bmatrix} A_{II,\varepsilon}^{(i)} & B_{II}^{(i)^T} & A_{\Gamma I,\varepsilon}^{(i)^T} & 0 \\ B_{II}^{(i)} & -1/\lambda\, C_{II}^{(i)} & B_{I\Gamma}^{(i)} & 0 \\ A_{\Gamma I,\varepsilon}^{(i)} & B_{I\Gamma}^{(i)^T} & A_{\Gamma\Gamma,\varepsilon}^{(i)} & B_0^{(i)^T} \\ 0 & 0 & B_0^{(i)} & -1/\lambda\, C_0^{(i)} \end{bmatrix},$$

where

$$A_\varepsilon^{(i)} = A^{(i)} + \varepsilon M^{(i)}.$$

Here $M^{(i)}$ is the local displacement mass matrix.

In variational form, the local problems are defined as follows. Since we need to apply local solvers only to elements $\mathbf{w} = \begin{bmatrix} \mathbf{u}_\Gamma \\ p_0 \end{bmatrix}$ of $Range(I - T_0)$ and $(I - T_0)$ is a projection on its range, we have $\mathbf{w} = (I - T_0) \begin{bmatrix} \mathbf{u}_\Gamma \\ p_0 \end{bmatrix} = (I - T_0) \begin{bmatrix} \mathbf{u}_\Gamma \\ 0 \end{bmatrix}$. By construction, the pair $\mathbf{u}_\Gamma = (I - T_0^u)\mathbf{u}_\Gamma$ and $p_0 = -T_0^p \mathbf{u}_\Gamma$ is benign and we can define $T_i \mathbf{w} = \begin{bmatrix} T_i^u \mathbf{u}_\Gamma \\ T_i^p \mathbf{u}_\Gamma \end{bmatrix} \in \mathbf{V}_{\Gamma,i} \times U_{0,i}$ as the solution of a local saddle point problem with natural boundary conditions: $\forall \mathbf{v}_i \in \mathbf{V}_{\Gamma,i}, \forall q_i \in U_{0,i}$,

$$\begin{cases} s_{\lambda,\varepsilon,i}(\delta_i T_i^u \mathbf{u}_\Gamma, \delta_i \mathbf{v}_i) + b_{0,i}(\delta_i \mathbf{v}_i, T_i^p \mathbf{u}_\Gamma) = s_\lambda(\mathbf{u}_\Gamma, \mathbf{v}_i) + b_0(\mathbf{v}_i, p_0) \\[2mm] b_{0,i}(\delta_i T_i^u \mathbf{u}_\Gamma, q_i) - 1/\lambda\, c_{0,i}(T_i^p \mathbf{u}_\Gamma, q_i) = b_0(\mathbf{u}_\Gamma, q_i) - 1/\lambda\, c_0(p_0, q_i) = 0. \end{cases} \tag{20}$$

Here,

$$s_{\lambda,\varepsilon,i}(\mathbf{u}_\Gamma, \mathbf{v}_\Gamma) = a_{\varepsilon,i}(\mathcal{SH}_{i,\varepsilon,\lambda}(\mathbf{u}_\Gamma), \mathcal{SH}_{\lambda,\varepsilon,i}(\mathbf{v}_\Gamma)),$$

and

$$a_{\varepsilon,i}(\mathbf{u}, \mathbf{v}) = a_i(\mathbf{u}, \mathbf{v}) + \varepsilon \int_{\Omega_i} \mathbf{u} \cdot \mathbf{v}\, dx, \tag{21}$$

where $\mathcal{SH}_{\lambda,\varepsilon,i}$ is the displacement component of the discrete saddle point extension operator defined in terms of the regularized $a_{\varepsilon,i}(\cdot,\cdot)$ displacement bilinear form instead of the standard $a(\cdot,\cdot)$ form.

5   Main Result

On the benign subspace $(\mathbf{V}_\Gamma \times U_0)_B$, the bilinear form defined by S_λ

$$\left\langle \begin{bmatrix} \mathbf{u}_\Gamma \\ p_0 \end{bmatrix}, \begin{bmatrix} \mathbf{v}_\Gamma \\ q_0 \end{bmatrix} \right\rangle_{S_\lambda} = \left\langle S_\lambda \begin{bmatrix} \mathbf{u}_\Gamma \\ p_0 \end{bmatrix}, \begin{bmatrix} \mathbf{v}_\Gamma \\ q_0 \end{bmatrix} \right\rangle,$$

coincides with the $\lambda-$inner product

$$\left\langle \begin{bmatrix} \mathbf{u}_\Gamma \\ p_0 \end{bmatrix}, \begin{bmatrix} \mathbf{v}_\Gamma \\ q_0 \end{bmatrix} \right\rangle_\lambda = s_\lambda(\mathbf{u}_\Gamma, \mathbf{v}_\Gamma) + 1/\lambda\, c_0(p_0, q_0),$$

since $B_0 \mathbf{u}_\Gamma - 1/\lambda\, C_0 p_0 = B_0 \mathbf{v}_\Gamma - 1/\lambda\, C_0 q_0 = 0$.

Theorem 1. *On the benign subspace* $(\mathbf{V}_\Gamma \times U_0)_B$, *cf. (8), the balancing Neumann-Neumann operator* T_λ *is symmetric positive definite with respect to the* $\lambda-$*inner product and*

$$cond(T_\lambda) \le C\Big(\frac{1}{\sqrt{\beta_0^2 + \frac{\mu}{\lambda}}}\Big)\Big(\frac{1}{\beta^2 + \frac{\mu}{\lambda}}\Big)\, \alpha,$$

where

$$\alpha = \begin{cases} (1 + \log(H/h))^2 & \textit{for finite elements} \\[2mm] (1 + \log n)^2 & \textit{for spectral elements,} \end{cases}$$

and β_0 *and* β *are the inf-sup constants of the coarse problem and the original discrete saddle point problem, respectively.*

6   Variable Coefficient and Composite Materials

Our algorithm can be extended to composite materials with different Lamé constants λ_i, μ_i in each subdomain Ω_i:

$$\begin{cases} 2\sum_{i=1}^N \int_{\Omega_i} \mu_i\, \varepsilon(\mathbf{u}) : \varepsilon(\mathbf{v})\, dx - \int_\Omega \operatorname{div}\mathbf{v}\, p\, dx = <\mathbf{F}, \mathbf{v}> \forall \mathbf{v} \in \mathbf{V} \\[4mm] -\int_\Omega \operatorname{div}\mathbf{u}\, q\, dx - \sum_{i=1}^N \int_{\Omega_i} 1/\lambda_i\, pq\, dx = 0 \qquad \forall q \in U; \end{cases}$$

$$(22)$$

Using the convention of padding local vectors by zeros, when they are needed as global vectors, the discrete problem can now be written as

$$K \begin{bmatrix} \mathbf{u} \\ p \end{bmatrix} = \sum_{i=1}^N \begin{bmatrix} A^{(i)} & B^{(i)^T} \\ B^{(i)} & -1/\lambda_i\, C^{(i)} \end{bmatrix} \begin{bmatrix} \mathbf{u}^{(i)} \\ p^{(i)} \end{bmatrix},$$

and the saddle point Schur complement obtained by static condensation as

$$ S \begin{bmatrix} \mathbf{u}_\Gamma \\ p_0 \end{bmatrix} = \sum_{i=1}^N \begin{bmatrix} S_\Gamma^{(i)} & B_0^{(i)^T} \\ B_0^{(i)} & -1/\lambda_i \, C_0^{(i)} \end{bmatrix} \begin{bmatrix} \mathbf{u}_\Gamma^{(i)} \\ p_0^{(i)} \end{bmatrix}. $$

The balancing Neumann-Neumann preconditioner Q for S has the same matrix form as before, but with modified local and coarse spaces. As in the scalar elliptic case, the jumps in the coefficients μ_i are taken care of by appropriately scaling the special counting functions δ_i and their pseudo inverses $\delta_i^\dagger$. At any node x on $\partial\Omega_i$, we use the definition:

$$ \delta_i^\dagger(x) = \frac{\mu_i^\gamma(x)}{\sum_{j \in N_x} \mu_j^\gamma(x)}, $$

where $\gamma \in [1/2, \infty)$ and N_x is the set of indices of all the subdomains that have x on their boundaries. Both functions δ_i and $\delta_i^\dagger$ vanish at each interface node outside $\partial\Omega_{i,h}$ and are extended inside each subdomain as discrete saddle point harmonic extensions. The pseudo inverses $\delta_i^\dagger$ still form a partition of unity.

The local and coarse problems are then defined formally as before but using these modified functions δ_i and $\delta_i^\dagger$.

7 Numerical Experiments

We report on two sets of numerical experiments.

7.1 Parallel Results with $Q_2 - Q_0$ Finite Elements

We report here on some results of parallel numerical experiments on the Beowulf cluster Chiba City at Argonne National Laboratory (with 256 Dual Pentium III processors). The algorithm has been implemented by Paulo Goldfeld in C, using the PETSc library; see [4], [5]. We report on results for the incompressible Stokes equation and for compressible elasticity only, although similar results have been obtained for generalized Stokes and incompressible elasticity problems. The domain considered is the unit square and the boundary conditions are of Dirichlet type and consistent with the incompressibility condition, when so required. The problem is discretized with $Q_2 - Q_0$ finite elements and the domain Ω divided into $\sqrt{N} \times \sqrt{N}$ square subdomains. The saddle point Schur complement (13) is solved iteratively by PCG with our balancing Neumann-Neumann preconditioner and the third choice of coarse space $\mathbf{V}_0^2 = \{\text{scaled rigid body motions}\} + Q_1^H$. The initial guess is a random vector modified so that the initial error is in the range of $(I - T_0)$, the right hand side is a random, uniformly distributed vector projected onto the range of S, and the stopping criterion is $\|r_k\|_2/\|r_0\|_2 \leq 10^{-6}$, where r_k is the residual at the k-th iterate.

Stokes problem on a homogeneous medium. We consider first the incompressible Stokes problem with constant coefficients, which is closely related to mixed elasticity models. In the upper half of Table 1, we show the results for increasing mesh sizes, always partitioned into 64 subdomains.

Table 1. Parallel results for Stokes system (homogeneous medium) and $Q_2 - Q_0$ finite elements: PCG iteration counts and extremal eigenvalues of T_λ for the balancing Neumann-Neumann preconditioner with coarse space $\mathbf{V}_0^2$

Fixed number of subdomains $N = 8 \times 8$					
mesh size	local size	# unknowns	iterations	eig min	eig max
80×80	10×10	58,243	15	1.04	5.93
160×160	20×20	231,683	18	1.06	8.31
240×240	30×30	520,323	20	1.06	9.88
320×320	40×40	924,163	21	1.07	11.08
400×400	50×50	1,443,203	22	1.07	12.06
480×480	60×60	2,077,443	22	1.07	12.89

Fixed local size 60×60 elements (32,883 unknowns)					
mesh size	# subdomains	# unknowns	iterations	eig min	eig max
120×120	2×2	130,563	14	1.07	9.83
180×180	3×3	293,043	19	1.07	10.06
240×240	4×4	520,323	20	1.07	11.36
360×360	6×6	1,169,283	22	1.07	12.43
480×480	8×8	2,077,443	22	1.07	12.89
720×720	12×12	4,671,363	23	1.07	13.32
840×840	14×14	6,357,123	23	1.07	13.44

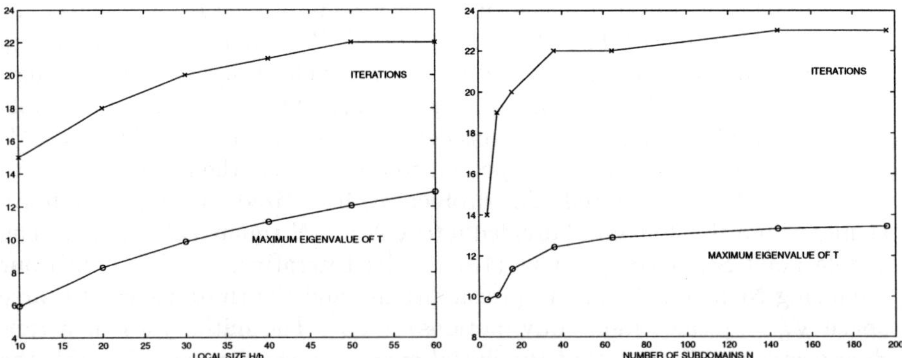

Fig. 1. Parallel results for Stokes system (homogeneous medium) and $Q_2 - Q_0$ finite elements: PCG iteration counts and maximum eigenvalue of T_λ vs. local size H/h (left) and number of subdomains N (right), from Table 1

The condition number and the iteration count grow weakly as we increase the size of the local problems, as can also be observed in the left part of Figure 7.1. The lower part of Table 1 shows results for an increasing number of subdomains of a fixed size. The corresponding graph, on the right in Figure 7.1, shows an almost horizontal tail, indicating independence of the condition number and the iteration count on the number of subdomains. These results are in perfect accordance with the theory, since in the finite element case we have inf-sup stability of both the original and coarse problems and therefore both inf-sup constants β and β_0 are bounded away from zero independently of h and H.

Elasticity problem on a heterogeneous medium. Here we consider an elasticity problem defined in a heterogeneous medium, which is composed of an arrangement of three different materials in the following pattern:

$$
\begin{array}{|c|c|c|c|c|c|c|}
\hline
s & r & s & r & \cdots & s & r \\
\hline
r & a & r & a & \cdots & r & a \\
\hline
s & r & s & r & \cdots & s & r \\
\hline
r & a & r & a & \cdots & r & a \\
\hline
\vdots & \vdots & \vdots & \vdots & \ddots & \vdots & \vdots \\
\hline
s & r & s & r & \cdots & s & r \\
\hline
r & a & r & a & \cdots & r & a \\
\hline
\end{array} \, ,
$$

where

$$
\begin{array}{llll}
s = \text{steel-like:} & \mu_s = 8.20 & \lambda_s = 10.00 & \nu_s = 0.275 \\
a = \text{aluminium-like:} & \mu_a = 2.60 & \lambda_a = 5.60 & \nu_a = 0.341 \\
r = \text{rubber-like:} & \mu_r = 0.01 & \lambda_r = 0.99 & \nu_r = 0.495
\end{array}
$$

Note that the material r is almost incompressible, with a Poisson ratio close to 0.5.

As in the previous example, we show, in the upper half of Table 2, the results for meshes of increasing sizes partitioned into 64 subdomains. Again, the condition number and the iteration count grow slowly with the size of the local problem, as in the homogeneous case; see also the left part of Figure 7.1. The last two columns of this table display CPU-time for these runs. The last column gives the total time for the code to run, while the column labeled "fact." gives the time spent on LU factorizations; there are three of them: two local, namely a Dirichlet and a Neumann subdomain-level problem, and one global coarse problem. We note that the cost of the factorizations grows rapidly and dominates the cost of the computation. The lower part of Table 2 shows results for an increasing number of subdomains of fixed size (about 58 thousand dofs). The corresponding graph is on the right in Figure 7.1. Again, we observe no influence of the number of subdomains on the condition number or iteration count. This is numerical evidence that our main result on Section 5 remains valid in the case of discontinuous coefficients. The fact that the factorization time remained constant for the entire range of problem sizes tested (from 16 to 169 subdomains) indicates that the cost associated with the factorization of the coarse problem is still tiny compared with that of the local problems. One can expect this scenario to change if the number of subdomains increases significantly.

Table 2. Parallel results for elasticity system (heterogeneous medium) and $Q_2 - Q_0$ finite elements: PCG iteration counts and extremal eigenvalues of T_λ for the balancing Neumann-Neumann preconditioner with coarse space $\mathbf{V}_0^2$

		Fixed number of subdomains $N = 8 \times 8$			CPU time (sec.)	
mesh size	local size	# unkn.	iter.	eig max	fact.	total
160×160	20×20	230,000	12	4.06	1.4	18.0
320×320	40×40	920,000	13	4.65	18.2	40.9
480×480	60×60	2,080,000	14	4.99	84.2	126.3
640×640	80×80	3,690,000	14	5.22	260.8	345.3

		Fixed local size 80×80 elements (58,242 unknowns)			CPU time (sec.)	
mesh size	# subdom.	# unkn.	iter.	eig max	fact.	total
320×320	4×4	920,000	12	5.18	258.0	321.4
480×480	6×6	2,080,000	13	5.21	253.7	317.4
640×640	8×8	3,690,000	14	5.22	260.8	345.3
800×800	10×10	5,770,000	14	5.14	262.8	356.7
1040×1040	13×13	9,740,000	14	4.93	261.2	363.9

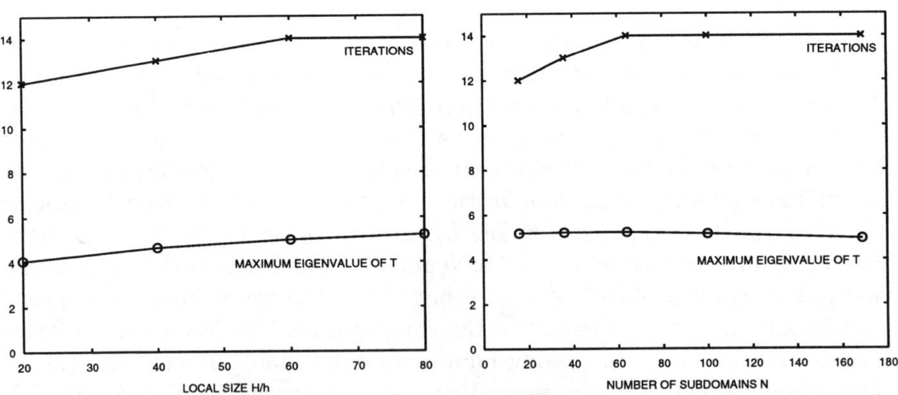

Fig. 2. Parallel results for elasticity system (heterogeneous medium) and $Q_2 - Q_0$ finite elements: PCG iteration counts and maximum eigenvalue of T_λ vs. local size H/h (left) and number of subdomains N (right), from Table 2

7.2   Serial Results for $Q_n - Q_{n-2}$ Spectral Elements

We report here on some results of serial numerical experiments, carried out in Matlab 5.3 on Unix workstations, for elasticity problems, this time discretized with $Q_n - Q_{n-2}$ spectral elements. Again, the saddle point Schur complement (13) is solved iteratively by PCG with our balancing Neumann-Neumann preconditioner. The initial guess is always zero, the right hand side is random and uniformly distributed, and the stopping criterion is as before.

Homogeneous materials. We consider first the case of homogeneous materials with fixed Lamé constants over the whole domain Ω. We report the results in two tables corresponding to two of the coarse spaces introduced in Section 4, $\mathbf{V}_0^0$ in Table 3 and $\mathbf{V}_0^1$ in Table 4. Some results are also plotted in Figures 7.2 and 7.2, respectively. In the upper half of each table the number of subdomains, $N = 3 \times 3$, is fixed, while the spectral degree n is increased from 2 to 10; in the lower half the spectral degree $n = 4$ is fixed and the number of subdomains N is increased from 2×2 to 10×10. Each table reports the PCG iteration counts and, in brackets, the maximum eigenvalue of the preconditioned operator T_λ (the minimum eigenvalue is always very close to 1 and therefore not reported). In each table, we have considered in four different columns the four cases $\nu = 0.3, 0.4, 0.49, 0.5$, ranging from compressible to incompressible materials. The results, in agreement with the theory, show that our balancing Neumann-Neumann algorithm is quasi-optimal (i.e., there is only a weak dependence on the spectral degree n) and scalable (i.e., there is no dependence on the number of subdomains N) for both coarse spaces, except with the first coarse space $\mathbf{V}_0^0$ in the incompressible and almost incompressible case (see the last two columns of Table 3, lower part). This is due to the fact that $\mathbf{V}_0^0$ is not uniformly inf-sup stable with respect to H and therefore β_0 approaches zero with increasing N. In the compressible case, the term μ/λ in Theorem 1 allows us to still obtain an upper bound independent of N and therefore scalability, but in the incompressible case μ/λ vanishes and we lose scalability. On the other hand, the use of the inf-sup stable coarse space $\mathbf{V}_0^1$ yields a scalable algorithm independently of the compressibility of the material; see Table 4.

Composite materials. We now consider the case of a composite material occupying a domain Ω divided into 4×4 subdomains and with the following distribution of Lamé coefficients:

$$
\lambda = \begin{array}{|c|c|c|c|}
\hline
1 & \infty & 1 & \infty \\
\hline
\infty & 1 & 10 & 0.1 \\
\hline
1 & 10 & \infty & 1 \\
\hline
0.1 & \infty & 0.1 & 10 \\
\hline
\end{array}, \quad
\mu = \begin{array}{|c|c|c|c|}
\hline
1 & 10^t & 1 & 10^t/3 \\
\hline
3 \cdot 10^t & 1/2 & 10^t/5 & 2 \\
\hline
1 & 10^t/2 & 1/3 & 10^t \\
\hline
10^t & 2/5 & 5 \cdot 10^t & 3 \\
\hline
\end{array},
$$

where the exponent t assumes the values $t = -3, -2, \cdots, 5, 6$. This example does not reflect any physical model, but it is considered in order to illustrate the robustness of our algorithm with respect to the variation of the Lamé coefficients, that are allowed here to have jumps of many orders of magnitude

Table 3. Serial results for elasticity system (homogeneous medium) and $Q_n - Q_{n-2}$ spectral elements: PCG iteration counts and maximum eigenvalue of T_λ (in brackets) for the balancing Neumann-Neumann preconditioner with first coarse space $\mathbf{V}_0^0$

Fixed number of subdomains $N = 3 \times 3$				
spectral degree n	Poisson ratio ν			
	$\nu = 0.3$	$\nu = 0.4$	$\nu = 0.49$	$\nu = 0.5$
3	8 (2.40)	8 (2.38)	9 (2.86)	9 (3.51)
4	9 (2.72)	9 (2.51)	10 (3.51)	10 (4.56)
5	10 (3.43)	10 (3.17)	11 (4.18)	11 (5.84)
6	11 (4.04)	11 (3.69)	12 (4.69)	12 (6.67)
7	12 (4.64)	11 (4.24)	13 (5.11)	13 (7.62)
8	12 (5.21)	12 (4.75)	13 (5.66)	14 (8.44)
9	13 (5.74)	12 (5.24)	14 (5.93)	15 (9.20)
10	13 (6.21)	13 (5.70)	15 (6.56)	15 (10.03)
Fixed spectral degree $n = 4$				
# of subdomains N	Poisson ratio ν			
	$\nu = 0.3$	$\nu = 0.4$	$\nu = 0.49$	$\nu = 0.5$
3×3	9 (2.72)	9 (2.51)	10 (3.51)	10 (4.56)
4×4	10 (2.75)	10 (2.52)	11 (3.92)	12 (5.23)
5×5	10 (2.77)	10 (2.52)	13 (6.07)	14 (10.82)
6×6	10 (2.67)	10 (2.53)	14 (6.20)	15 (11.11)
7×7	11 (2.79)	10 (2.53)	15 (8.05)	17 (20.22)
8×8	11 (2.71)	10 (2.54)	16 (7.98)	19 (19.42)
9×9	11 (2.80)	10 (2.54)	17 (9.40)	21 (32.76)
10×10	11 (2.79)	10 (2.54)	18 (9.24)	22 (30.12)

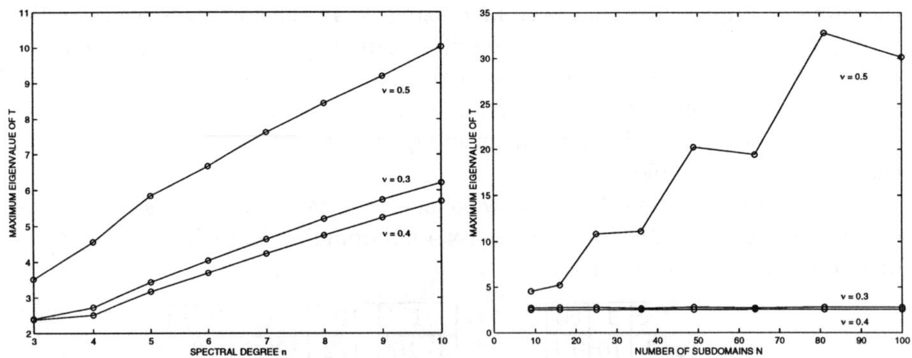

Fig. 3. Serial results for elasticity system (homogeneous medium) and $Q_n - Q_{n-2}$ spectral elements: maximum eigenvalue of T_λ vs. spectral degree n (left) and number of subdomains N (right), from Table 3

Table 4. Serial results for elasticity system (homogeneous medium) and $Q_n - Q_{n-2}$ spectral elements: PCG iteration counts and maximum eigenvalue of T_λ (in brackets) for the balancing Neumann-Neumann preconditioner with second coarse space $\mathbf{V}_0^1$

Fixed number of subdomains $N = 3 \times 3$			
spectral degree n	Poisson ratio ν		
	$\nu = 0.3$	$\nu = 0.4$	$\nu = 0.49$	$\nu = 0.5$
3	8 (2.40)	8 (2.38)	8 (2.40)	9 (2.40)
4	9 (2.71)	9 (2.49)	9 (2.39)	9 (2.39)
5	10 (3.42)	10 (3.14)	10 (2.94)	10 (2.92)
6	11 (4.02)	11 (3.66)	11 (3.45)	11 (3.46)
7	11 (4.62)	11 (4.21)	12 (3.93)	12 (3.94)
8	12 (5.19)	12 (4.72)	13 (4.45)	12 (4.48)
9	13 (5.71)	12 (5.20)	13 (4.86)	13 (4.90)
10	12 (6.23)	13 (5.66)	14 (5.35)	14 (5.41)
Fixed spectral degree $n = 4$			
# of subdomains N	Poisson ratio ν		
	$\nu = 0.3$	$\nu = 0.4$	$\nu = 0.49$	$\nu = 0.5$
9	9 (2.71)	9 (2.49)	9 (2.39)	9 (2.39)
16	10 (2.75)	10 (2.52)	10 (2.45)	10 (2.45)
25	10 (2.76)	10 (2.51)	10 (2.45)	10 (2.46)
36	11 (2.77)	10 (2.51)	10 (2.46)	10 (2.46)
49	10 (2.78)	10 (2.50)	10 (2.42)	10 (2.42)
64	11 (2.75)	10 (2.51)	10 (2.45)	10 (2.45)
81	11 (2.79)	10 (2.51)	10 (2.45)	10 (2.45)
100	11 (2.80)	10 (2.52)	10 (2.45)	10 (2.45)

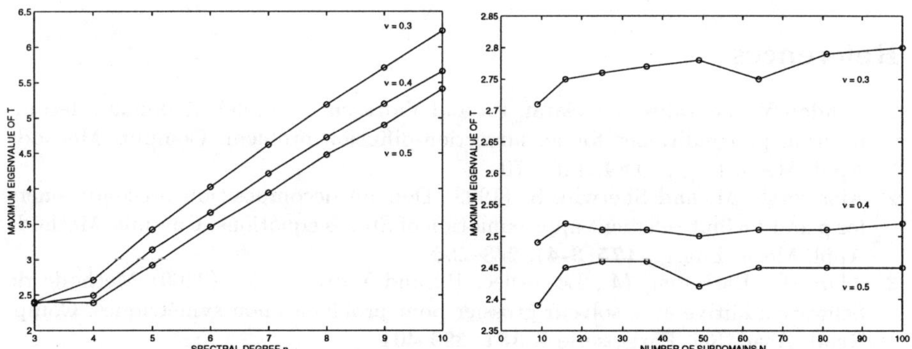

Fig. 4. Serial results for elasticity system (homogeneous medium) and $Q_n - Q_{n-2}$ spectral elements: maximum eigenvalue of T_λ vs. spectral degree n (left) and number of subdomains N (right), from Table 4, coarse space $\mathbf{V}_0^1$

across subdomain boundaries. We have tried many possible combinations of compressible and incompressible materials and have found our algorithm to be practically insensitive to these variations. This is clearly shown in the results of Table 5, where variations of up to ten orders of magnitude in the Lamé coefficients cause a comparable increase in the condition number of the discrete problem (computed by the Matlab function `condest`) but do not affect the performance of our algorithm or the spectrum of the preconditioned operator.

Table 5. Serial results for elasticity system (heterogeneous medium) and $Q_n - Q_{n-2}$ spectral elements: PCG iteration counts and extremal eigenvalues of T_λ for the balancing Neumann-Neumann preconditioner with first coarse space $\mathbf{V}_0^0$. Fixed number of subdomains $N = 4 \times 4$ and spectral degree $n = 5$

exponent t in μ	# iterations	eig max	eig min	estim. cond(K)
-3	11	3.48	1.0012	$2.4 \cdot 10^5$
-2	11	3.53	1.0024	$3.3 \cdot 10^4$
-1	12	3.73	1.0026	$5.3 \cdot 10^3$
0	12	3.47	1.0017	$1.7 \cdot 10^4$
1	12	3.46	1.0017	$8.3 \cdot 10^5$
2	12	3.35	1.0016	$7.2 \cdot 10^7$
3	11	3.47	1.0017	$7.1 \cdot 10^9$
4	10	3.48	1.0003	$7.1 \cdot 10^{11}$
5	10	3.49	1.0001	$7.1 \cdot 10^{13}$
6	9	3.49	1.0000	$7.1 \cdot 10^{15}$

References

1. Achdou Y., Le Tallec, P., Nataf, F., and Vidrascu, M..(2000) A domain decomposition preconditioner for an advection-diffusion problem. Comput. Methods Appl. Mech. Engrg, **184**, 145–170
2. Ainsworth, M. and Sherwin, S. (1999) Domain decomposition preconditioners for p and hp finite element approximation of Stokes equations. Comput. Methods Appl. Mech. Engrg., **175(3-4)**, 243–266
3. Alart, P., Barboteu, M., Le Tallec, P., and Vidrascu, M. (2000) Méthode de Schwarz additive avec solveur grossier pour problèmes non symétriques. Comp. Rend. Acad. Sci. Paris, serie I, **331**, 399–404
4. Balay, S., Buschelman, K., Gropp, W.D., Kaushik, D., Curfman McInnes, L., and Smith, B.F. (2001) PETSc home page, http://www.mcs.anl.gov/petsc
5. Balay, S., Gropp, W.D., Curfman McInnes, L., and Smith, B.F. (2001) PETSc users manual, Technical Report ANL-**95/11** - Revision 2.1.0, Argonne National Laboratory
6. Bernardi, C. and Maday, Y. (1999) Uniform inf-sup conditions for the spectral discretization of the Stokes problem. Math. Models Methods Appl. Sci, **9(3)**, 395–414

7. Bramble, J.H. and Pasciak, J.E. (1989/90) A domain decomposition technique for Stokes problems. Appl. Numer. Math., **6**, 251–261

8. Brezzi, F. and M. Fortin, M. (1991) Mixed and Hybrid Finite Element Methods. Springer-Verlag, Berlin

9. Casarin, M.A. (2001) Schwarz Preconditioners for the Spectral Element Discretization of the Steady Stokes and Navier-Stokes Equations. Numer. Math. **89:2**, 307–339

10. Cowsar, L.C., Mandel, J., and Wheeler, M.F. (1995) Balancing domain decomposition for mixed finite elements. Math. Comp., **64(211)**, 989–1015

11. Dryja, M. and Widlund, O.B. (1995) Schwarz methods of Neumann-Neumann type for three-dimensional elliptic finite element problems. Comm. Pure Appl. Math., **48(2)**, 121–155

12. Fischer, P.F. (1997) An overlapping Schwarz method for spectral element solution of the incompressible Navier-Stokes equations. J. Comp. Phys., **133(1)**, 84–101

13. Fischer, P.F., Miller, N.I., and Tufo, H.M. (2000) An overlapping Schwarz method for spectral element simulation of three-dimensional incompressible flows. In: Bjørstad, P.E. and Luskin, M. (Eds.) Parallel Solution of PDE, IMA Volumes in Mathematics and Its Applications, **120** Springer-Verlag, 1–30

14. Fischer, P.F. and Rønquist, E. (1994) Spectral element methods for large scale parallel Navier-Stokes calculations. Comp. Meths. Appl. Mech. Eng., **116**, 69–76

15. Gervasio, P. (1995) Risoluzione delle equazioni alle derivate parziali con metodi spettrali in regioni partizionate in sottodomini. PhD thesis, Università di Milano

16. Klawonn, A. and Pavarino, L.F. (1998) Overlapping Schwarz methods for mixed linear elasticity and Stokes problems. Comp. Meths. Appl. Mech. Eng., **165**, 233–245

17. Klawonn, A. and Widlund, O.B. (2000) A domain decomposition method with Lagrange multipliers and inexact solvers for linear elasticity. SIAM J. Sci. Comp., **22(4)**, 1199–1219

18. Klawonn, A. and Widlund, O.B. (2001) FETI and Neumann-Neumann iterative substructuring methods: connections and new results. Comm. Pure Appl. Math., **54(1)**, 57–90

19. Le Tallec, P. and Patra, A. (1997) Non-overlapping domain decomposition methods for adaptive *hp* approximations of the Stokes problem with discontinuous pressure fields. Comp. Meths. Appl. Mech. Eng., **145:361–379**

20. Le Tallec, P., Mandel, J., and Vidrascu, M. (1998) A Neumann-Neumann domain decomposition algorithm for solving plate and shell problems. SIAM J. Numer. Anal., **35(2)**, 836–867,

21. Maday, Y., Meiron, D., Patera, A., and Rønquist, E. (1993) Analysis of iterative methods for the steady and unsteady Stokes problem: application to spectral element discretizations. SIAM J. Sci. Comp., **14(2)**, 310–337

22. Mandel, J. (1993) Balancing domain decomposition. Comm. Numer. Methods Engrg., **9(3)**, 233–241

23. Mandel, J. and Brezina, M. (1996) Balancing domain decomposition for problems with large jumps in coefficients. Math. Comp., **65(216)**, 1387–1401

24. Marini, D. and Quarteroni, A. (1989) A relaxation procedure for domain decomposition methods using finite elements. Numer. Math., **55(5)**, 575–598

25. Pasciak, J.E. (1989) Two domain decomposition techniques for Stokes problems. In: Chan, T., Glowinski, R., Périaux, J., and Widlund, O (Eds.) Domain Decomposition Methods, SIAM, Philadelphia 419–430

26. Pavarino, L.F. (1997) Neumann-Neumann algorithms for spectral elements in three dimensions. RAIRO M^2AN, **31(4)**, 471–493
27. Pavarino, L.F. and Widlund, O.B. (1999) Iterative substructuring methods for spectral element discretizations of elliptic systems. II: Mixed methods for linear elasticity and Stokes flow. SIAM J. Numer. Anal., **37(2)**, 375–402
28. Pavarino, L.F. and Widlund, O.B. (2002) Balancing Neumann-Neumann methods for incompressible Stokes equations. Comm. Pure Appl. Math., **55(3)**, 302–335
29. Quarteroni, A. (1989) Domain decomposition algorithms for the Stokes equations. In: Chan, T., Glowinski, R., Périaux, J., and Widlund, O (Eds.) Domain Decomposition Methods, SIAM, Philadelphia, 431–442
30. Quarteroni, A. and Valli, A.(1999) Domain Decomposition Methods for Partial Differential Equations. Oxford Science Publications
31. Rønquist, E. (1996) A domain decomposition solver for the steady Navier-Stokes equations. In: Ilin, A.V. and Scott, L.R. (Eds.) Proc. of ICOSAHOM.95, published by the Houston Journal of Mathematics, 469–485
32. Rønquist, E. (1999) Domain decomposition methods for the steady Stokes equations. In: Lai, C.-H., Bjørstad, P.E., Cross, M., and Widlund, O. (Eds.) Proceedings of DD11. DDM.org, 326–336
33. Smith, B.F., Bjørstad, P., and Gropp, W.D. (1996) Domain Decomposition: Parallel Multilevel Methods for Elliptic Partial Differential Equations. Cambridge University Press
34. Stenberg, R. and Suri, M. (1996). Mixed hp finite element methods for problems in elasticity and Stokes flow. Numer. Math., **72(3)**, 367–390
35. Toselli., A. (2000) Neumann–Neumann methods for vector field problems. ETNA, **11**, 1–24

Partition of Unity Coarse Spaces and Schwarz Methods with Harmonic Overlap

Marcus Sarkis[1]

Mathematical Sciences Department, Worcester Polytechnic Institute, Worcester, MA 01609, USA and Instituto de Matemática Pura e Aplicada, Est. Dona Castorina, 110, Rio de Janeiro, RJ, CEP 22420-320, Brazil

Abstract. Coarse spaces play a crucial role in making Schwarz type domain decomposition methods scalable with respect to the number of subdomains. In this paper, we consider coarse spaces based on a class of partition of unity (PU) for some domain decomposition methods including the classical overlapping Schwarz method, and the new Schwarz method with harmonic overlap. PU has been used as a very powerful tool in the theoretical analysis of Schwarz type domain decomposition methods and meshless discretization schemes. In this paper, we show that PU can also be used effectively in the numerical construction of coarse spaces. PU based coarse spaces are easy to construct and need less communication than the standard finite-element-basis-function-based coarse space in distributed memory parallel implementations. We prove the new result that the condition number of the algorithms grows only linearly with respect to the relative size of the overlap. We also introduce the additive Schwarz method (AS) with harmonic overlap (ASHO), where all functions are made harmonic in part of the overlapping regions. As a result, the communication cost and condition number of ASHO is smaller than that of AS. Numerical experiments and a conditioning theory are presented in the paper.

1 Introduction

Fast domain decomposition algorithms for elliptic problems are typically two-level methods. In this paper, we introduce and analyze two-level overlapping Schwarz methods for unstructured meshes. These methods are based on partition of unity (PU) coarse spaces and/or on the concept of harmonic overlap introduced in [4]. Work on two-level methods on unstructured meshes is not new. Several different approaches have been introduced and some can be found in [1,2,6–14,16] and papers cited therein. Related works, based on two-level agglomeration techniques, can be found in [1,10]. These works use a class of partition of unity coarse spaces based on agglomeration smoothing techniques and prove an upper bound for the condition number which depends quadratically on the relative overlap. In this paper, we consider coarse basis functions based on the kind of partition of unity used in the theoretical analysis of Schwarz methods. These coarse basis functions have been around for some time and can be found in some of the numerical experiments of [10]. This kind of partition of unity has a controlled decaying property on the overlapping region. Instead of using two distinct partitions of unity, one for designing the algorithm and another one for analyzing the algorithm, as used in [10], we use here the same partition of unity for both except near the boundary. To accomplish that, we use a key argument based on a combination of Neumann-Neumann [12,9] and small overlap techniques [8] to establish a new result in which an upper bound for the condition number of the preconditioners depends only linearly on the relative overlap. We remark that this key argument for the analysis was directly inspired by the analysis developed for the RASHO method [3,4].

We note that partition of unity functions do not vanish on the boundary of the original domain. Hence, they cannot be used straightforwardly as coarse basis functions since they should satisfy zero Dirichlet boundary conditions for Dirichlet boundary problems. We have tested two approaches. In the first approach, we do not include in the coarse spaces the coarse basis functions that touch $\partial\Omega$, i.e. the boundary coarse functions. In the second approach, we modify the boundary coarse functions so that they have a controlled decay to zero near $\partial\Omega$ and include them in the coarse spaces. We show that coarse spaces with boundary coarse functions are much more effective than coarse spaces without them. In this paper, we focus on analyzing the second approach. The analysis for coarse spaces without the boundary coarse basis functions will then follow easily. We refer to [11] for a modified proof of the first approach and for discussions on the choice of coarse basis functions and related rate of convergence.

The preconditioners to be discussed below are applicable for general symmetric positive definite problems. They are algebraic in the sense that the notions of subdomains, harmonic overlaps, the classification of the nodal points, etc., can all be defined in terms of graphs and sparse matrices. In order to provide a complete and simple mathematical analysis, we restrict our discussion to a finite element discretization of the Poisson problem with zero Dirichlet boundary condition. Find $u \in H_0^1(\Omega)$, such that

$$a(u,v) = f(v), \quad \forall \, v \in H_0^1(\Omega), \tag{1}$$

where

$$a(u,v) = \int_\Omega \nabla u \cdot \nabla v \, dx \quad \text{and} \quad f(v) = \int_\Omega fv \, dx \text{ for } f \in L^2(\Omega).$$

For simplicity, let Ω be a bounded polygonal region in $\Re^2$ with a diameter of size $O(1)$. The extension of the results to $\Re^3$ can be carried out easily. Let $\mathcal{T}^h(\Omega)$ be a shape regular, quasi-uniform triangulation, of size $O(h)$, of Ω and $\mathcal{V} \subset H_0^1(\Omega)$ the finite element space consisting of continuous piecewise linear functions associated with the triangulation. The extension of the theory for the case of local quasi-uniform triangulation is also straightforward. We are interested in solving the following discrete problem associated with (1): Find $u^* \in \mathcal{V}$ such that

$$a(u^*,v) = f(v), \quad \forall \, v \in \mathcal{V}. \tag{2}$$

Using the standard basis functions, (2) can be rewritten as a linear system
of equations

$$Au^* = f. \tag{3}$$

For simplicity, we understand u^* and f both as functions and vectors depending on the situation. Throughout this paper, C and C_0, are positive
generic constants that are independent of any of the mesh parameters and
the number of subdomains. All the domains and subdomains are assumed to
be open; i.e., boundaries are not included in their definitions.

2   Notations

Given the domain Ω and triangulation $\mathcal{T}^h(\Omega)$, we assume that a domain
partition has been applied and resulted in N non-overlapping subdomains
$\Omega_i, i = 1, \ldots N$ of size $O(H)$, such that

$$\overline{\Omega} = \cup_{i=1}^{N} \overline{\Omega}_i, \quad \Omega_i \cap \Omega_j = \emptyset, \quad \text{for} \quad j \neq i.$$

We define the overlapping subdomains Ω_i^δ as follows. Let Ω_i^1 be the one-
overlap element extension of Ω_i, where $\Omega_i^1 \supset \Omega_i$ is obtained by including
all the immediate neighboring elements $\tau_h \in \mathcal{T}^h(\Omega)$ of Ω_i such that $\overline{\tau}_h \cap$
$\overline{\Omega}_i \neq \emptyset$. Using the idea recursively, we can define a δ-extension overlapping
subdomains Ω_i^δ

$$\Omega_i \subset \Omega_i^1 \subset \cdots \Omega_i^\delta.$$

Here the integer $\delta \geq 1$ indicates the level of element extension and δh is
the approximate length of the extension. We note that this extension can be
coded easily through the knowledge of the adjacent matrix associated to the
mesh.

In this paper we consider the case of Dirichlet boundary condition on
the whole $\partial\Omega$. To introduce coarse basis functions of low energy near $\partial\Omega$ we
introduce a Dirichlet boundary treatment. Let Ω_B^1 be one layer of elements
near the Dirichlet boundary $\partial\Omega$ and then define recursively,

$$\Omega_B^1 \subset \Omega_B^2 \cdots \Omega_B^\delta$$

with δ levels of extension by adding recursively neighboring elements. We
note that we can choose δ extensions to obtain Ω_i^δ and $\hat{\delta}$ to obtain $\Omega_B^{\hat{\delta}}$.
If $O(\delta) \leq \hat{\delta} \leq O(H/h)$, we obtain similar upper bounds in the analysis.
Numerically, $\hat{\delta} = \delta$ gives good results.

To define and analyze the new preconditioners, we subdivide Ω_i^δ as follow.
Let $\gamma_i^\delta = \partial\Omega_i^\delta \backslash \partial\Omega, i = 1, \cdots, N$; i.e., the part of the boundary of Ω_i^δ that
does not belong to the physical boundary of Ω. Let $\gamma_B^\delta = \partial\Omega_B^\delta \backslash \partial\Omega$. We define
the interface overlapping boundary Γ^δ as the union of all the γ_i^δ and γ_B^δ; i.e.,
$\Gamma^\delta = \cup_{i=B,1}^{N} \gamma_i^\delta$. We also need the following subsets of Ω_i^δ

- $\Gamma_i^\delta = \Gamma^\delta \cap \Omega_i^\delta$ (local interface)

- $\mathcal{N}_i^\delta = \Omega_i^\delta \backslash (\cup_{j \neq i} \Omega_j^\delta \cup \Omega_B^\delta \cup \Gamma_i^\delta)$ (non-overlapping region)

- $\mathcal{O}_i^\delta = \Omega_i^\delta \backslash (\mathcal{N}_i^\delta \cup \Gamma_i^\delta)$ (overlapping region)

We note that $\Omega_i^\delta = \mathcal{N}_i^\delta \cup \Gamma_i^\delta \cup \mathcal{O}_i^\delta$. The region $\mathcal{O}_i^\delta$ is the overlapping region of Ω_i^δ excluding the Γ_i^δ. The region $\mathcal{N}_i^\delta$ is the subregion of Ω_i^δ which does not overlap any neighboring extended subdomain $\overline{\Omega}_j^\delta$ and $\overline{\Omega}_B^\delta$. We recall that the regions $\mathcal{O}_i^\delta$ and $\mathcal{N}_i^\delta$ are open sets.

3 Overlapping Additive Schwarz (AS) Methods

We next review the AS method and then introduce a coarse space based on a partition of unity.

3.1 The AS Method without a Coarse Space

We introduce the space $\mathcal{V}_i^\delta = \mathcal{V} \cap H_0^1(\Omega_i^\delta)$ extended by zero to $\Omega \backslash \Omega_i^\delta$. It is easy to verify that
$$\mathcal{V} = \mathcal{V}_1^\delta + \mathcal{V}_2^\delta + \cdots + \mathcal{V}_N^\delta.$$

This decomposition is used in defining the classical one-level additive Schwarz algorithm [15]. Note that this decomposition is not a direct sum. Let us define $P_i^\delta : \mathcal{V} \to \mathcal{V}_i^\delta$ by: for any $u \in \mathcal{V}$

$$a(P_i^\delta u, v) = a(u, v), \quad \forall v \in \mathcal{V}_i^\delta. \tag{4}$$

Then, the classical one-level additive Schwarz operator has the form

$$P^\delta = P_1^\delta + \cdots + P_N^\delta.$$

We next introduce a partition of unity (PU) coarse space $\mathcal{V}_0^\delta$ to the AS method.

3.2 A PU Coarse Space for the AS Method

We next construct a partition of unity θ_i^δ such that $\theta_i^\delta \in \mathcal{V}_i^\delta$, $0 \leq \theta_i^\delta(x) \leq 1$, $|\nabla \theta_i^\delta(x)| \leq C/(\delta h)$, and $\sum_{i=B,1}^{N} \theta_i^\delta \equiv 1$. We next give one possible construction of the θ_i^δ. We first construct the function $\hat{\theta}_B^\delta \in \mathcal{V}_i^\delta$ as follows. We let $\hat{\theta}_B^\delta(x) = 1$ and $\hat{\theta}_B^\delta(x) = 0$ for nodes x on $\partial \Omega$ and $\Omega \backslash \Omega_B^\delta$, respectively. For the first layer of neighboring nodes x of $\partial \Omega$ we let $\hat{\theta}_B^\delta(x) = (\delta - 1)/\delta$, For the second layer of neighboring nodes x of $\partial \Omega$ we let $\hat{\theta}_B^\delta(x) = (\delta - 2)/\delta$, and recursively, we let $\hat{\theta}_B^\delta(x) = (\delta - k)/\delta$ for the $(k)st$ layer of neighboring nodes x of $\partial \Omega$. Similarly, for $i = 1, \cdots, N$, we let $\hat{\theta}_i^\delta(x) = 1$ and $\hat{\theta}_i^\delta(x) = 0$ for nodes

x of $\overline{\Omega}_i\backslash\Omega_B^\delta$ and $\overline{\Omega}\backslash\Omega_i^\delta$, respectively. For the first layer of neighboring nodes x of $\overline{\Omega}_i\backslash\Omega_B^\delta$ we let $\hat{\theta}_i^\delta(x) = (\delta-1)/\delta$, and recursively, we let $\hat{\theta}_i^\delta(x) = (\delta-k)/\delta$ for the $(k)st$ layer of neighboring nodes x of $\overline{\Omega}_i\backslash\Omega_B^\delta$. It is easy to verify that $0 \le \hat{\theta}_i^\delta(x) \le 1$, and for quasi-uniform triangulation $|\nabla\hat{\theta}_i^\delta(x)| \le C/(\delta h)$. The partition of unity θ_i^δ is defined as

$$\theta_i^\delta = I_h\Big(\frac{\hat{\theta}_i^\delta}{\sum_{j=B,1}^N \hat{\theta}_j^\delta}\Big).$$

It is easy to verify that $\sum_{i=B,1}^N \theta_i^\delta(x) = 1, \quad 0 \le \theta_i^\delta(x) \le 1$, and $|\nabla\theta_i^\delta(x)| \le C/(\delta h), \forall x \in \bar{\Omega}$.

The PU coarse space $\mathcal{V}_0^\delta$ is given as the linear combination of the coarse basis functions $\theta_i^\delta, i = 1, \cdots, N$. Note that we do not include the θ_B^δ to the PU coarse space $\mathcal{V}_0^\delta$. Let us define $P_0 : \mathcal{V} \to \mathcal{V}_0^\delta$ by: for any $u \in \mathcal{V}$

$$a(P_0^\delta u, v) = a(u, v), \quad \forall v \in \mathcal{V}_0^\delta.$$

Then, the two-level additive Schwarz operator with the PU coarse problem P_0^δ has the form

$$P_C^\delta = \sum_{i=0}^N P_i^\delta.$$

4   AS Methods with Harmonic Overlap (ASHO)

We next introduce two ASHO methods: One without coarse space and a second one with a coarse space based on a partition of unity.

4.1   ASHO Method without a Coarse Space

We define $\widetilde{\mathcal{V}}_i^\delta$ as a subspace of $\mathcal{V}_i^\delta$ consisting of functions that are discrete harmonic at all nodes interior to $\mathcal{O}_i^\delta$, i.e. $u \in \widetilde{\mathcal{V}}_i^\delta$, if for all nodes $x_k \in \mathcal{O}_i^\delta$,

$$a(u, \phi_{x_k}) = 0.$$

Here, $\phi_{x_k} \in \mathcal{V}$ is the regular finite element basis function associated with node x_k, i.e. $\phi_{x_k}(x_k) = 1$, and $\phi_{x_k}(x_j) = 0, j \neq k$.

We define $\widetilde{\mathcal{V}}^\delta$ as a subspace of $\mathcal{V}$ defined as

$$\widetilde{\mathcal{V}}^\delta = \widetilde{\mathcal{V}}_1^\delta + \widetilde{\mathcal{V}}_2^\delta + \cdots + \widetilde{\mathcal{V}}_N^\delta.$$

We note that the above sum is not a direct sum and $\widetilde{\mathcal{V}}_i^\delta \neq \mathcal{V}_i^\delta$. We define $\widetilde{P}_i^\delta : \widetilde{\mathcal{V}}^\delta \to \widetilde{\mathcal{V}}_i^\delta$ to be the projection operators such that, for any $u \in \widetilde{\mathcal{V}}^\delta$

$$a(\widetilde{P}_i^\delta u, v) = a(u, v), \quad \forall v \in \widetilde{\mathcal{V}}_i^\delta.$$

Then, the one-level additive overlapping Schwarz method (ASHO) with harmonic overlap is defined as

$$\widetilde{P}^\delta = \sum_{i=1}^{N} \widetilde{P}_i^\delta.$$

We next introduce a PU coarse space $\widetilde{\mathcal{V}}_0^\delta$ for the ASHO method.

4.2 A PU Coarse Space for ASHO Method

To define the PU coarse space $\widetilde{\mathcal{V}}_0^\delta \subset \widetilde{\mathcal{V}}^\delta$, we simply modify our basis functions θ_i^δ to $\widetilde{\theta}_i^\delta$. The $\widetilde{\theta}_i^\delta$ is defined to be equal to θ_i^δ except on $\mathcal{O}_i^\delta$. On $\mathcal{O}_i^\delta$ we make $\widetilde{\theta}_i^\delta$ discrete harmonic.

The PU coarse space $\widetilde{\mathcal{V}}_0^\delta$ is given as the linear combination of the coarse basis functions $\widetilde{\theta}_i^\delta, i = 1, \cdots, N$. We introduce $\widetilde{P}_0 : \widetilde{\mathcal{V}}^\delta \to \widetilde{\mathcal{V}}_0^\delta$ as the operator such that, for any $u \in \widetilde{\mathcal{V}}^\delta$,

$$a(\widetilde{P}_0^\delta u, v) = a(u, v), \quad \forall v \in \widetilde{\mathcal{V}}_0^\delta \tag{5}$$

Then, the two-level ASHO with the PU coarse problem $\widetilde{P}_0^\delta$ is defined as

$$\widetilde{P}_C^\delta = \sum_{i=0}^{N} \widetilde{P}_i^\delta. \tag{6}$$

5 Hybrid Methods with PU Coarse Spaces

In this paper we also consider the special hybrid Schwarz operator (see [15,13]) with the error propagation operator given by

$$(I - \sum_{i=1}^{N} P_i^\delta)(I - P_0^\delta),$$

or after an additional coarse solve,

$$(I - P_0^\delta)(I - \sum_{i=1}^{N} P_i^\delta)(I - P_0^\delta). \tag{7}$$

This is a symmetric operator with which we can work essentially without any extra cost, since when forming powers of the operator (7), we can use the fact that $I - P_0^\delta$ is a projection, and therefore $(I - P_0^\delta)^2 = I - P_0^\delta$. Subtracting the operator (7) from the identity operator I, we obtain the operator

$$P_{hyb}^\delta = P_0^\delta + (I - P_0^\delta)(\sum_{i=1}^{N} P_i^\delta)(I - P_0^\delta).$$

We also consider the harmonic overlap version for the hybrid algorithm defined by

$$\widetilde{P}^{\delta}_{hyb} = \widetilde{P}^{\delta}_0 + (I - \widetilde{P}^{\delta}_0)(\sum_{i=1}^{N} \widetilde{P}^{\delta}_i)(I - \widetilde{P}^{\delta}_0).$$

6   Remarks about ASHO Methods

We next show that the explicit elimination of the variables associated with the overlapping nodes is not needed in order to apply $\widetilde{P}^{\delta}$ to any given vector $v \in \widetilde{\mathcal{V}}^{\delta}$.

Lemma 1. *For any* $u \in \widetilde{\mathcal{V}}^{\delta}$*, we have*

$$\widetilde{P}^{\delta}_i u = P^{\delta}_i u, \quad i = 1, \cdots, N.$$

Hence, $\widetilde{P}^{\delta} u = P^{\delta} u, \, u \in \widetilde{\mathcal{V}}^{\delta}$.

Proof. If $u \in \widetilde{\mathcal{V}}^{\delta}$ then

$$a(P^{\delta}_i u, \phi_{x_k}) = a(u, \phi_{x_k}) = 0, \quad \forall x_k \in \mathcal{O}^{\delta}_i.$$

Hence, $P^{\delta}_i u \in \widetilde{\mathcal{V}}^{\delta}_i$. Here, $\phi_{x_k} \in \mathcal{V}^{\delta}_i$ are the regular basis functions associated to the nodes x_k. To complete the proof of the lemma, we just need to verify that

$$a(P^{\delta}_i u, v) = a(u, v), \quad \forall v \in \widetilde{\mathcal{V}}^{\delta}_i. \tag{8}$$

To verify (8), we use the definition of P^{δ}_i (4) and that $\widetilde{\mathcal{V}}^{\delta}_i$ is a subset of $\mathcal{V}^{\delta}$.

We note that the solution u^* of (3) is not in the subspace $\widetilde{\mathcal{V}}^{\delta}$, therefore, the operators $\widetilde{P}^{\delta}, \widetilde{P}^{\delta}_C$, and $\widetilde{P}^{\delta}_{hyb}$ cannot be used to solve the linear system (3) directly. We will need to modify the right-hand side of the system (3). A reformulated (3) will be presented in Lemma 2 below. Using the matrix notations, the next lemma shows how to modify the system (3) so that its solution belongs to $\widetilde{\mathcal{V}}^{\delta}$. Let $\mathcal{O}^{\delta} = \cup_i \mathcal{O}^{\delta}_i$. Let $W^{\delta}_{\mathcal{O}}$ be the set of nodes associated to the degree of freedom of $\mathcal{V}^{\delta}$ in $\mathcal{O}^{\delta}$. We define the restriction operator, or a matrix, $R_{\mathcal{O}^{\delta}} \colon W \to W$ as follows

$$(R_{\mathcal{O}^{\delta}} v)(x_k) = \begin{cases} v_k & \text{if } x_k \in W_{\mathcal{O}^{\delta}} \\ \\ 0 & \text{otherwise.} \end{cases}$$

The matrix representation of $R^{\delta}_{\mathcal{O}}$ is given by a diagonal matrix with 1 for nodal points in the interior of $\mathcal{O}^{\delta}$ and zero for the remaining nodal points. Using this restriction operator, we define the subdomain stiffness matrix as

$$A_{\mathcal{O}^{\delta}} = R_{\mathcal{O}^{\delta}} A R^{T}_{\mathcal{O}^{\delta}},$$

which can also be obtained by the discretization of the original finite element problem on $\mathcal{O}^\delta$ with zero Dirichlet data on $\partial\mathcal{O}^\delta$ and extended by zero outside of $\mathcal{O}^\delta$. We remark that $\mathcal{O}$ is a disconnected region where $\partial\mathcal{O} = \Gamma_i^\delta \cup \partial\Omega$. Therefore, $A_{\mathcal{O}^\delta} w = f$ can be solved locally and inexpensively.

It is easy to see that the following lemma holds; see [4].

Lemma 2. *Let u^* and f be the exact solution and the right-hand side of (3), and*

$$w = R_{\mathcal{O}^\delta}^T A_{\mathcal{O}^\delta}^+ R_{\mathcal{O}^\delta} f. \tag{9}$$

Then $\widetilde{u}^* = u^* - w \in \widetilde{\mathcal{V}}^\delta$ and satisfies the following modified linear system of equations

$$A\widetilde{u}^* = f - Aw = \widetilde{f}.$$

7   Theoretical Analysis

The algorithms presented in the previous section are applicable for general sparse, symmetric positive definite linear systems. The notions of subdomains, harmonic overlaps, the classification of the regions $\mathcal{O}_i^\delta$ and $\mathcal{N}_i^\delta$ and the interfaces Γ_i^δ, etc., can all be defined in terms of the graph of the sparse matrix. In this section we provide a sharp and nearly optimal estimate for a Poisson equation discretized with a piecewise linear finite element method. We estimate the condition numbers of the operators P^δ, $\widetilde{P}^\delta$, P_C^δ, $\widetilde{P}_C^\delta$, P_{hyb}^δ, and $\widetilde{P}_{hyb}^\delta$ in terms of the fine mesh size h, the subdomain size H, and the overlapping factor δ. We shall follow the abstract additive Schwarz theory [15] to analyze the additive versions, where three assumptions have to be checked and three parameters C_0, ω and $\rho(\mathcal{E})$ estimated. Two assumptions are trivial to check: $\omega = 1$ since we use exact solvers, and $\rho(\mathcal{E}) \leq C$ since we use two-level algorithms. So our focus in the rest of the paper is in bounding C_0 for each of the preconditioned operators P^δ, P_C^δ, $\widetilde{P}_C^\delta$, and $\widetilde{P}^\delta$. To analyze the hybrid algorithms we use a result due to Mandel (Lemma 3.2 [13]) which in our context is given by

Lemma 3.

$$\lambda_{min}(P_{hyb}^\delta) \geq \lambda_{min}(P_C^\delta), \qquad \lambda_{min}(\widetilde{P}_{hyb}^\delta) \geq \lambda_{min}(\widetilde{P}_C^\delta)$$

$$\lambda_{max}(P_{hyb}^\delta) \leq \lambda_{max}(P^\delta), \quad and \quad \lambda_{max}(\widetilde{P}_{hyb}^\delta) \leq \lambda_{max}(\widetilde{P}^\delta).$$

7.1   Additive Schwarz with a PU Coarse Space

Let $\mathcal{V}_{i,0}^\delta, i = 1, \cdots, N$ be the one-dimensional spaces generated by the θ_i^δ coarse basis functions. We introduce the interpolation-like operator $I_0^\delta = \sum_{i=1}^N I_{i,0}^\delta$, where the $I_{i,0}^\delta : \mathcal{V} \to \mathcal{V}_{i,0}^\delta$ is defined as follows:

$$(I_{i,0}^\delta u)(x) = \bar{u}_i^\delta \theta_i^\delta(x),$$

where

$$\bar{u}_i^\delta = \frac{\int_{\Omega_i^\delta} u\, dx}{\int_{\Omega_i^\delta} 1\, dx},$$

is the average of u on the extended region Ω_i^δ. Here $|\Omega_i^\delta|$ is the area of the region Ω_i^δ.

We next we give a direct proof of the H_1-stability of I_0^δ. We remark that the H_1-stability is not new and an alternative proof can be found in [10].

Lemma 4. *For any $u \in \mathcal{V}$ we have*

$$a(I_0^\delta u, I_0^\delta u) \le C \frac{H}{\delta h} a(u, u). \tag{10}$$

Proof. Let c be an arbitrary constant. Using Cauchy-Schwarz inequality and the fact $|\Omega_i^\delta| = O(H^2)$ we have

$$|\bar{u}_i^\delta - c| = \frac{1}{|\Omega_i^\delta|} \left| \int_{\Omega_i^\delta} (u - c)dx \right| \le C \frac{1}{H} \|u - c\|_{L^2(\Omega_i^\delta)}. \tag{11}$$

Let Ω_j^δ be a neighbor subdomain of Ω_i^δ; i.e., $\Omega_j^\delta \cap \Omega_i^\delta \ne \emptyset$. Using a triangular inequality and the estimate (11), we obtain

$$|\bar{u}_i^\delta - \bar{u}_j^\delta| \le |\bar{u}_i^\delta - c| + |\bar{u}_j^\delta - c| \le C \frac{1}{H} \|u - c\|_{L^2(\Omega_i^\delta \cup \Omega_j^\delta)}.$$

Applying a Bramble-Hilbert argument, we have

$$|\bar{u}_i^\delta - \bar{u}_j^\delta| \le \frac{C}{H} \inf_c \|u - c\|_{L^2(\Omega_i^\delta \cup \Omega_j^\delta)}$$

$$\le \frac{C}{H} \inf_c \|u - c\|_{L^2(\Omega_i^{\delta, ext})} \tag{12}$$

$$\le C\, |u|_{H^1(\Omega_i^{\delta, ext})}.$$

Here $\Omega_i^{\delta, ext}$ is the union of Ω_i^δ and all of its neighbors Ω_k^δ. We note that we use here that $\Omega_i^{\delta, ext}$ has size $O(H)$ and has nice shape in order to obtain the last inequality through Friedrichs inequality. For the case when Ω_i^δ is distant $O(H)$ from the boundary $\partial\Omega$, we use another Friedrichs inequality to obtain

$$|\bar{u}_i^\delta| \le C \frac{1}{H} \|u\|_{L^2(\Omega_i^\delta)} \le C \frac{1}{H} \|u\|_{L^2(\Omega_i^{\delta, ext})} \le C|u|_{H^1(\Omega_i^{\delta, ext})}, \tag{13}$$

where $\Omega_i^{\delta, ext}$ is the union of Ω_i^δ and all of its neighbors Ω_k^δ and possibly a few more $\Omega_{k'}^\delta$, so that $\Omega_i^{\delta, ext}$ has size $O(H)$ and intersects the boundary $\partial\Omega$ on a set of measure of $O(H)$.

The rest of the proof is devoted to bound $|I_0^\delta u|^2_{H^1(\mathcal{N}_i^\delta)}$ and $|I_0^\delta u|^2_{H^1(\mathcal{O}_i^\delta)}$. For $x \in \mathcal{N}_i^\delta$, we have $\theta_j^\delta(x) = 0$, $j \neq i$, and $\theta_i^\delta(x) = 1$. Therefore, we have

$$|I_0^\delta u|^2_{H^1(\mathcal{N}_i^\delta)} = |\bar{u}_i^\delta|^2_{H^1(\mathcal{N}_i^\delta)} = 0.$$

For the region $\mathcal{O}_i^\delta$ we have

$$|u_0|^2_{H^1(\mathcal{O}_i^\delta)} = |I_0^\delta u|^2_{H^1(\mathcal{O}_i^\delta)} = |\sum_{j=B,1}^{N} \bar{u}_j^\delta \theta_j^\delta|^2_{H^1(\mathcal{O}_i^\delta)},$$

where here we have introduced, artificially, the constant $\bar{u}_B^\delta = 0$.

We next use that

$$\sum_{j=B,1}^{N} \theta_j^\delta(x) = \theta_i^\delta(x) + \sum_{j \neq i} \theta_j^\delta(x) = 1, \quad \forall x \in \bar{\Omega}$$

to obtain

$$|I_0^\delta u|^2_{H^1(\mathcal{O}_i^\delta)} = \left| \sum_{j \neq i} (\bar{u}_j^\delta - \bar{u}_i^\delta) \theta_j^\delta \right|^2_{H^1(\mathcal{O}_i^\delta)}.$$

By a triangular inequality we have

$$|I_0^\delta u|^2_{H^1(\mathcal{O}_i^\delta)} \leq C \sum_{j \neq i} |(\bar{u}_j^\delta - \bar{u}_i^\delta) \theta_j^\delta|^2_{H^1(\mathcal{O}_i^\delta)} \leq C \sum_{j \neq i} |\bar{u}_j^\delta - \bar{u}_i^\delta|^2 |\theta_j^\delta|^2_{H^1(\mathcal{O}_i^\delta)}.$$

We next use that $|\nabla \theta_k^\delta(x)| \leq C/(\delta h)$, the fact the area of $\mathcal{O}_i^\delta$ is at most of $O(H\delta h)$, and the estimates (12) (case $j \neq B$) and (13) (case $j = B$), to obtain

$$|\bar{u}_j^\delta - \bar{u}_i^\delta|^2 |\theta_j^\delta|^2_{H^1(\mathcal{O}_i^\delta)} \leq C \frac{H}{\delta h} |u|^2_{H^1(\Omega_i^{\delta,ext})}. \tag{14}$$

The estimate (10) follows by summing the contributions of all extended domains $\Omega_i^{\delta,ext}$ and using a coloring argument.

For the case when the boundary coarse basis functions θ_i^δ are not included in the coarse space, we just set $\bar{u}_i^\delta = 0$ and the proof above also holds. In this case, we use (12) for subdomains that are neighbors of boundary domains. We note that the last constant C in (13) grows when the size of $\Omega_i^{\delta,ext}$ gets larger, or equivalently, when Ω_i^δ gets more distant from $\partial\Omega$. That explain why coarse spaces without the boundary coarse basis functions are less effective than with them.

We next prove the main result of the paper.

Theorem 1. *There exists a constant $C > 0$, independent of h, δ, and H, such that*

$$\kappa(P^\delta) \leq C \frac{1}{H^2}\left(1 + \frac{H}{\delta h}\right) \tag{15}$$

and

$$\kappa(P_{hyb}^{\delta}) \le \kappa(P_C^{\delta}) \le C(1 + \frac{H}{\delta h}). \tag{16}$$

Proof. The bound (15) is well known and it is proved in Dryja and Widlund [8]. The first inequality of (16) follows directly from the Lemma 3. What remains to complete the proof is to derive a bound for C_0; i.e., to find C_0 such that for any given $u \in \mathcal{V}$, there exist $u_i \in \mathcal{V}_i^{\delta}$, such that

$$u = \sum_{i=0}^{N} u_i, \tag{17}$$

and

$$\sum_{i=0}^{N} a(u_i, u_i) \le C_0 a(u, u). \tag{18}$$

We define the decomposition $u = \sum_{i=0}^{N} u_i$ as follows. Let $u_0 \in \mathcal{V}_0^{\delta}$ be defined as

$$u_0 = I_0^{\delta} u = \sum_{i=1}^{N} \bar{u}_i^{\delta} \theta_i^{\delta},$$

and let $u_i \in \mathcal{V}_i^{\delta}$ be defined as

$$u_i = I_h(\vartheta_i^{\delta} u) - \bar{u}_i^{\delta} \theta_i^{\delta}, i = 1, \ldots, N.$$

Here, I_h is the standard pointwise interpolator. The piecewise linear functions $\vartheta_i^{\delta} \in H^1(\Omega_i^{\delta})$ are defined below and form a partition of unity $\sum_{i=1}^{N} \vartheta_i \equiv 1$ on $\bar{\Omega}$. It is easy to see (17) holds.

We next modify the coarse basis functions θ_i^{δ} on $(\bar{\Omega}_i^{\delta} \cap \bar{\Omega}_B^{\delta})$ to define the partition of unity ϑ_i^{δ}. We first construct the function $\hat{\vartheta}_i^{\delta} \in H^1(\Omega_i^{\delta})$. Let $\hat{\vartheta}_i^{\delta}(x) = 1$ and $\hat{\vartheta}_i^{\delta}(x) = 0$ for nodes x of $\overline{\Omega}_i$ and $\overline{\Omega} \backslash \Omega_i^{\delta}$, respectively. For the first layer of neighboring nodes x of $\overline{\Omega}_i$ we let $\hat{\vartheta}_i^{\delta}(x) = (\delta - 1)/\delta$, and recursively, we let $\hat{\vartheta}_i^{\delta}(x) = (\delta - k)/\delta$ for the $(k)st$ layer of neighboring nodes x of $\overline{\Omega}_i$. The partition of unity ϑ_i^{δ} is defined as

$$\vartheta_i^{\delta} = I_h\left(\frac{\hat{\vartheta}_i^{\delta}}{\sum_{j=1}^{N} \hat{\vartheta}_j^{\delta}}\right).$$

It is easy to verify that $\sum_{i=1}^{N} \vartheta_i^{\delta}(x) = 1$, $\quad 0 \le \vartheta_i^{\delta}(x) \le 1$, and $|\nabla \vartheta_i^{\delta}(x)| \le C/(\delta h)$, when $x \in \bar{\Omega}$, and also that $\vartheta_i^{\delta}(x) = \theta_i^{\delta}(x), i = 1, \cdots, N$, when $x \in \Omega \backslash \Omega_B^{\delta}$.

We decompose u_i as $u_i = u_i^0 + u_i^B$ where

$$u_i^0 = I_h(\theta_i^{\delta}(u - \bar{u}_i^{\delta})), \quad \text{and} \quad u_i^B = I_h((\vartheta_i^{\delta} - \theta_i^{\delta})u)), \quad i = 1, \cdots, N. \tag{19}$$

The next step is to bound $\sum_{i=0}^{N} a(u_i, u_i)$. In order to bound $a(u_0, u_0)$ we use Lemma 4. The remaining of the proof is to bound $a(u_i^0, u_i^0)$ and $a(u_i^B, u_i^B)$ since

$$a(u_i, u_i) \leq 2a(u_i^0, u_i^0) + 2a(u_i^B, u_i^B), i = 1, \cdots, N.$$

We denote $w_i = u - \bar{u}_i^\delta$. Estimating $a(u_i^0, u_i^0)$ as in the standard additive Schwarz method [8], we obtain

$$a(u_i^0, u_i^0) = |I_h(\theta_i^\delta w_i)|_{H^1(\Omega_i^\delta)}^2 \leq C \left(|w_i|_{H^1(\Omega_i^\delta)}^2 + \frac{1}{(\delta h)^2} \|w_i\|_{L^2(\mathcal{O}_i^\delta)} \right).$$

Using Lemma 3 in Dryja and Widlund [8]; i.e.,

$$\frac{1}{(\delta h)^2} \|w_i\|_{L^2(\mathcal{O}_i^\delta)}^2 \leq C \left((1 + \frac{H}{\delta h}) |w_i|_{H^1(\Omega_i^\delta)}^2 + \frac{1}{H(\delta h)} \|w_i\|_{L^2(\Omega_i^\delta)}^2 \right),$$

we obtain

$$|u_i^0|_{H^1(\Omega_i^\delta)}^2 \leq C \left((1 + \frac{H}{\delta h}) |w_i|_{H^1(\Omega_i^\delta)}^2 + \frac{1}{H(\delta h)} \|w_i\|_{L^2(\Omega_i^\delta)}^2 \right).$$

Combining these estimates and that $|w_i|_{H^1(\Omega_i^\delta)}^2 = |u|_{H^1(\Omega_i^\delta)}^2$ and a Friedrichs inequality

$$\|w_i\|_{L^2(\Omega_i^\delta)}^2 \leq C H^2 |u|_{H^1(\Omega_i^\delta)}^2,$$

we obtain

$$a(u_i^0, u_i^0) \leq C(1 + \frac{H}{\delta h}) |u|_{H^1(\Omega_i^\delta)}^2.$$

We next bound $a(u_i^B, u_i^B)$. We note that for $i = 1, \cdots, N$

$$\vartheta_i^\delta(x) = \theta_i^\delta(x), \forall x \in \Omega \backslash \bar{\Omega}_B^\delta,$$

and therefore the support of u_i^B is $\bar{\Omega}_B^\delta \cap \bar{\Omega}_i^\delta$.

Using standard additive Schwarz methods arguments, we obtain

$$a(u_i^B, u_i^B) = |u_i^B|_{H^1(\Omega_i^\delta \cap \Omega_B^\delta)}^2 \leq C \left(|u|_{H^1(\Omega_i^\delta \cap \Omega_B^\delta)}^2 + \frac{1}{(\delta h)^2} \|u\|_{L^2(\Omega_i^\delta \cap \Omega_B^\delta)}^2 \right).$$

Using a Friedrichs inequality for the case when the width of $\Omega_i^\delta \cap \Omega_B^\delta$ is $O(\delta h)$ and u vanishes on the larger side (u vanishes on $\partial \Omega$); i.e.,

$$\|u\|_{L^2(\Omega_i^\delta \cap \Omega_B^\delta)}^2 \leq C(\delta h)^2 |u|_{H^1(\Omega_i^\delta \cap \Omega_B^\delta)}^2,$$

we obtain

$$a(u_i^B, u_i^B) \leq C|u|_{H^1(\Omega_i^\delta \cap \Omega_B^\delta)}^2.$$

And the proof is complete by summing all the contributions and using a coloring argument.

The proof also holds for the case when the boundary coarse basis functions θ_i^δ are not included in the coarse space. In this case, we set $\bar{u}_i^\delta = 0$, and use that

$$\|w_i\|_{L^2(\Omega_i^\delta)}^2 \leq C H^2 |u|_{H^1(\Omega_i^{\delta, ext})}^2.$$

7.2   Additive Schwarz with Harmonic Overlap Methods

We next find upper bounds for the condition numbers of the preconditioners with harmonic overlap.

Theorem 2.

$$\kappa(\widetilde{P}^\delta) \leq C \frac{1}{H^2}(1 + \frac{H}{\delta h}) \tag{20}$$

and

$$\kappa(\widetilde{P}^\delta_{hyb}) \leq \kappa(\widetilde{P}^\delta_C) \leq C(1 + \frac{H}{\delta h}). \tag{21}$$

Proof. The first inequality of (21) follows directly from the Lemma 3. We next show the second inequality of (21). Let $u \in \widetilde{\mathcal{V}}^\delta$, and let the u_i be the decomposition (17) introduced in Theorem 1. We next define $\tilde{u}_i$ equals to u_i on $(\cup_{j=1}^N \mathcal{N}_j^\delta) \cup \Gamma_i^\delta \cup \partial\Omega$, and discrete harmonic on $\mathcal{O} = \cup_{j=1}^N \mathcal{O}_j^\delta$. We recall that $\mathcal{O}$ is a disconnected region and therefore discrete harmonic extension on $\mathcal{O}$ can be obtained locally. Using that $u \in \widetilde{\mathcal{V}}^\delta$, it is easy to see that

$$u = \sum_{i=0}^N \tilde{u}_i, \quad \text{and the} \quad u_i \in \widetilde{\mathcal{V}}_i^\delta.$$

Since the discrete harmonic extensions have the minimal semi-energy norm, we have

$$a(\tilde{u}_i, \tilde{u}_i) \leq a(u_i, u_i), \quad i = 0, \cdots, N,$$

and using the bound (18) in Theorem 1 we obtain

$$\sum_{i=0}^N a(\tilde{u}_i, \tilde{u}_i) \leq \sum_{i=0}^N a(u_i, u_i) \leq C(1 + \frac{H}{\delta h})a(u, u).$$

The proof of (20) follows the same lines as above; i.e., we first introduce a decomposition using the subspaces $\mathcal{V}_i^\delta$ (see [8]), and then we modify this decomposition on $\mathcal{O}$ to obtain a decomposition with functions on $\widetilde{\mathcal{V}}_i^\delta$.

8   Numerical Experiments and Final Remarks

In this section, we present some numerical results for solving the Poisson's equation on the unit square with zero Dirichlet boundary conditions. We compare the performance of the preconditioned Conjugate Gradient methods. As preconditioners, we consider the ASHO and AS with PU coarse spaces (given by $\widetilde{P}_C$ and P_C) and without coarse spaces (given by $\widetilde{P}$ and P), HybridHO and Hybrid with PU coarse spaces (given by $\widetilde{P}_{hyb}$ and P_{hyb}), and the AS and Hybrid with the standard non-nested coarse space; see [2,6]. In the numerical experiments below, we pay particular attention to the dependence on the number of subdomains, mesh size, and the size of overlap.

We first discuss a few implementation issues related to the preconditioner based on harmonic overlap. In order to apply the ASHO/CG and HybridHO/CG methods, it is necessary to force the solution to belong to $\widetilde{\mathcal{V}}^\delta$. To do so, a pre-CG-computation is needed, and it is done through the formula (9). We note, by Lemma 2, that $u = u^* - w \in \widetilde{\mathcal{V}}^\delta$. Hence, we can apply the regular PCG to the $\widetilde{P}^\delta$, $\widetilde{P}^\delta_C$, and $\widetilde{P}^\delta_{Hyb}$ preconditioned systems.

The exact solution of the equation is $u(x,y) = e^{5(x+y)} \sin(\pi x) \sin(\pi y)$. All subdomain problems are solved exactly. The stopping condition for CG is to reduce the initial residual by a factor of 10^{-6}. The iteration counts (iter), condition numbers (cond), maximum (max) and minimum (min) eigenvalues of the preconditioned matrix are summarized in Tables 1, 2, 3, 4, and 5.

From all the Tables below, it is clear that ASHO/CG (HybridHO/CG) is always better than the classical AS/CG (Hybrid/CG) in terms of the condition numbers, while they have similar behavior in terms of iteration numbers. This is an advantage because the AS based preconditioners can be modified to ASHO ones with a large saving in communications on a parallel computer with distributed memory. From Tables 1, 2, 6, and 7, it is clear that the hybrid versions with PU coarse spaces are always much better than the additive versions with PU coarse spaces. This is an important result since the extra computational cost of the hybrid ones over the additive ones is the calculation of one residual per iteration. From Tables 1 and 6 we can see the effective of the PU coarse spaces. It is clear that the partition of unity coarse space makes the algorithms scalable with respect to the number of subdomains. The hybrid formulations attains the asymptotic behavior of scalability faster. From Table 3 we have the AS/CG and ASHO/CG without a coarse space. We can see a dramatic grow of iterations and condition numbers and also the ASHO is slightly better than AS preconditioner. From Tables 2 and 7 we can see that the condition number of the preconditioners all grow linearly with the size of the overlap. This is an important result since we can recover the same linear behavior as obtained for regular coarse spaces on structured meshes or non-nested coarse spaces. This results agree with the theory developed in this paper. From Tables 4 and 5 we compare the PU coarse spaces with standard non-nested coarse spaces. It is clear that the hybrid versions using the partition of unity coarse spaces can perform the same or better than non-nested coarse spaces. This is an advantage since the partition of unity is easy to implement for unstructured meshes and on parallel computers. a coarse space. Finally we see, by comparing Tables 1 and 6, and by comparing with Tables 2 and 7, that all the algorithms perform considerably better if the boundary coarse functions are included in the coarse space.

Acknowledgements: The author is indebted to Xiao-Chuan Cai and Maksymilian Dryja for helpful comments and suggestions to improve this paper. The work was supported in part by the NSF grant CCR-9984404.

Table 1. Two-level HybridHO/CG (Hybrid/CG) and ASHO/CG (AS/CG) using PU coarse spaces for solving the Poisson's equation on a $16 * DOM \times 16 * DOM$ mesh decomposed into $DOM \times DOM$ subdomains with overlapping size $\delta = 2$.

HybridHO/CG (Hybrid/CG)

$DOM \times DOM$	iter	cond	max	min
2×2	13 (13)	5.89 (9.71)	2.19 (4.00)	0.372 (0.412)
4×4	17 (18)	6.32 (11.4)	2.20 (4.00)	0.347 (0.345)
8×8	18 (19)	6.48 (11.8)	2.20 (4.00)	0.340 (0.340)
16×16	17 (19)	6.54 (11.9)	2.20 (4.00)	0.337 (0.340)

ASHO/CG (AS/CG)

$DOM \times DOM$	iter	cond	max	min
2×2	14 (15)	7.71 (11.2)	2.52 (4.00)	0.326 (0.356)
4×4	24 (24)	12.7 (16.6)	2.79 (4.00)	0.278 (0.241)
8×8	30 (31)	16.3 (22.0)	2.85 (4.00)	0.175 (0.182)
16×16	32 (34)	17.5 (24.0)	2.85 (4.00)	0.164 (0.166)

Table 2. Two-level HybridHO/CG (Hybrid/CG) and ASHO/CG (AS/CG) using PU coarse spaces for solving the Poisson's equation on a 256×256 mesh decomposed into 16×16 subdomains with different overlapping sizes δ.

HybridHO/CG (Hybrid/CG)

δ	iter	cond	max	min
1	25 (26)	12.7 (23.5)	2.17 (4.00)	0.170 (0.170)
2	17 (19)	6.54 (11.9)	2.20 (4.00)	0.337 (0.340)
3	15 (16)	4.53 (8.07)	2.24 (4.00)	0.495 (0.495)
4	13 (14)	3.48 (6.19)	2.26 (4.00)	0.649 (0.646)

ASHO/CG (AS/CG)

δ	iter	cond	max	min
1	47 (48)	36.5 (49.7)	2.94 (4.00)	0.081 (0.081)
2	32 (34)	17.5 (24.0)	2.85 (4.00)	0.164 (0.166)
3	25 (26)	11.0 (15.4)	2.76 (4.00)	0.251 (0.260)
4	21 (22)	7.67 (11.0)	2.68 (4.00)	0.351 (0.363)

Table 3. One-level ASHO/CG (AS/CG) without coarse spaces for solving the Poisson's equation on a $16*DOM \times 16*DOM$ mesh decomposed into $DOM \times DOM$ subdomains with overlapping sizes $\delta = 2$.

	ASHO/CG (AS/CG)			
$DOM \times DOM$	iter	cond	max	min
2×2	13 (14)	9.05 (16.4)	2.21 (4.00)	0.2455 (0.2445)
4×4	25 (27)	28.7 (51.8)	2.22 (4.00)	0.0772 (0.0772)
8×8	46 (48)	108. (195.)	2.22 (4.00)	0.0205 (0.0205)
16×16	89 (93)	426. (768.)	2.22 (4.00)	0.0052 (0.0052)

Table 4. Two-level Hybrid/CG (AS/CG) using non-nested coarse spaces for solving the Poisson's equation on a $16*DOM \times 16*DOM$ mesh decomposed into $DOM \times DOM$ subdomains with overlapping size $\delta = 2$.

	Hybrid/CG (AS/CG)			
$DOM \times DOM$	iter	cond	max	min
2×2	13 (14)	7.18 (7.26)	4.04 (4.00)	0.557 (0.557)
4×4	18 (18)	7.32 (7.53)	3.98 (4.10)	0.554 (0.544)
8×8	18 (18)	7.35 (7.62)	3.98 (4.12)	0.543 (0.541)
16×16	18 (19)	7.40 (7.65)	3.99 (4.11)	0.540 (0.538)

Table 5. Two-level Hybrid/CG (AS/CG) using non-nested coarse spaces for solving the Poisson's equation on a 256×256 mesh decomposed into 16×16 subdomains with different overlapping sizes δ.

	Hybrid/CG (AS/CG)			
δ	iter	cond	max	min
1	23 (23)	12.9 (13.0)	4.00 (4.02)	0.310 (0.310)
2	18 (19)	7.40 (7.65)	4.00 (4.11)	0.540 (0.538)
3	16 (17)	5.65 (6.06)	4.00 (4.24)	0.708 (0.700)
4	15 (16)	4.83 (5.41)	4.00 (4.38)	0.825 (0.809)

Table 6. Two-level HybridHO/CG (Hybrid/CG) and ASHO/CG (AS/CG) using the PU coarse space without the boundary coarse basis functions. We solve the Poisson's equation on a $16*DOM \times 16*DOM$ mesh decomposed into $DOM \times DOM$ subdomains with overlapping size $\delta = 2$.

Hybrid/CG (AS/CG)

$DOM \times DOM$	iter	cond	max	min
2×2	13 (14)	9.06 (16.4)	2.21 (4.00)	0.245 (0.245)
4×4	22 (23)	13.7 (24.7)	2.21 (4.00)	0.162 (0.162)
8×8	27 (29)	14.9 (26.9)	2.21 (4.00)	0.149 (0.149)
16×16	29 (30)	15.3 (27.6)	2.21 (4.00)	0.145 (0.145)

ASHO/CG (AS/CG)

$DOM \times DOM$	iter	cond	max	min
2×2	13 (14)	9.05 (16.4)	2.21 (4.00)	0.245(0.245)
4×4	26 (27)	22.5 (32.6)	2.76 (4.00)	0.123 (0.123)
8×8	37 (38)	28.1 (39.5)	2.84(4.00)	0.101 (0.101)
16×16	41 (42)	29.6 (41.3)	2.86 (4.00)	0.097 (0.097)

Table 7. Two-level HybridHO/CG (Hybrid/CG) and ASHO/CG (AS/CG) using the PU coarse space without boundary coarse basis functions. We solve the Poisson's equation on a 256×256 mesh decomposed into 16×16 subdomains with different overlapping sizes δ.

HybridHO/CG (Hybrid/CG)

δ	iter	cond	max	min
1	41 (43)	30.5(56.3)	2.17 (4.00)	0.071 (0.071)
2	29 (30)	13.7 (27.6)	2.21 (4.00)	0.162 (0.145)
3	24 (26)	10.2 (18.1)	2.25 (4.00)	0.220 (0.220)
4	21 (21)	7.62 (13.4)	2.27 (4.00)	0.298 (0.298)

ASHO/CG (AS/CG)

δ	iter	cond	max	min
1	59 (60)	62.5 (84.8)	2.94 (4.00)	0.047 (0.047)
2	41 (42)	29.6 (41.3)	2.86 (4.00)	0.097 (0.097)
3	32 (34)	18.6 (26.9)	2.77 (4.00)	0.149 (0.149)
4	28 (29)	13.2 (19.6)	2.70 (4.00)	0.204 (0.204)

References

1. Brezina M. and Vaněk P. (1999) A Black-box Iterative Solvers Based on a Two-level Schwarz Method. Computing. **63**, 233–363
2. Cai X.-C. (1993) An Optimal Two-level Overlapping Domain Decomposition Method for Elliptic Problems in Two and Three Dimensions. SIAM J. Sci. Comp. **14**, 239–247
3. Cai X.-C., Dryja M., and Sarkis M. RASHO: A Restricted Additive Schwarz Preconditioner with Harmonic Overlap. Proceedings of the Thirteen International Symposium on Domain Decomposition Methods, Lyon, France, Oct, 2000, (Submitted)
4. Cai X.-C., Dryja M., and Sarkis M. A Restricted Additive Schwarz Preconditioner with Harmonic Overlap for Symmetric Positive Definite Linear Systems. SIAM J. Sci. Comp. May, 2001, (Submitted)
5. Cai X.-C. and Sarkis M. (1999) A Restricted Additive Schwarz Preconditioner for General Sparse Linear systems. SIAM J. Sci. Comput. **21**, 792–797
6. Chan T., Smith B., and Zou J. (1996) Overlapping Schwarz Methods on Unstructured Meshes using Non-matching Coarse Grids. Numer. Math. **73**, 149–167
7. Dryja M., Smith B., and Widlund O. (1994) Schwarz Analysis of Iterative Substructuring Algorithms for Elliptic Problems in Three Dimensions. SIAM J. Numer. Anal. **31(6)**, 1662–1694
8. Dryja M. and Widlund O. (1994) Domain Decomposition Algorithms with Small Overlap. SIAM J. Sci. Comp. **15**, 604–620
9. Dryja M. and Widlund O. (1995) Schwarz Methods of Neumann-Neumann Type for Three-dimensional Elliptic Finite Elements Problems. Comm. Pure Appl. Math. **48**, 121–155
10. Jenkins E., Kees C., Kelley C., and Miller C. (2001). An Aggregation-based Domain Decomposition Preconditioner for Groundwater Flow. SIAM J. Sci. Comp. **25**, 430–441
11. Lasser C. and Toselli A. Convergence of some Two-level Overlapping Decomposition Preconditioners with Smoothed Aggregation Coarse Spaces. The Proceedings of the Workshop on Domain Decomposition, ETH, Zurich, June, 2001
12. Mandel J. (1993) Balancing Domain Decomposition. Communications in Numerical Methods in Engineerings. **9**, 233–241.
13. Mandel J. (1994) Hybrid Domain Decomposition with Unstructured Subdomains. Contemporary Mathematics. **157**, 103–112
14. Bjørstad P., Dryja M., and Vainikko E. (1996) Additive Schwarz Methods without Subdomain Overlap and with New Coarse Spaces. Applied Parallel Computing in Industrial Problems and Optimization. Springer, Lecture Notes in Computer Science, **1184**
15. B. Smith, Bjørstad P., and Gropp W. (1995) Domain Decomposition: Parallel Multilevel Methods for Elliptic Partial Differential Equations. Cambridge University Press
16. Tezaur R. Vaněk P., and Brezina M. (1995) Two-Level Method for Solids on Unstructured Meshes. Center for Computational Mathematics Report CCM TR **73**, University of Colorado at Denver

Convergence of Some Two-Level Overlapping Domain Decomposition Preconditioners with Smoothed Aggregation Coarse Spaces

Caroline Lasser[1] and Andrea Toselli[2]

[1] Center for Mathematical Sciences, TU München, Arcisstr. 21,
   D-80290 München, Germany
[2] Seminar for Applied Mathematics, ETH Zürich, Rämisstr. 101, CH-8092 Zürich,
   Switzerland

Abstract. We study two-level overlapping preconditioners with smoothed aggregation coarse spaces for the solution of sparse linear systems arising from finite element discretizations of second order elliptic problems. Smoothed aggregation coarse spaces do not require a coarse triangulation. After aggregation of the fine mesh nodes, a suitable smoothing operator is applied to obtain a family of overlapping subdomains and a set of coarse basis functions. We consider a set of algebraic assumptions on the smoother, that ensure optimal bounds for the condition number of the resulting preconditioned system. These assumptions only involve geometrical quantities associated to the subdomains, namely the diameter of the subdomains and the overlap. We first prove an upper bound for the condition number, which depends quadratically on the relative overlap. If additional assumptions on the coarse basis functions hold, a linear bound can be found. Finally, the performance of the preconditioners obtained by different smoothing procedures is illustrated by numerical experiments for linear finite elements in two dimensions.

1 Introduction

We consider the scalar Poisson problem

$$-\Delta u = f, \quad \text{in } \Omega, \\ u = 0, \quad \text{on } \partial\Omega, \tag{1}$$

where Ω is a bounded polyhedral domain in $\mathbb{R}^d$ with $d = 2, 3$.

The discretization of this equation by finite element methods results in a sparse linear system, which is typically too large to be solved directly by Gaussian elimination. Therefore, an iterative solver like the Conjugate Gradient algorithm has to be used. The condition number of the linear system is usually very large and grows quadratically with h^{-1}, where h is the mesh size of the triangulation, thus making convergence very slow. A preconditioner needs to be employed.

Here, we consider a class of two-level overlapping Schwarz preconditioners. These preconditioners consist of two components: the solution of *local* problems associated to an overlapping partition of Ω into subdomains and the solution of a *coarse* problem defined on a low dimensional global space. Local components typically ensure that convergence is independent of h and thus of the size of the original problem (optimality). The coarse level ensures independence of the number of local problems (scalability). Convergence is expected to improve when the relative overlap between the subdomains is increased.

A typical choice for the coarse problem is a finite element approximation on a coarse mesh. For structured meshes, finding a coarse triangulation, such that the fine mesh is a refinement of this coarse one, is relatively easily achieved; see [13] and the references therein. For unstructured meshes, a more general coarse mesh can be employed, but only as long as an interpolation operator from the coarse to the fine finite element space can be found and efficiently implemented; see, e.g., [3]. However, this is not always a trivial task, especially in three dimensions. An alternative approach is realized by smoothed aggregation techniques or partition of unity coarse spaces (see [12]), which provide efficient coarsening procedures without the need of introducing coarse triangulations.

The basic ideas of smoothed aggregation are fairly simple and natural: in a first step, the fine mesh points are aggregated to an initial non-overlapping partition of the domain Ω, and the characteristic functions associated to this non-overlapping partition are considered (aggregation). In a second step, these characteristic functions, whose values typically decrease from one to zero in a layer of width $O(h)$, are smoothed out by the application of a suitable smoothing operator (smoothing). The supports of the smoothed functions define an overlapping partition and corresponding local problems, while their linear span provides a low dimensional coarse problem. This procedure can be applied recursively in order to obtain additional coarse levels for the construction of a multilevel method. The smoother employs the stencil of the finite element matrix and is typically chosen a polynomial of degree $q > 0$ of the original stiffness matrix. The overlap is thus $\delta \sim qh$. The property, that the coarse space represents constant functions, is ensured by exploiting the kernel of the original problem.

An aggregation technique was first introduced in [8] and then quite extensively used for the solution of problems arising in Economics; see [9] and the references therein. Smoothed aggregation techniques have been considered in [17,1] for two-level methods and in [14,16,4,15] for multi-level methods. There, extensive work has been reported on the study of certain smoothers, and practical procedures have been proposed for the initial aggregation, i.e. the initial partition into subdomains. Numerical tests on a large class of scalar and vector problems have been performed. We also mention [7], where smoothed aggregation techniques are applied to discontinuous Galerkin approximations of advection-diffusion problems.

Our assumptions on the smoothers considered here are essentially the same as those already proposed in [17,1]. There, the authors ensure that an optimal preconditioner can be found for the case of generous overlap, i.e., if the overlap δ is comparable to the diameter of the subdomains H. In [6], the case of small overlap is considered, and the condition number of the resulting two-level method is shown to be bounded, if more general assumptions on the overlapping partition and the set of coarse basis functions hold. All these bounds grow quadratically with the inverse of the relative overlap, H/δ. However, a link between the *algebraic* properties of the smoother and optimal bounds for the condition number given in terms of *geometrical* quantities of the overlapping partition is still missing for the case of small overlap. It is the purpose of this paper to bridge this gap. A similar set of assumptions as those given in [17,1] allows us to find the same quadratic bound as in [6]. Using the arguments originally proposed in [12], we also show that if additional assumptions are verified, then a linear bound can be found – as in the case of two-level methods with standard coarse spaces or with partition of unity coarse spaces; see [13, Ch. 5] and [12], respectively. These additional assumptions on the coarse basis functions, however, do not seem to translate into simple algebraic properties on the smoother.

In addition, we enclose some numerical tests in two-dimensions for different choices of the smoother. Although not all of the smoothers tested satisfy the proposed assumptions, our tests do not show any appreciable difference in their numerical performance, i.e., in the number of iterations or the condition number of the preconditioned operator. Our numerical results are consistent with the linear bound on the condition number, even if the additional assumptions on the coarse space required for the proof could not be verified for any of the smoothers considered.

We note that aggregation and smoothed aggregation techniques and partition of unity coarse spaces (see [12]) rely on a similar idea: the coarse basis functions are associated to an overlapping partition into subdomains and no coarse mesh needs to be introduced. However, while in [12] such functions are constructed by assigning explicit nodal values inside the subdomains, in smoothed aggregation techniques they are found by applying a smoother to some initial functions.

The rest of the paper is organized as follows:
In Section 2, we introduce the problem setting and two-level overlapping preconditioners. Section 3 contains the convergence result with quadratic growth in the relative overlap. It is valid, if a suitable set of coarse functions and a proper overlapping partition into subdomains are given. Section 4 deals with their construction by smoothed aggregation techniques. In Section 5, we provide the improved convergence result with linear growth. Section 6 contains the discussion of some smoothing operators, and finally, we present numerical results for a two dimensional problem in Section 7.

2   Problem Setting and Two-Level Overlapping Preconditioners

We consider the Poisson problem (1). We note, that homogeneous Dirichlet conditions have been chosen just for simplicity, and that more general boundary conditions can be dealt with.

For $u, v \in H^1(\Omega)$, we define the bilinear form

$$A(u, v) = \int_\Omega \nabla u \cdot \nabla v \, dx.$$

To approximate the solution of (1), we introduce a shape-regular, quasi-uniform triangulation $\mathcal{T}_h$ of Ω, consisting of triangles or tetrahedra. Let h be the maximum of the diameters of its elements. We define

$$V = \{u \in H^1(\Omega) \mid \ u_{|_\kappa} \in \mathbb{P}_1(\kappa), \ \kappa \in \mathcal{T}_h\},$$

where $\mathbb{P}_1(\kappa)$ is the space of polynomials of maximum degree 1 on κ, and

$$V^0 = V \cap H_0^1(\Omega).$$

Finite element spaces built on quadrilaterals or hexahedra can also be considered, and higher order spaces of piecewise polynomial functions of degree $k > 1$ are possible. The results in this paper remain valid in these cases, with bounds that in general depend on k.

The approximate solution of (1) is then defined as the unique $u \in V^0$ such that

$$A(u, v) = (f, v)_{L^2(\Omega)}, \quad v \in V^0. \tag{2}$$

Problem (2) can be written in matrix form as

$$Au = f. \tag{3}$$

Here, we have used the same notation for a function $u \in V^0$ and the corresponding vector of degrees of freedom, and for a bilinear form $A(\cdot, \cdot)$ and its matrix representation in the space V^0. Similarly, we will use the same notation for functional spaces and the corresponding vector spaces of degrees of freedom.

Next, we introduce a class of two-level overlapping Schwarz preconditioners. Again for simplicity, we only consider additive preconditioners, though multiplicative or hybrid methods can also be devised and analyzed; see [13]. We always assume that we employ exact solvers for the local and coarse problems but approximate solvers could be considered as well. Our theory is easily adjusted to this case; see [13, Ch. 5].

Our preconditioner is uniquely determined by two components:

- an overlapping partition of Ω into subdomains

$$\mathcal{F} = \{\Omega_i' \subset \Omega \mid 1 \leq i \leq N\},$$

  which determines the local solvers and ensures optimality;
- a set of coarse basis functions $\{\Phi_i \mid 1 \leq i \leq N\} \subset V^0$, which determines the coarse solver and ensures scalability.

We note that we consider coarse basis functions that are associated to the subdomains. We will make further assumptions in the following.

Given the partition $\mathcal{F}$, the local spaces are defined by

$$V_i = H_0^1(\Omega_i') \cap V^0, \quad 1 \leq i \leq N. \tag{4}$$

Let $R_i^T : V_i \to V^0$ be the natural injection operator from the subspace V_i into V^0, which extends a local function by zero to the whole of Ω. We recall, that the restriction operator $R_i : V^0 \to V_i$, defined as the transpose of R_i^T with respect to the Euclidean scalar product, extracts the degrees of freedom inside Ω_i'. The matrix block corresponding to the space V_i is obtained by extracting the degrees of freedom relative to the nodes contained in Ω_i' and is equal to

$$A_i = R_i A R_i^T : V_i \longrightarrow V_i.$$

Our coarse space is defined by

$$V_0 = \operatorname{span}\{\Phi_i, \, 1 \leq i \leq N\}.$$

If $R_0^T : V_0 \to V^0$ is the natural injection operator from the subspace V_0 into V^0, then our coarse solver is

$$A_0 = R_0 A R_0^T.$$

The operators $\{A_i, \, i \geq 0\}$ are symmetric and positive-definite.

The additive Schwarz preconditioner is thus defined as

$$\hat{A}^{-1} = \sum_{i=0}^{N} R_i^T A_i^{-1} R_i,$$

and the corresponding preconditioned operator is

$$P = \hat{A}^{-1} A.$$

3   A Convergence Result

Optimality and scalability of the Schwarz algorithms introduced in the previous section are not guaranteed for general partitions and coarse spaces without further assumptions. In this section we introduce two sets of sufficient conditions on $\mathcal{F}$ and the coarse functions. They ensure that the resulting additive preconditioner is optimal and scalable and allow to derive quantitative bounds which only involve the relative overlap between the subdomains, as in the case of two-level methods with standard coarse space; see [13, Ch. 5]. We note that most of the content of this section can be found in [1] for the case of generous overlap, and in [6] for the case of small overlap.

Here, we consider functions

$$\{\Phi_i \mid \ 1 \leq i \leq N\} \subset V^0,$$

such that $\operatorname{supp}\{\Phi_i\} \subset \overline{\Omega'_i}$, $1 \leq i \leq N$. Every function Φ_i vanishes on $\partial\Omega$, and the $\{\Phi_i\}$ are required to form a partition of unity, but only within a proper subset of Ω.

The following two sets of assumptions for the coarse functions $\{\Phi_i\}$ and the partition $\mathcal{F}$ are given in terms of H and δ, $H > \delta > 0$, which reflect the size of the subdomains and the overlap, respectively.

Property 1 (Coarse space I).

1. $|\Phi_i|_1^2 \leq CH^{(d-1)}/\delta$;
2. $\|\Phi_i\|_0^2 \leq CH^d$;
3. There exists $\Omega_{int} \subset \Omega$, such that $\sum_{i=1}^{N} \Phi_i(x) = 1$ for $x \in \Omega_{int}$, and $\operatorname{dist}(x, \partial\Omega) \leq C\delta$ for $x \in \Omega \setminus \Omega_{int}$;
4. $\operatorname{supp}\{\Phi_i\} \subset \overline{\Omega'_i}$.

We note, that a non-negative function Φ_i, which is constant in the interior of Ω'_i and decreases to zero in a layer of width δ around $\partial\Omega'_i$, satisfies the given bounds for the energy $|\Phi_i|_1^2$ as well as for $\|\Phi_i\|_0^2$. The additional property, that coarse functions must reproduce the constants everywhere except on a layer of width δ around the boundary, will translate into an error estimate for a suitably defined interpolation operator; see Lemma 1.

Property 2 (Partition).

1. $\operatorname{diam}(\Omega'_i) \leq CH$;
2. For every $x \in \Omega$, there exists $\Omega'_i \in \mathcal{F}$, such that $x \in \Omega'_i$ and $\operatorname{dist}(x, \partial\Omega'_i \setminus \partial\Omega) \geq c\delta$;
3. There exists C_1 and C_2, such that, for $x \in \Omega$, the ball

$$B(x, rH) = \{y \in \Omega \mid \ \operatorname{dist}(y, x) \leq rH\}$$

intersects at most $C_1 + C_2 r^d$ subdomains in $\mathcal{F}$;
4. $\operatorname{meas}(\Omega_i) \geq CH^d$.

The first and the last property together ensure that the subdomains have diameter of comparable size H and are shape-regular. According to the second property, δ is a measure of the overlap between the subdomains. The third property is equivalent to the finite covering property, which is standard in overlapping methods; see, e.g., [13, Ch. 5].

The following lemma and its proof can be found in [6, Lem. 2.2].

Lemma 1 (Coarse Interpolant I). *Let Property 1 hold. Then there exists an operator $Q_0 : H_0^1(\Omega) \to V_0$, such that*

$$|Q_0 u|_1^2 \leq C \frac{H}{\delta} |u|_1^2,$$

$$\|u - Q_0 u\|_0^2 \leq C H^2 |u|_1^2.$$

To prove a bound for the lowest eigenvalue of the additive operator we need to find a stable decomposition into subspaces, which is given in the following lemma.

Lemma 2. *Let Properties 1 and 2 hold. Then there exists a decomposition $\{u_i \in V_i,\ 0 \leq i \leq N\}$ such that*

$$\sum_{i=0}^{N} A(u_i, u_i) \leq C \left(1 + \frac{H}{\delta}\right)^2 A(u, u), \quad u \in V^0.$$

Proof. Given $u \in V^0$, we define

$$u_0 = Q_0 u,$$
$$u_i = I_h(\theta_i(u - u_0)), \quad 1 \leq i \leq N,$$

where I_h is the nodal interpolation operator into the fine mesh, and the family $\{\theta_i\} \subset V$ is a continuous piecewise linear partition of unity relative to $\mathcal{F}$. We recall, that we can find partitions of unity such that

$$\sum \theta_i(x) = 1 \quad x \in \Omega, \qquad \|\theta_i\|_{0,\infty} \leq C, \qquad |\theta_i|_{1,\infty} \leq C/\delta; \qquad (5)$$

see, e.g., [13, Pg. 166].

Standard arguments, see [13, Pg. 168], give

$$\sum_{i=1}^{N} |u_i|_{1,\Omega}^2 \leq C \left(\left(1 + \frac{H}{\delta}\right) |u - Q_0 u|_{1,\Omega}^2 + \frac{1}{H\delta} \|u - Q_0 u\|_{0,\Omega}^2 \right).$$

The quadratic bound is then found by applying Lemma 1.

We note, that the previous lemma and its proof have already been given in [6, Lem. 2.3], and that their reformulation here is only meant to motivate the additional assumptions on the coarse basis functions, which will be made in Section 5.

Given Lemma 2 and a coloring argument, we can prove a bound for the condition number of the additive operator; see, e.g., [13, Ch. 5].

Theorem 1. *Let Properties 1 and 2 hold. Then there exist constants $c_1 > 0$, $C_2 > 0$, such that for all $u \in V^0$*

$$c_1 \left(1 + \frac{H}{\delta}\right)^{-2} A(u, u) \leq A(u, Pu) \leq C_2 A(u, u).$$

Remark 1. We note that the assumption on the triangulation $\mathcal{T}_h$ being quasi-uniform is not employed in the proofs of this section. Indeed, Theorem 1 is a consequence of Properties 1 and 2 for any arbitrary shape-regular mesh $\mathcal{T}_h$.

4 Smoothed Aggregation

We now consider the task of finding an overlapping partition and a set of coarse functions that satisfy Properties 1 and 2. We start from an initial partition into non-overlapping subdomains $\mathcal{F}_0 = \{\Omega_i \mid 1 \le i \le N\}$. We always assume that these non-overlapping subdomains are shape-regular, and that the diameter of each subdomain is of order H. While algorithms that generate overlapping partitions starting from $\mathcal{F}_0$ can be easily found and implemented, coarse functions that satisfy Property 1 cannot be constructed easily for subdomains of general shape. The method we consider will generate both an overlapping partition and coarse basis functions starting from $\mathcal{F}_0$.

A first choice is to build a coarse space by *aggregation*. We define a set of 'characteristic' functions relative to the initial non-overlapping partition $\mathcal{F}_0$, $\{\Psi_i\} \subset V^0$ and consider the span of these functions. For every node x of $\mathcal{T}_h$ we set

$$\Psi_i(x) = \begin{cases} 0, & x \in \Omega \setminus \Omega_i \text{ or } x \in \partial\Omega, \\ \operatorname{card}(\{j \mid x \in \partial\Omega_j\})^{-1}, & x \in \partial\Omega_i \setminus \partial\Omega, \\ 1, & x \in \Omega_i, \end{cases}$$

where $\operatorname{card}(M)$ denotes the cardinality of a finite set M.

We note, that, if the subdomain boundaries do not contain nodes of the fine mesh $\mathcal{T}_h$, the value of these functions at the nodes is either zero or one, and that they decrease from one to zero in a strip of width h. In the general case they assume values between zero and one, and they decrease from one to zero in a strip of width at most $2h$. Furthermore, the non-vanishing nodal values of Ψ_i cannot be arbitrarily small, since the partition $\mathcal{F}_0$ is shape-regular. These functions form a partition of unity for Ω except in a strip of width $O(h)$ along $\partial\Omega$.

The set $\{\Psi_i\}$ thus satisfies Property 1 with $\delta = h$, and the corresponding coarse space can be analyzed within the framework introduced in the previous section. However, the corresponding additive preconditioner would result in an unsatisfactory bound for the condition number that increases quadratically with H/h. Therefore, the coarse functions $\{\Psi_i\}$ need to be 'smoothed out' to decrease their energy. In order to do so, we apply a suitable operator, called smoother,

$$\Phi_i = S\Psi_i, \quad 1 \le i \le N.$$

This smoothing process shall have the effect of increasing the support of the original functions and of creating additional overlap between their supports. We then define the overlapping subdomains by

$$\overline{\Omega_i'} = \operatorname{supp}\{\Phi_i\}, \tag{6}$$

and obtain an overlapping partition $\mathcal{F} = \{\Omega_i' \subset \Omega \mid 1 \le i \le N\}$.

The smoothing shall also exploit the stencil of the operator A. If $S = p_q(DA)$, where p_q is a polynomial of degree $q \ge 0$ and D a diagonal matrix, then the support of the initial function Ψ_i is increased by q layers of fine elements, which gives an overlap of order $\delta = qh$. In addition, we need to preserve the property that the modified coarse functions $\{\Phi_i\}$ reproduce the constants. This property is guaranteed by the null space of the original differential operator which consists of constant functions. We note that A, the representation of $A(\cdot, \cdot)$ on V^0, is not singular since homogeneous Dirichlet conditions are imposed on $\partial\Omega$, but that when applied to a constant vector, it produces a vector that vanishes everywhere except in a strip around $\partial\Omega$ of width $O(h)$. If $p_q(0) = 1$, then we can write

$$\sum_{i=1}^{N} \Phi_i = S \sum_{i=1}^{N} \Psi_i = S\mathbf{1} = p_q(DA)\,\mathbf{1} = \mathbf{0}' + p_q(0)\mathbf{1} = \mathbf{1}',$$

where $\mathbf{1}$ is the vector of all ones $(1, \ldots, 1)^T$, while $\mathbf{0}'$ and $\mathbf{1}'$ are vectors of zeros and ones, respectively, except for entries relative to nodes inside a neighborhood of width $O(qh)$ around $\partial\Omega$. The smoothed coarse functions thus satisfy Property 1.3 with $\delta = qh$.

In view of these remarks, we consider the following assumptions on the initial partition $\mathcal{F}_0$ and the smoother S.

Property 3 (Initial partition and smoother).

1. The initial partition $\mathcal{F}_0$ satisfies

$$cH^d \leq \text{meas}(\Omega_i) \leq CH^d. \tag{7}$$

2. S is equal to $p_q(DA)$, where p_q is a polynomial of degree q and D a diagonal matrix, such that
   (a) $c\delta \leq qh \leq C\delta \leq C'H$;
   (b) $p_q(0) = 1$;
   (c) $\|S\|_2 \leq 1$;
   (d) $\varrho(S^T AS) \leq Cq^{-2}\varrho(A)$,
   where $\|\cdot\|_2$ and $\varrho(\cdot)$ denote the spectral norm and the spectral radius of a matrix, respectively.

We note that Properties 3.2.c and 3.2.d have already been considered in [17, Lem. 2.8] and [1, Lem. 4.2] for the case when S is a polynomial in A. A similar property to 3.2.a has been stated in [1, Ass. 4.1], but in terms of the graph corresponding to the initial partition $\mathcal{F}_0$.

Lemma 3. *Let S satisfy Property 3. Then we have for the functions $\Phi_i = S\Psi_i$, $1 \leq i \leq N$,*

$$\|\Phi_i\|_0^2 \leq CH^d,$$
$$|\Phi_i|_1^2 \leq CH^{d-1}/\delta. \tag{8}$$

Proof. By construction, the functions Ψ_i and the corresponding column vectors of degrees of freedom, also denoted by Ψ_i, consist of zeros for nodes that belong to elements outside Ω_i. In addition, each non-vanishing entry can be bounded from above and below by a constant. Therefore, $\Psi_i^t \Psi_i$ is bounded from above and below by a constant times the number of nodes inside Ω_i. Since $\mathcal{T}_h$ is shape-regular and quasi-uniform, we have

$$c(\Psi_i^T \Psi_i)\, h^d \leq \mathrm{meas}(\Omega_i) \leq C(\Psi_i^T \Psi_i)\, h^d.$$

If M is the mass matrix, we have

$$\|\Phi_i\|_0^2 = \Psi_i^T S^T M S \Psi_i \leq \varrho(M)\|S\|_2^2 \Psi_i^T \Psi_i \leq C h^d (\Psi_i^T \Psi_i) \leq C \,\mathrm{meas}(\Omega_i),$$

where we have used the property that $\varrho(M)$ is bounded from above by $C h^d$; see, e.g., [11, Sect. 6.3.2].

We next consider the second inequality of (8). A trivial unsatisfactory bound can be easily derived, using the fact that $\varrho(A)$ can be bounded from above by $C h^{d-2}$ (see [11, Sect. 6.3.2]):

$$\|\Phi_i\|_A^2 = \Psi_i^T S^T A S \Psi_i \leq \varrho(S^T A S)(\Psi_i^T \Psi_i) \leq C \frac{h^d\,(\Psi_i^T \Psi_i)}{q^2 h^2} \leq C \frac{H^d}{\delta^2}.$$

To prove a sharper bound, we need to take into account that $A S \Psi_i$ vanishes except in a strip along $\partial \Omega_i'$ of width $O(\delta)$.

For $\kappa \in \mathcal{T}_h$, we denote by R_κ the restriction operator which extracts the degrees of freedom relative to κ and by A_κ the stiffness matrix relative to κ. We note, that if $\overline{\kappa}$ does not intersect $\partial \Omega$, A_κ has a one-dimensional null space consisting of constant functions on κ. We have

$$\|\Phi_i\|_A^2 = \sum_{\kappa \in \mathcal{T}_h} (R_\kappa \Phi_i)^T A_\kappa (R_\kappa \Phi_i).$$

We next define $\Gamma_{i,q}$ as the region of elements, where Φ_i is not constant:

$$\overline{\Gamma_{i,q}} = \bigcup \{\overline{\kappa} \mid \quad \kappa \in \mathcal{T}_h,\ A_\kappa(R_\kappa \Phi_i) \neq 0\}.$$

For $q = 0$, the region $\Gamma_{i,q}$ consists of at most two layers of fine elements, and every application of A adds one additional fine layer in each direction. Therefore, $\Gamma_{i,q}$ is a strip of elements along $\partial \Omega_i'$ of width $O(2qh)$ and thus of measure $O(\delta H^{d-1})$. We clearly have

$$\|\Phi_i\|_A^2 = \sum_{\kappa \subset \Gamma_{i,q}} (R_\kappa \Phi_i)^T A_\kappa (R_\kappa \Phi_i). \tag{9}$$

We next need to relate $R_\kappa \Phi_i = R_\kappa S \Psi_i$ to Ψ_i. We consider an element κ lying in $\Gamma_{i,q}$ and define recursively the regions ω_κ^j, $j \geq 0$. We set $\omega_\kappa^0 = \kappa$ and define ω_κ^j, $j \geq 1$, by taking the union of ω_κ^{j-1} and the neighboring elements that share at least a vertex with ω_κ^{j-1}. Since S is a polynomial of degree q in DA and D is diagonal, the vector $R_\kappa \Phi_i$ is determined only by the values of Ψ_i in ω_κ^q. We set

$$\overline{\Gamma'_{i,q}} = \bigcup_{\kappa \subset \Gamma_{i,q}} \overline{\omega_\kappa^q} \supset \overline{\Gamma_{i,q}},$$

and note that $\Gamma'_{i,q}$ is a layer of elements along $\partial \Omega'_i$ of width $O(4qh)$. Consequently, the expression on the right hand side of (9) is independent of the degrees of freedom of Ψ_i outside the closure of $\Gamma'_{i,q}$. For each node x of $\mathcal{T}_h$, we define

$$\Psi_i^q(x) = \begin{cases} \Psi_i(x), & x \in \overline{\Gamma'_{i,q}}, \\ 0, & \text{otherwise.} \end{cases}$$

Then we can write

$$\|\Phi_i\|_A^2 = (S\Psi_i)^T A (S\Psi_i) = (S\Psi_i^q)^T A (S\Psi_i^q) \le \varrho(S^T A S)(\Psi_i^q)^T \Psi_i^q.$$

Since $(\Psi_i^q)^T \Psi_i^q$ is bounded from above by the number of nodes contained in the closure of $\Gamma'_{i,q}$, we finally find

$$\|\Phi_i\|_A^2 \le C \frac{\varrho(A)}{q^2}(\Psi_i^q)^T \Psi_i^q \le C \frac{h^d (\Psi_i^q)^T \Psi_i^q}{q^2 h^2} \le C \frac{H^{d-1}\delta}{\delta^2} = C \frac{H^{d-1}}{\delta}. \quad (10)$$

Given these bounds on the energy and the norm, respectively, we just need to verify that the functions Φ_i form a partition of unity inside Ω, and that the overlapping partition $\mathcal{F}$ fulfills Property 2. This will be done in the following two lemmas:

Lemma 4. *Let S satisfy Property 3. Then the coarse functions $\Phi_i = S\Psi_i$, $1 \le i \le N$, satisfy Property 1.*

Proof. We only need to prove Property 1.3. The function

$$u(x) = \sum_{i=1}^N \Psi_i(x)$$

is equal to one at every node x that does not belong to $\partial\Omega$, and consequently at every $x \in \Omega$ outside a strip of width $O(h)$ around $\partial\Omega$. For $k \le q$, we have

$$((DA)^k u)(x) = 0,$$

except in a strip of width $O(kh)$ around $\partial\Omega$. We thus obtain

$$\sum_{i=1}^N \Phi_i(x) = \left(S \sum_{i=1}^N \Psi_i \right)(x) = (Su)(x) = 1,$$

at every $x \in \Omega$ outside a strip of width $O(qh) = O(\delta)$ around $\partial\Omega$.

Lemma 5. *Let the initial partition $\mathcal{F}_0$ satisfy Property 3. Then the overlapping partition $\mathcal{F}$ satisfies Property 2.*

Proof. Since the fine mesh $\mathcal{T}_h$ is quasi-uniform, we have

$$\mathrm{diam}(\Omega_i') \leq C(H + qh) \leq CH.$$

Since the original partition is shape-regular, also $\mathcal{F}$ is, and thus we have

$$\mathrm{meas}(\Omega_i') \geq cH^d.$$

We next consider Property 2.2. We first note that for every $x \in \Omega$, there exists i such that $x \in \overline{\Omega_i}$. Since the overlapping subdomains are obtained by adding layers of width $O(qh)$, we have

$$\mathrm{dist}(x, \partial\Omega_i') \geq cqh \geq c\delta.$$

Property 3.1 ensures that, for every i, there is a ball $\mathcal{B}_i \subset \Omega_i$ with $\mathrm{diam}(\mathcal{B}_i) \geq cH$. Since in addition $\mathrm{diam}(\Omega_i') \leq CH$, Property 2.3 holds.

Lemmas 3, 4, and 5, set the stage to apply Theorem 1, and we have:

Theorem 2. *Let Property 3 hold. Then Properties 1 and 2 hold. Therefore, there exist constants c_1, C_2, such that for all $u \in V^0$*

$$c_1 \left(1 + \frac{H}{\delta}\right)^{-2} A(u, u) \leq A(u, Pu) \leq C_2 A(u, u).$$

Remark 2. A closer look at the proofs of this section reveals that the assumption on the global mesh $\mathcal{T}_h$ being quasi-uniform can be relaxed. Indeed it is enough to assume that only the local meshes on the subdomains $\{\Omega_i\}$ are quasi-uniform.

5   An Improved Convergence Result

In this section we find a sharper bound for the condition number of P, which is linear in H/δ. We recall that Schwarz preconditioners with coarse solvers built on coarse triangulations or partition of unity coarse spaces satisfy a linear bound as well. Here, we employ the tools developed in [12]. We recall that there the coarse basis functions are also associated to the subdomains, but are not obtained through smoothed aggregation.

To improve the quadratic bounds of Theorems 1 and 2, we need to impose that our coarse basis functions satisfy additional properties, originally proposed in [12]. In particular they involve bounds on the L^∞-norm of the coarse functions and of their gradients. These additional conditions, however, do not appear to translate into simple conditions on the smoother S, though our numerical results confirm the linear bound, see Section 7.

The algorithm remains the same as before. Only the proof of the decomposition lemma changes, see (5) in Lemma 2 and Lemma 7, and employs the coarse basis functions themselves as the partition of unity $\{\theta_i\}$. Once additional assumptions are satisfied on the coarse basis functions, the proof of the decomposition lemma can be carried out exactly as in [12] and is not presented here.

We consider the same set of coarse functions as before

$$\{\Phi_i \mid \quad 1 \le i \le N\} \subset V^0,$$

and the same coarse space

$$V_0 = \operatorname{span}\{\Phi_i,\ 1 \le i \le N\}.$$

However, we also need an additional function $\Phi_B \in V$ associated to the boundary $\partial\Omega$, so that the augmented set of functions forms a partition of unity on the entire $\overline{\Omega}$. This additional function is only needed for the proof and need not be implemented in practice. Given a partition $\mathcal{F}$, we consider the following assumptions; see [12]:

Property 4 (Coarse space II).

1. $\|\Phi_i\|_{0,\infty} \le C$ for $1 \le i \le N$ and $i = B$;
2. $|\Phi_i|_{1,\infty} \le C/\delta$ for $1 \le i \le N$ and $i = B$;
3. $\displaystyle\sum_{i=B,1}^{N} \Phi_i(x) = 1$ for $x \in \overline{\Omega}$;
4. $\operatorname{supp}\{\Phi_i\} \subset \overline{\Omega_i'},\ 1 \le i \le N$.

The same interpolation operator defined in Lemma 1 can be considered here and the same bounds in Lemma 1 can be proven by noting that Properties 4.1 and 4.2 imply Properties 2.1 and 2.2, respectively.

Lemma 6 (Coarse Interpolant II). *Let Property 4 hold. Then there exists an operator $Q_0 : H_0^1(\Omega) \to V_0$, such that*

$$|Q_0 u|_1^2 \le C \frac{H}{\delta} |u|_1^2,$$

$$\|u - Q_0 u\|_0^2 \le C H^2 |u|_1^2.$$

A stable decomposition can then be found by using the same proof as in [12, Th. 1], since our coarse functions Φ_B and Φ_i satisfy the same properties as θ_B^δ and θ_i^δ in [12]. We thus have

Lemma 7. *Let Properties 4 and 2 hold. Then there exists a decomposition $\{u_i \in V_i,\ 0 \le i \le N\}$ such that*

$$\sum_{i=0}^{N} A(u_i, u_i) \le C \left(1 + \frac{H}{\delta}\right) A(u, u), \quad u \in V^0.$$

Our final result is the following theorem:

Theorem 3. *Let Properties 4 and 2 hold. Then there exist constants $c_1, C_2 > 0$, such that, for all $u \in V^0$,*

$$c_1 \left(1 + \frac{H}{\delta}\right)^{-1} A(u, u) \le A(u, Pu) \le C_2 A(u, u).$$

Remark 3.

1. If $\|S\|_\infty < C$, then Property 4.1 holds. This is easily seen using an analogous argument as in the proof for the corresponding L^2-bound of Lemma 3. However, we were not able to find a suitable condition on S that ensures Property 4.2.
2. The error bound in Lemma 6 is not needed for the proof of Lemma 7.

6   Some Choices for the Smoother

Most of the choices for the smoother S presented in this section have already been proposed in the literature; see [1], [17], [14], and [2]. Not all the Properties 3.2 can be proved for them, except for the recursive Richardson smoother. However, all of them show comparable iteration counts and condition numbers in our numerical experiments; see Section 7. This is due to the fact that the inequalities in Properties 3.2.c and 3.2.d are asymptotic bounds for large q, which is roughly the number of fine mesh layers of overlap between the subdomains, while in practice q is usually not large at all.

Both, the simple and the recursive Richardson smoother, rely on a known upper bound $\hat{\varrho}$ for the spectral radius $\varrho(A)$, that satisfies

$$\varrho(A) \le \hat{\varrho} \le C_\varrho \, \varrho(A). \tag{11}$$

6.1   A Simple Richardson Smoother

We define

$$S = S_q = (Id - \omega \hat{\varrho}^{-1} A)^q, \quad \omega \in (0, 1], \quad q \ge 0.$$

S is the smoother for Richardson's method with relaxation parameter $\omega \hat{\varrho}^{-1}$, similar to a smoother proposed in [14]. It involves q applications of A and it can be defined recursively by the relation

$$S_q = (Id - \omega \hat{\varrho}^{-1} A) S_{q-1}.$$

Consequently, functions and overlapping subdomains with larger overlap can be calculated from the previous ones. By construction, S is a polynomial p_q of degree q in A with $p_q(0) = 1$. It satisfies Property 3 except 3.2.d. We can only prove a somewhat weaker bound for $\varrho(S^T A S)$ that is of order q^{-1}:

Lemma 8. *Let $S = (Id - \omega\hat{\varrho}^{-1}A)^q$, $\omega \in \mathbb{R}$, and $q \geq 0$. Then we have*

$$\|S\|_2 \leq (\max\{|1 - \omega|, 1\})^q .$$

If in addition $\omega \in (0, 1]$ and $q > 0$, then

$$\varrho(S^T A S) \leq C_\varrho \frac{\varrho(A)}{2q\omega} .$$

Proof. For the first bound it is enough to show that

$$\|S_1\|_2 \leq \max\{|1 - \omega|, 1\}.$$

Since S_1 is symmetric, we have to examine its eigenvalues. They satisfy $\lambda(S_1) = 1 - \omega\hat{\varrho}^{-1}\lambda(A)$. Since A is positive definite, we have $\lambda(S_1) \leq 1$. Furthermore,

$$\omega\hat{\varrho}^{-1}\lambda(A) \leq \omega\hat{\varrho}^{-1}\varrho(A) \leq \omega ,$$

and therefore $\lambda(S_1) \geq 1 - \omega$.

We now consider the second bound. We have $S^T A S = S^2 A$, and obtain for $\omega \in (0, 1]$

$$\varrho(S^T A S) \leq \max_{t \in [0, \varrho(A)]} \left|(1 - \omega\hat{\varrho}^{-1}t)^{2q} t\right| \leq \max_{t \in [0, \hat{\varrho}\omega^{-1}]} \left|(1 - \omega\hat{\varrho}^{-1}t)^{2q} t\right| .$$

The function $f(t) := (1 - \omega\hat{\varrho}^{-1}t)^{2q} t$ is non-negative in $[0, \hat{\varrho}\omega^{-1}]$ and satisfies $f(0) = f(\hat{\varrho}\omega^{-1}) = 0$. Its maximum is attained for

$$\bar{t} = \frac{\hat{\varrho}}{(2q + 1)\omega} .$$

Evaluating $f(\bar{t})$ and using $\hat{\varrho} \leq C_\varrho \varrho(A)$ yields the upper bound.

With the same arguments as in the proof of Lemma 3, we can prove, that for a fixed ω the coarse basis functions Φ_i only satisfy

$$|\Phi_i|_1^2 \leq C \frac{H^{d-1}}{h} .$$

As already discussed in Section 4, one would expect that the condition number of the corresponding preconditioner increases quadratically with H/h. However, our numerical experiments show comparable behavior for this simple smoother and the recursive one introduced later, for which the desired sharper bound holds; see Sections 6.2 and 7. This is due to the fact, that in practice q is not large, and thus the following remark applies.

Remark 4. For every $q_0 \geq 0$ there exists a constant $C_0 = C(q_0) > 0$, such that for $0 \leq q \leq q_0$ and $\omega \in (0, 1]$

$$\varrho(S^T A S) \leq C_\varrho C_0 \frac{\varrho(A)}{q^2\omega} .$$

This can be easily seen from the proof of the previous lemma. Indeed we have

$$C_0 = \max_{0 \leq q \leq q_0} \left\{ \left(\frac{2q}{2q+1} \right)^{2q} \frac{q^2}{(2q+1)} \right\}. \tag{12}$$

The key point here is that C_0 remains small for high values of q. We have for instance $C_0 \leq 19$ for $q_0 = 100$, which is a value for q that is far larger than those employed in practice.

6.2 A Recursive Richardson Smoother

We now recall the smoother that was introduced and studied in [1,17]. We define

$$S = S_k = \prod_{i=0}^{k} \left(Id - \frac{4}{3} \varrho_i^{-1} A_i \right), \quad k \geq 0 \tag{13}$$

where $\varrho_i = 9^{-i} \hat{\varrho}$, $i \geq 0$, and

$$A_0 = A, \quad A_i = \left(Id - \frac{4}{3} \varrho_{i-1}^{-1} A_{i-1} \right)^2 A_{i-1}, \quad i \geq 1.$$

S is a recursive Richardson smoother with prefixed relaxation parameters $4/3 \varrho_i^{-1}$. It is a polynomial p_q in A with $p_q(0) = 1$. For the polynomial degree q only certain values are possible. Since $\deg(A_0) = 1$ and $\deg(A_i) = 3 \deg(A_{i-1})$, $i \geq 1$, we have $\deg(A_i) = 3^i$, $i \geq 0$. Therefore, $\deg(S_0) = 1$ and

$$\deg(S_k) = \deg(A_k) + \deg(S_{k-1}) = 3^k + \deg(S_{k-1}), \quad k \geq 1.$$

Consequently,

$$q = \deg(S_k) = \sum_{i=0}^{k} 3^i = \frac{3^{k+1} - 1}{2}, \quad k \geq 0.$$

The following lemma and its proof can be found in [17, Lem. 2.8]. It shows that S satisfies Property 3.

Lemma 9. *Let S be the smoother defined in (13), and $q = \deg(S)$. Then we have*

$$\|S\|_2 \leq 1, \quad \varrho(S^T A S) \leq \frac{C_\varrho}{4} q^{-2} \varrho(A).$$

6.3   A SPAI Smoother

The Richardson smoothers previously introduced depend on a relaxation parameter, which is prefixed in the case of the recursive smoother. However, a preconditioner that does not involve parameters that need to be tuned to the particular type of problem and its coefficients, is preferable; see [7] for the application of smoothed aggregation techniques to an advection-diffusion problem. A parameter-free smoothing operator built with a so-called sparse approximate inverse (SPAI) of the stiffness matrix A is given by

$$S = S_q = (Id - DA)^q, \quad q \geq 0, \tag{14}$$

where the SPAI-matrix D minimizes the Frobenius norm $\|Id - DA\|_F$ over the set of diagonal matrices.

Let n be the size of A. If the column vectors of a matrix M are denoted by m_k, and if e_k is the k-th column of the identity matrix, we write

$$\|Id - DA\|_F^2 = \sum_{k=1}^{n} \|Ad_k - e_k\|_2^2 = \sum_{k=1}^{n} \|d_{kk}a_k - e_k\|_2^2,$$

which is minimized by $D = \text{diag}\{d_{kk}\}$ with

$$d_{kk} = \frac{a_{kk}}{\|a_k\|_2^2}, \quad 1 \leq k \leq n;$$

see [2]. If A is the discrete Laplacian with periodic boundary conditions, which results from a standard second-order finite difference discretization, the SPAI smoother of (14) is the standard damped Jacobi smoother with optimal damping parameter ω; see [2, Proposition 1].

The norm of S satisfies a weaker bound than that stated in Property 3.2.c.

Lemma 10. *Let S be the SPAI smoother defined in (14). Let n denote the size of A, and p_i, $1 \leq i \leq n$, the number of nonzero off-diagonal entries in the i-th row of A. Then we have*

$$\|S\|_2 \leq C(p)^q,$$

where $p = \max_{1 \leq i \leq n} p_i$, and $C(p) = \sqrt{(1 + p)(1 + \sqrt{p})}$.

Proof. It is enough to show $\|S_1\|_2 \leq C(p)$. We will establish bounds for the row-sum-norm of S and S^T, since

$$\|S\|_2 \leq \|S\|_\infty^{\frac{1}{2}} \|S^T\|_\infty^{\frac{1}{2}};$$

see [5, Exercise 2.9.6].

$$\|S\|_\infty = \max_{1\le i\le n} \sum_{j=1}^{n} |s_{ij}| = \max_{1\le i\le n} \sum_{j=1}^{n} \left| \delta_{ij} - \frac{a_{ii}a_{ij}}{\|a_i\|_2^2} \right|$$

$$= \max_{1\le i\le n} \left(1 - \frac{a_{ii}^2}{\|a_i\|_2^2} + \sum_{j\ne i} \frac{a_{ii}|a_{ij}|}{\|a_i\|_2^2} \right)$$

$$\le \max_{1\le i\le n} \left(1 - \frac{a_{ii}^2}{\|a_i\|_2^2} + \frac{a_{ii}}{\|a_i\|_2^2} \cdot \|a_i\|_2 \cdot \sqrt{p_i} \right)$$

$$\le 1 + \sqrt{p},$$

where we have employed Cauchy-Schwarz for the last but one inequality, and used the symmetry of A.

Furthermore,

$$\|S^T\|_\infty = \max_{1\le i\le n} \sum_{j=1}^{n} |s_{ji}| = \max_{1\le i\le n} \sum_{j=1}^{n} \left| \delta_{ji} - \frac{a_{jj}a_{ji}}{\|a_j\|_2^2} \right|$$

$$= \max_{1\le i\le n} \left(1 - \frac{a_{ii}^2}{\|a_i\|_2^2} + \sum_{j\ne i} \frac{a_{jj}|a_{ij}|}{\|a_j\|_2^2} \right)$$

$$\le \max_{1\le i\le n} \left(1 - \frac{a_{ii}^2}{\|a_i\|_2^2} + \sum_{j\ne i} \frac{|a_{ij}|}{\|a_j\|_2} \right) \le \max_{1\le i\le n} \left(1 - \frac{a_{ii}^2}{\|a_i\|_2^2} + 1 \cdot p_i \right)$$

$$\le 1 + p.$$

Here p is equal to the maximal number of neighbors that a mesh point of $\mathcal{T}_h$ has, and for, e.g., $p = 7$, we have $C(p) \sim 5.4$. We note that $C(p) > 1$ and that consequently Property 3.2.c cannot be derived from Lemma 10.

If one does not assume A to be weakly diagonal dominant, only triangulations with $p \le 7$ give an upper bound for $\varrho(S^T AS)$ which asymptotically decays to zero for $q \to \infty$; see [2, Theorem 1]) for the proof of the following lemma.

Lemma 11. *Assume $p \le 7$, and denote with*

$$\hat{C} = \max_{1\le i\le n} \frac{\|a_i\|_2^2}{a_{ii}^2}, \quad \Gamma = \frac{1}{2}(1 + \sqrt{1+p}) < 2.$$

Then we have

$$\varrho(S^T AS) \le \hat{C}\, \eta(2q)\, \varrho(A), \tag{15}$$

where

$$\eta(2q) = \max\left\{ \left(\frac{2q}{2q+1}\right)^{2q} \frac{1}{2q+1}, \ \Gamma(\Gamma-1)^{2q} \right\}.$$

Though the bound in the previous lemma is even weaker than the one obtained for the simple Richardson smoother and, of course, fails to satisfy Property 3.2.d, the numerical performance of the SPAI smoother is comparable to the results obtained by the other choices for S, see Section 7.

7 Numerical Results

We have tested the performance of our Schwarz preconditioner for the Poisson problem

$$-\Delta u = xe^y, \quad \text{in } \Omega = (0,1)^2 \,,$$
$$u = -xe^y, \quad \text{on } \partial\Omega \,. \tag{16}$$

This choice of Dirichlet conditions also numerically confirms our claim that inhomogeneous boundary conditions can be dealt with.

The mesh is built by dividing Ω into n^2 equal fine squares and cutting them into halves. Thus, we obtain a triangulation $\mathcal{T}_h$ with $h = \frac{1}{n}$, $h \in \left\{\frac{1}{16}, \frac{1}{32}, \frac{1}{64}, \frac{1}{128}\right\}$. The maximum number of neighbors, which a mesh point in $\mathcal{T}_h$ has, is $p = 6$. The aggregation routine partitions Ω into non-overlapping subsquares Ω_i of area H^2 with $H \in \left\{\frac{1}{2}, \frac{1}{4}, \frac{1}{8}, \frac{1}{16}, \frac{1}{32}\right\}$.

Depending on the polynomial degree q of the smoother, which varies between 0 and 14 in our experiments, we obtain an overlapping partition $\mathcal{F}$ with overlap $\delta = (q + 2)h$. The sum $q + 2$ results from the fact, that the boundaries $\partial\Omega_i$ of the non-overlapping subsquares contain fine mesh points, which causes the support of the initial, unsmoothed coarse basis functions Ψ_i overlap in a strip of width $2h$.

For the simple and the recursive Richardson smoother, we replace the upper bound for the spectral radius of the stiffness matrix by an estimate for the spectral radius, which is given by the *Matlab* built-in function *normest*. We use linear finite elements, and solve the resulting linear system by Conjugate Gradient. The Conjugate Gradient routine employed belongs to the NetLib-linalg package, see [10]. It provides an estimate for the condition number of the preconditioned operator by dividing the maximal and the minimal eigenvalue of a suitable tridiagonal matrix, which approximates the preconditioned operator. Our stopping criterion is the reduction of the residual norm by a factor of 10^{-6} or the exceeding of 100 iterations.

Effectiveness of the coarse solve: Tables 1 and 2 show iteration counts and estimated condition numbers for the one- and two-level preconditioners built with the SPAI smoother. As expected, iteration counts and condition numbers for the one-level algorithm increase rapidly with the number of subdomains, while the coarse solve in the corresponding two-level algorithm, see Table 2, only allows a moderate increase or keeps the iterations counts bounded for the case of generous overlap.

Comparison of some smoothers: Tables 2, 3 and 4 show iteration counts and estimated condition numbers of the two-level preconditioned operators which were obtained with the SPAI, the simple Richardson, and the recursive Richardson smoother, respectively. The relaxation parameter for the simple Richardson smoother was chosen as $\omega = 2/3$. Though only the recursive Richardson smoother meets the required theoretical bounds, there is just a slight difference in the numerical performance of the three smoothers. As previously remarked, this can be explained by the fact that only small values of the smoother polynomial degree q are employed in practice, while our theory requires bounds which hold for large q as well.

Linear dependence on the relative overlap: Figure 1 shows a linear and quadratic least-squares fit for the estimated condition number of the preconditioner built with the recursive Richardson smoother versus the relative overlap H/δ. Since the coefficient for the quadratic term is small and the least-square relative error is the same for the linear and the quadratic fit, our experiments confirm a linear dependence.

Table 1. Iteration counts and estimated condition numbers (in parenthesis) for Conjugate Gradient with one-level preconditioner versus h and the relative overlap.

		H/δ			
h^{-1}	H^{-1}	16	8	4	2
16	2	-	-	13 (16.7)	12 (6.0)
16	4	-	-	-	17 (25.4)
32	2	-	17 (32.6)	14 (11.1)	11 (5.3)
32	4	-	-	24 (52.8)	16 (15.3)
32	8	-	-	-	31 (93.2)
64	2	22 (64.6)	17 (21.7)	13 (9.6)	11 (5.0)
64	4	-	34 (107.4)	22 (33.6)	15 (12.5)
64	8	-	-	42 (198.5)	27 (54.9)
64	16	-	-	-	57 (365.5)
128	4	46 (216.7)	30 (70.2)	21 (28.3)	14 (11.4)
128	8	-	59 (408.8)	40 (125.2)	25 (44.2)
128	16	-	-	80 (782.1)	52 (214.1)
128	32	-	-	-	100 (1454.8)

References

1. Brezina, M., Vaněk, P. (1999) A black–box iterative solver based on a two–level Schwarz method. Computing **63**, 233–263
2. Bröker, O., Grote. M. J., Mayer, C., Reusken, A. (2000) Robust Parallel Smoothing for Multigrid Via Sparse Approximate Inverses. Submitted to SIAM J. Sc. Comp.
3. Chan, T. F., Smith. B. F., Zou. J. (1996) Overlapping Schwarz Methods on Unstructured Meshes using Non-matching Coarse Grids. Numer. Math. **73(2)**, 149–167

Table 2. Iteration counts and estimated condition numbers (in parenthesis) for Conjugate Gradient with the two–level preconditioner, using the SPAI smoother.

h^{-1}	H^{-1}	H/δ 16	8	4	2
16	2	-	-	15 (13.8)	13 (5.1)
16	4	-	-	-	16 (10.0)
32	2	-	18 (27.4)	15 (9.7)	13 (4.8)
32	4	-	-	22 (21.3)	16 (7.8)
32	8	-	-	-	19 (11.8)
64	2	24 (54.6)	19 (19.2)	14 (8.8)	12 (4.8)
64	4	-	30 (43.6)	21 (17.5)	15 (8.0)
64	8	-	-	26 (25.3)	19 (10.5)
64	16	-	-	-	19 (12.3)
128	4	41 (88.3)	28 (36.4)	19 (17.8)	15 (8.7)
128	8	-	38 (51.9)	27 (23.8)	19 (12.3)
128	16	-	-	27 (26.2)	21 (11.6)
128	32	-	-	-	19 (12.4)

Table 3. Iteration counts and estimated condition numbers (in parenthesis) for Conjugate Gradient with the two-level preconditioner, using the simple Richardson smoother.

h^{-1}	H^{-1}	H/δ 16	8	4	2
16	2	-	-	15 (13.8)	13 (5.4)
16	4	-	-	-	16 (10.0)
32	2	-	18 (27.4)	15 (10.1)	13 (5.0)
32	4	-	-	22 (21.3)	16 (9.4)
32	8	-	-	-	19 (11.8)
64	2	24 (54.6)	19 (19.9)	14 (9.1)	13 (5.0)
64	4	-	30 (43.6)	22 (20.4)	16 (9.6)
64	8	-	-	26 (25.3)	22 (13.9)
64	16	-	-	-	19 (12.3)
128	4	41 (88.3)	30 (42.0)	20 (20.9)	15 (10.0)
128	8	-	38 (51.9)	30 (30.6)	21 (16.9)
128	16	-	-	27 (26.2)	25 (16.1)
128	32	-	-	-	19 (12.4)

4. Guillard, H., Vaněk, P. (1998) An aggregation multigrid solver for convection-diffusion problems on unstructured meshes. TR **130**, Center for Computational Mathematics, University of Colorado, Denver

5. Hackbusch, W. (1994) Iterative Solution of Large Sparse Systems of Equations. Springer-Verlag

6. Jenkins, E. W., Kelley, C. T., Miller, C. T., Kees, C. E. (2000) An Aggregation-based Domain Decomposition Preconditioner for Groundwater Flow. TR **00–13**, Department of Mathematics, North Carolina State University

Table 4. Iteration counts (first rows) and estimated condition numbers (second rows, in parenthesis) of the two-level preconditioner with recursive Richardson smoothing.

h^{-1}	H^{-1}	H/δ							
		16	$10\frac{2}{3}$	8	$5\frac{1}{3}$	4	$2\frac{2}{3}$	$2\frac{2}{15}$	2
16	2	-	-	-	-	15	14	-	-
		-	-	-	-	(13.8)	(7.3)	-	-
16	4	-	-	-	-	-	-	-	16
		-	-	-	-	-	-	-	(10.0)
32	2	-	-	18	17	-	13	-	-
		-	-	(27.4)	(14.3)	-	(5.9)	-	-
32	4	-	-	-	-	22	18	-	-
		-	-	-	-	(21.3)	(12.0)	-	-
32	8	-	-	-	-	-	-	-	19
		-	-	-	-	-	-	-	(11.8)
64	2	24	20	-	15	-	-	11	-
		(54.6)	(28.5)	-	(11.4)	-	-	(4.4)	-
64	4	-	-	30	25	-	16	-	-
		-	-	(43.6)	(25.6)	-	(9.1)	-	-
64	8	-	-	-	-	26	23	-	-
		-	-	-	-	(25.3)	(15.7)	-	-
64	16	-	-	-	-	-	-	-	19
		-	-	-	-	-	-	-	(12.3)
128	4	41	33	-	-	-	22	-	13
		(88.3)	(52.5)	-	-	-	(20.0)	-	(5.8)
128	8	-	-	38	31	-	20	-	-
		-	-	(51.9)	(33.6)	-	(11.9)	-	-
128	16	-	-	-	-	27	24	-	-
		-	-	-	-	(26.2)	(16.9)	-	-
128	32	-	-	-	-	-	-	-	19
		-	-	-	-	-	-	-	(12.4)

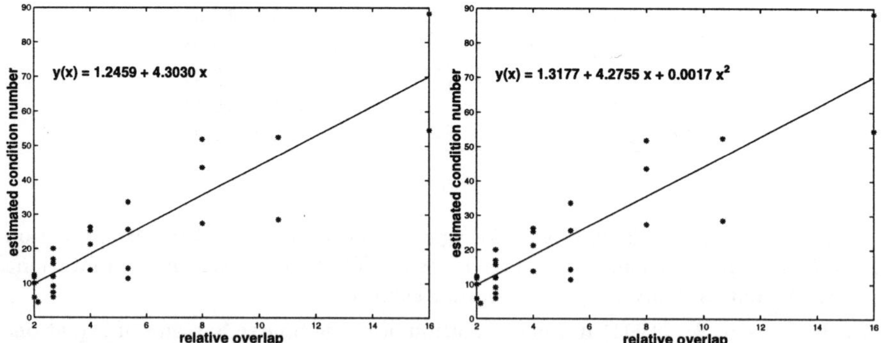

Fig. 1. Linear and quadratic least-squares fit for the estimated condition number of the two-level preconditioned operator versus the relative overlap H/δ. The preconditioner is built using the recursive Richardson smoother.

7. Lasser, C., Toselli, A. Overlapping preconditioners for discontinuous Galerkin approximations of second order problems. To appear in the proceedings of the 13th International Conference of Domain Decomposition, Lyon, October 2000. http://www.sam.math.ethz.ch/ toselli/dd13.ps.gz

8. Leontief, W. (1951) The structure of the American Economy. Oxford University Press, New York

9. Mandel, J., Sekerka, B. (1983) A local convergence proof for the iterative aggregation method. Lin. Alg. and Its Applic. **51**, 163–172

10. NetLib (1993) Univ. of Tennessee and Oak Ridge National Laboratory. http://www.netlib.org/linalg/

11. Quarteroni, A., Valli, A. (1994) Numerical approximation of partial differential equations. Springer-Verlag, Berlin

12. Sarkis, M. (2001) Partition of Unity coarse spaces and Schwarz with harmonic overlap. Submitted to the proceedings of the Workshop on Domain Decomposition, ETH Zürich

13. Smith, B. F., Bjørstad, P. E., Gropp, W. D. (1996) Domain Decomposition: Parallel Multilevel Methods for Elliptic Partial Differential Equations. Oxford University Press

14. Vaněk, P., Mandel, J., Brezina, M. (1996) Algebraic multigrid by smooth aggregation for second and fourth order elliptic problems. Computing **56**, 179–196

15. Vaněk, P., Brezina, M., Mandel, J. (2001) Convergence of algebraic multigrid based on smoothed aggregation. Numerische Mathematik **88**, 559–679

16. Vaněk, P., Mandel, J., Brezina, M. (1997) Solving a two-dimensional Helmholtz problem using algebraic multigrid TR **110**, Center for Computational Mathematics, University of Colorado, Denver

17. Vaněk, P., Brezina, M., Tezaur, R. (1999) Two-grid method for linear elasticity on unstructured meshes. SIAM J. Sci. Comput. **21**, 900–923

Wavelet/FEM Coupling by the Mortar Method

Silvia Bertoluzza[1], Silvia Falletta[2], and Valérie Perrier[3]

[1] I.A.N.-C.N.R., v. Ferrata 1, 27100 Pavia, Italy
[2] Dipartimento di Matematica, Universitá di Pavia, v. Ferrata 1, 27100 Pavia, Italy
[3] Laboratoire de Modélisation et Calcul de l'IMAG, BP 53 - 38 041 Grenoble cedex 9, France

Abstract. We propose and analyze in an abstract framework a mortar method with approximate constraint, based on replacing the "master side" function appearing in the weak continuity condition with its projection on a suitably chosen auxiliary space. We show how to choose the auxiliary space in such a way that such a technique can be applied for computing the weak continuity constraints arising in the framework of the wavelet/FEM coupling.

1 Introduction

One of the advantages of non-conforming domain decomposition methods, is that they allow in principle to couple discretizations of different type. This feature is particularly attractive when considering discretizations whose applicability is, by nature, limited to tensor product domains. This is the case of wavelet type methods. Wavelets perform in a very promising way on academic examples. For instance they allow the design of simple and efficient adaptive schemes and the stiffness matrices resulting from their application are well conditioned (after a suitable rescaling). Their application to real life problems is however seriously limited by the issue of geometry, among other things. Coupling (in a domain decomposition approach) with finite elements allows in principle to overcome this limitation.

In non-conforming domain decomposition methods, and in particular in the mortar method ([1]) which we are considering here, continuity across the interface is relaxed to a weak continuity constraint. In order to couple discretizations of different type, the need eventually arises for applying the constraint operator, which implies the need for computing integrals of products of functions belonging to the two different discretizations. In particular, in the Wavelet/FEM coupling the problem arises of computing the integral of a wavelet type function times a piecewise polynomial defined on an unstructured grid. Unfortunately, due to the particular nature of wavelets, which are not known in closed form, there is up to date no way of computing such an integral exactly, and it is therefore necessary to approximate it somehow. Clearly the approximation of such an integral in the constraint results in general in a different definition of the weak continuity constraint, and therefore in a different approximation space with respect to the one obtained when computing the integrals exactly. Such change has an effect on both consistency and approximation errors ([5]). Moreover it is well known that when considering wavelets, standard quadrature formulae are not well suited, since wavelets have lower regularity than the one needed to match the potential accuracy of the method.

In this paper we propose, in an abstract framework, a different way of approximating such integrals which is particularly well suited for the coupling of wavelet and finite elements. This consists in approximating, within the integral, a product of the form $u_h^+ \lambda_h$ (u_h^+ being the trace on the so called "master" or "mortar" side of the piecewise H^1 function u_h, and λ_h being the multiplier) with the product $P^-(u_h^+)\lambda_h$, where P^- is an $H_{00}^{1/2}$–bounded projector onto a suitable auxiliary space U_h^-. We provide an abstract error estimate for the resulting modified mortar method and we show how to choose the auxiliary space and the projector when considering the coupling of wavelets with finite elements.

2   The Mortar Method with Approximate Constraint

Let $\Omega \subset \mathbb{R}^2$ be a bounded polygonal domain and consider the model problem: given $f \in L^2(\Omega)$ find $u : \Omega \to \mathbb{R}$ s.t.

$$-\nabla \cdot (a\nabla u) = f \ \text{ in } \Omega, \qquad u = 0 \ \text{ on } \Gamma = \partial\Omega, \tag{1}$$

where for simplicity we assume that the matrix a is constant symmetric positive definite. We consider here a very simple example of non-conforming domain decomposition. More precisely consider a geometrically conforming splitting of Ω in two polygonal subdomains as

$$\bar\Omega = \bar\Omega_+ \cup \bar\Omega_-, \qquad \text{with} \qquad \gamma = \partial\Omega_+ \cap \partial\Omega_-,$$

in such a way that Ω_+ and Ω_- share one common edge γ.

Denote by V_h^+ and V_h^-

$$V_h^+ \subset H_\Gamma^1(\Omega_+) = \{u \in H^1(\Omega_+) : \ u = 0 \text{ on } \Gamma \cap \Omega_+\} \tag{2a}$$

$$V_h^- \subset H_\Gamma^1(\Omega_-) = \{u \in H^1(\Omega_-) : \ u = 0 \text{ on } \Gamma \cap \Omega_+\} \tag{2b}$$

the two discrete spaces chosen for approximating u in Ω_+ and Ω_- respectively. Without loss of generality, we can assume the side of γ coming from Ω_+ to be the master side of the side of γ coming from Ω_-. Let $M_h \subset H^{-1/2}(\gamma)$, with $dim(M_h) = dim(V_h^-|_\gamma)$, be a suitable multiplier space — which in the mortar method is obtained from a subspace of $V_h^-|_\gamma$ with suitable modifications at the vertices of γ ([1]), or which coincides, in a more general formulation, with a suitable "dual space" of $V_h^-|_\gamma$ ([3,9]). In the classical formulation of the mortar method, the approximation of the solution of (1) is sought in the constrained space $\mathcal{X}_h$ defined as

$$\mathcal{X}_h = \{u : \ u|_{\Omega_+} \in V_h^+, \ \ u|_{\Omega_-} \in V_h^-, \ \ \int_\gamma [u]\lambda \, ds = 0 \ \ \forall \lambda \in M_h\},$$

where, introducing the notation $u^+ = u|_{\Omega_+}$ and $u^- = u|_{\Omega_-}$, $[u] = u^+|_\gamma - u^-|_\gamma$ denotes the jump of the function u across the interface γ. The solution u to problem (1) is approximated by looking for u_h in $\mathcal{X}_h$ such that for all $v_h \in \mathcal{X}_h$ it holds

$$\int_{\Omega_+} a\nabla u_h \cdot \nabla v_h \, dx + \int_{\Omega_-} a\nabla u_h \cdot \nabla v_h \, dx = \int_\Omega f v_h.$$

In the solution of the linear system resulting from such problem the need arises eventually of computing the integrals appearing in the constraint

$$\int_\gamma (u_h^+ - u_h^-)\lambda \, ds = 0, \qquad \forall \lambda \in M_h. \tag{3}$$

Since in the mortar method the multiplier space M_h is strongly related to the "slave" space V_h^-, it is reasonable to assume that the integrals of the products $u_h^- \lambda$ are computable in practice. This is not necessarily the case of the products $u_h^+ \lambda$, where functions originating from totally unrelated spaces are involved. We will concentrate here on approximating this term in the constraint.

In order to do that, let us introduce two auxiliary spaces $U_\delta^- \subset L^2(\gamma)$ and $U_\delta^+ \subset L^2(\gamma)$ depending on a parameter δ, which we assume to have the same finite dimension and to satisfy

$$\inf_{\zeta \in U_\delta^-} \sup_{\eta \in U_\delta^+} \frac{\int_\gamma \zeta \eta \, ds}{\|\zeta\|_{H_{00}^{1/2}(\gamma)} \|\eta\|_{-1/2,\gamma}} \geq \alpha > 0. \tag{4}$$

Assume that the two auxiliary spaces are chosen in such a way that the integrals of the form $\int_\gamma \zeta \eta \, ds$ are computable provided either $\zeta \in V_h^+|_\gamma$ and $\eta \in U_\delta^+$ or $\zeta \in V_h^-|_\gamma$ and $\eta \in U_\delta^-$. For all $\zeta \in L^2(\gamma)$ let $P^-(\zeta) \in U_\delta^-$ be the unique element in U_δ^- such that

$$\int_\gamma P^-(\zeta)\eta \, ds = \int_\gamma \zeta \eta \, ds, \qquad \forall \eta \in U_\delta^+. \tag{5}$$

Remark that, thanks to property (4) the projector P^- is well defined and it verifies

$$\|P^-(\zeta)\|_{H_{00}^{1/2}(\gamma)} \lesssim \|\zeta\|_{H_{00}^{1/2}(\gamma)}.$$

We propose here to approximate the integral of the product $u_h^+ \lambda$ with the integral of $P^-(u_h^+)\lambda$ (where, by abuse of notation we will write u_h^+ instead of $u_h^+|_\gamma$). The constraint (3) is then replaced by the approximated constraint

$$\int_\gamma (P^-(u_h^+) - u_h^-)\lambda \, ds = 0, \qquad \forall \lambda \in M_h, \tag{6}$$

which corresponds to defining a new constrained space as

$$\mathcal{X}_h^* = \{u: \ u|_{\Omega_+} \in V_h^+, \ u|_{\Omega_-} \in V_h^-, \ \int_\gamma (P^-(u^+) - u^-)\lambda \, ds = 0 \ \forall \lambda \in M_h\},$$

and approximating the solution to (1) by the solution of the following discrete problem: find $u_h \in \mathcal{X}_h^*$ such that for all $v_h \in \mathcal{X}_h^*$ it holds

$$\int_{\Omega_+} a\nabla u_h \cdot \nabla v_h \, dx + \int_{\Omega_-} a\nabla u_h \cdot \nabla v_h \, dx = \int_\Omega f v_h. \tag{7}$$

Denoting by $\| \cdot \|_{1,*}$

$$\| \cdot \|_{1,*} = \| \cdot \|_{1,\Omega_+} + \| \cdot \|_{1,\Omega_-}$$

the broken H^1 norm, we can prove the following bound.

Theorem 1. *Let the multiplier space M_h be chosen in such a way that the following assumptions are satisfied:*

(A1) *there exists a bounded projection $\pi : L^2(\gamma) \longrightarrow V_h^-|_\gamma$, such that for all $\eta \in L^2(\gamma)$ and for all $\lambda \in M_h$*

$$\int_\gamma (\eta - \pi\eta)\lambda = 0, \tag{8}$$

and for all $\eta \in H_{00}^{1/2}(\gamma)$

$$\|\pi\eta\|_{H_{00}^{1/2}(\gamma)} \lesssim \|\eta\|_{H_{00}^{1/2}(\gamma)}. \tag{9}$$

(A2) *there exists a discrete lifting $R_h : V_h^-|_\gamma \longrightarrow V_h^-$ such that for all $\eta \in V_h^-|_\gamma$*

$$\|R_h\eta\|_{1,\Omega_-} \lesssim \|\eta\|_{H_{00}^{1/2}(\gamma)}. \tag{10}$$

Moreover let the two auxiliary spaces U_δ^+ and U_δ^- be chosen in such a way that the following Jackson type inequality holds for some $\tilde{R}, R \geq 1/2$: for all r, $1/2 \leq r \leq R$ (resp. for all $\tilde{r}$, $-1/2 \leq \tilde{r} \leq \tilde{R}$)

$$\eta \in H_0^r(\gamma) \quad \Longrightarrow \quad \inf_{\eta_\delta \in U_\delta^-} \|\eta - \eta_\delta\|_{1/2,\gamma} \lesssim \delta^{r-1/2}\|\eta\|_{r,\gamma}, \tag{11a}$$

$$\eta \in H^{\tilde{r}}(\gamma) \quad \Longrightarrow \quad \inf_{\eta_\delta \in U_\delta^+} \|\eta - \eta_\delta\|_{-1/2,\gamma} \lesssim \delta^{\tilde{r}+1/2}\|\eta\|_{\tilde{r},\gamma}. \tag{11b}$$

Then, if u_h is the solution of problem (7), and if the true solution u of problem (1) verifies $u \in H^s(\Omega)$ for some s, $2 \leq s \leq \min\{\tilde{R} + 3/2, R + 1/2\}$, the following error estimate holds:

$$\|u - u_h\|_{1,*} \lesssim \delta^{s-1}\|u\|_{s,\Omega} + \inf_{\lambda \in M_h} \|\partial_{\nu_a} u - \lambda\|_{H^{-1/2}(\gamma)}$$

$$+ \inf_{v_h \in V_h^+} \|u - v_h\|_{1,\Omega_+} + \inf_{v_h \in V_h^-} \|u - v_h\|_{1,\Omega_-} \tag{12}$$

where ∂_{ν_a} denotes the trace on γ of the outer co-normal derivative to the subdomain Ω_+.

Proof. Let

$$\mathcal{X}^* = \{u \in H^1_\Gamma(\Omega_+) \times H^1_\Gamma(\Omega_-) \text{ s.t. (6) holds }\},$$

and let $u_* \in \mathcal{X}^*$ be the unique element such that for all $v \in \mathcal{X}^*$ it holds that

$$\int_{\Omega_+} a\nabla u_* \cdot \nabla v \, dx + \int_{\Omega_-} a\nabla u_* \cdot \nabla v \, dx = \int_\Omega fv.$$

Let us denote by $a^* : H^1_\Gamma(\Omega_+) \times H^1_\Gamma(\Omega_-) \to \mathbb{R}$ the bilinear form

$$a^*(v, w) = \int_{\Omega_+} a\nabla v \cdot \nabla w \, dx + \int_{\Omega_-} a\nabla v \cdot \nabla w \, dx.$$

We first observe that, since both $\partial\Omega_+ \setminus \gamma$ and $\partial\Omega_- \setminus \gamma$ have strictly positive length (since γ coincides with just one edge of both Ω_+ and Ω_-), a^* is coercive on $H^1_\Gamma(\Omega_+) \times H^1_\Gamma(\Omega_-)$ with respect to the broken norm $\| \cdot \|_{1,*}$.

Since $\mathcal{X}^*_h \subset \mathcal{X}^*$, and since for all $w_h \in \mathcal{X}^*_h$ we have, by Galerkin orthogonality, $a^*(u_* - u, w_h) = 0$, it is not difficult to check that it holds for all $v_h \in \mathcal{X}^*_h$

$$\|u_* - u_h\|^2_{1,*} \lesssim a(u_* - u_h, u_* - u_h) = a(u_* - u_h, u_* - u + u - v_h)$$

which yields

$$\|u_* - u_h\|_{1,*} \lesssim \|u - u_*\|_{1,*} + \|u - v_h\|_{1,*}.$$

The arbitrariness of v_h implies

$$\|u - u_h\|_{1,*} \lesssim \|u - u_*\|_{1,*} + \inf_{v_h \in \mathcal{X}^*_h} \|u - v_h\|_{1,*}. \tag{13}$$

Since u is the solution of (1), for all $v \in H^1_\Gamma(\Omega_+) \times H^1_\Gamma(\Omega_-)$ it holds

$$\int_{\Omega_+} a\nabla u \cdot \nabla v \, dx + \int_{\Omega_-} a\nabla u \cdot \nabla v \, dx = \int_\Omega fv - \int_\gamma \partial_{\nu_a} u[v] \, ds.$$

Then for any $\tilde{\eta}_\delta \in U^+_\delta$ and $\lambda_h \in M_h$, using (6) and (5) we can bound the first term on the right hand side of (13) as:

$$\|u - u_*\|^2_{1,*} \lesssim \left| \int_\gamma \partial_{\nu_a} u[u_*] \, ds \right|$$

$$\lesssim \int_\gamma \left| \partial_{\nu_a} u(u^+_* - P^-(u^+_*)) \right| + \int_\gamma \left| \partial_{\nu_a} u(P^-(u^+_*) - u^-_*) \right|$$

$$= \int_\gamma \left| (\partial_{\nu_a} u - \tilde{\eta}_\delta)(u^+_* - P^-(u^+_*)) \right|$$

$$\quad + \int_\gamma \left| (\partial_{\nu_a} u - \lambda_h)(P^-(u^+_*) - u^-_*) \right|$$

$$\lesssim \|\partial_{\nu_a} u - \tilde{\eta}_\delta\|_{-1/2,\gamma} \|(u^+_* - P^-(u^+_*))\|_{H^{1/2}_{00}(\gamma)}$$

$$\quad + \|\partial_{\nu_a} u - \lambda_h\|_{-1/2,\gamma} \|P^-(u^+_*) - u^-_*\|_{H^{1/2}_{00}(\gamma)}.$$

By triangular inequality, we can bound

$$\|u_*^+ - P^-(u_*^+)\|_{H_{00}^{1/2}(\gamma)} \leq \|u_*^+ - u\|_{H_{00}^{1/2}(\gamma)}$$
$$+ \|u - P^-u\|_{H_{00}^{1/2}(\gamma)} + \|P^-(u - u_*^+)\|_{H_{00}^{1/2}(\gamma)}$$

and

$$\|P^-(u_*^+) - u_*^-\|_{H_{00}^{1/2}(\gamma)} \leq \|P^-(u_*^+ - u)\|_{H_{00}^{1/2}(\gamma)} + \|P^-u - u\|_{H_{00}^{1/2}(\gamma)}$$
$$+ \|u - u_*^-\|_{H_{00}^{1/2}(\gamma)},$$

from which, using the boundedness of P^- as well as the trace theorem we obtain

$$\|u_*^+ - P^-(u_*^+)\|_{H_{00}^{1/2}(\gamma)} \leq \|u - u_*\|_{1,*} + \|u - P^-u\|_{H_{00}^{1/2}(\gamma)},$$
$$\|P^-(u_*^+) - u_*^-\|_{H_{00}^{1/2}(\gamma)} \leq \|u - u_*\|_{1,*} + \|u - P^-u\|_{H_{00}^{1/2}(\gamma)}.$$

Then, using the arbitrariness of λ_h and $\tilde{\eta}_\delta$ as well as inequalities (11a) and (11b), it is not difficult to conclude by standard reasoning that

$$\|u - u_*\|_{1,*} \lesssim \delta^{s-1}\|u\|_{s,\Omega} + \inf_{\lambda \in M_h} \|\partial_{\nu_a} u - \lambda\|_{H^{-1/2}(\gamma)}. \tag{14}$$

In order to bound the second term in (13), let $v_h^+ \in V_h^+$ and $v_h^- \in V_h^-$ be respectively approximations to $u|_{\Omega_+}$ and $u|_{\Omega_-}$. The function $v_h \in L^2(\Omega)$ defined as $v_h|_{\Omega_+} = v_h^+$ and $v_h|_{\Omega_-} = v_h^-$ does not satisfy the constraint and therefore it does not belong to $\mathcal{X}_h^*$. Let $\check{v}_h \in \mathcal{X}_h^*$ be defined by

$$\check{v}_h|_{\Omega_+} = v_h^+, \qquad \check{v}_h|_{\Omega_-} = v_h^- - R_h\left[\pi\left(v_h^- - P^-(v_h^+)\right)\right].$$

It is not difficult to check that $\check{v}_h$ satisfies (6) and therefore $\check{v}_h \in \mathcal{X}_h^*$. We have

$$\|u - \check{v}_h\|_{1,*} \lesssim \|u - v_h\|_{1,*} + \|\pi\left(v_h^- - P^-(v_h^+)\right)\|_{H_{00}^{1/2}(\gamma)}.$$

Now, adding and subtracting $u|_\gamma$ and $P^-(u|_\gamma)$, by triangular inequality, using the boundedness of P^- and the trace theorem we obtain

$$\|\pi\left(v_h^- - P^-(v_h^+)\right)\|_{H_{00}^{1/2}(\gamma)} \lesssim \|u - v_h\|_{1,*} + \|u - P^-(u)\|_{H_{00}^{1/2}}$$
$$\lesssim \|u - v_h\|_{1,*} + \delta^{s-1}\|u\|_{1,\Omega}.$$

Again, by the arbitrariness of v_h we conclude that

$$\inf_{v_h \in \mathcal{X}_h^*} \|u - \check{v}_h\|_{1,*} \lesssim \inf_{v_h \in V_h^+ \times V_h^-} \|u - v_h\|_{1,*} + \delta^{s-1}\|u\|_{1,\Omega},$$

which, together with (14) and (13) yields (12).

Remark 1. The extremely simple configuration considered (only two subdomains), hides some of the issues related to the analysis of the mortar method in more general configurations — namely the treatment of cross points and the coercivity of the broken bilinear form on the constrained space. Moreover, in this simple case the solution obtained with the method proposed coincides with the one obtained using the three fields formulation (see [2]) with a suitable choice of the discretization spaces. However, the approach used and the results obtained in this paper carry over to more complex cases (with the presence of cross-points), where the method obtained does not fall into the three fields formulation framework, with, in the worse case, a loss of a logarithmic factor in the error estimate.

3   Wavelet/FEM Coupling

Let us now consider the case of the coupling of wavelets and finite elements. As already pointed out, the use of both kinds of discretization can be particularly useful in a domain decomposition framework, giving the possibility of using wavelet bases in some rectangular shaped subdomains, and finite element discretizations to catch more complicated geometries. Weak continuity across the interface requires the jump of the solution to be orthogonal to the multiplier space, which translates in requiring

$$\int_\gamma u_h^+ \lambda \, ds = \int_\gamma u_h^- \lambda \, ds \qquad \forall \lambda \in M_h,$$

where u_h^+ is a piecewise polynomial function and λ a wavelet type function or, conversely, u_h^+ is a wavelet and λ a piecewise polynomial. Unfortunately, according to the particular nature of wavelets (whose expression is not known directly but only via the so called *refinement equation*), in both cases it is not possible to compute the above integral $\int_\gamma u_h^+ \lambda$ exactly.

In this section we will describe how to overcome such a problem by building up the two auxiliary spaces introduced in Section 2 to design the approximate constraint operator in the particular case of Wavelet/FEM coupling.

3.1   Building the Auxiliary Spaces

We recall that in general a wavelet type function ζ on $(0,1)$ at level j ($h = 2^{-j}$ corresponding to the mesh-size) (be it an element of the trace space or of the corresponding multiplier space) can be written as the restriction to $(0,1)$ of a linear combination of the dilated and translated of a single *refinable* function φ

$$\zeta(x) = \sum_\ell \zeta_{j,\ell} \varphi(2^j x - \ell)|_{(0,1)}. \tag{15}$$

The *refinability* of the function φ is by definition the existence of coefficients $\{g_l\}_{l \in \mathbb{Z}}$ such that

$$\varphi(s) = \sum_{k=0}^{K_1} g_k \varphi(2s - k).$$

In order to construct suitable auxiliary spaces, we will need a couple of biorthogonal multiresolution analyses $\{V_j\}_{j \geq j_0}$ and $\{\tilde{V}_j\}_{j \geq j_0}$ of $L^2(\gamma)$ with the following characteristics ([6,8]):

- $V_j \subset H^1(\gamma)$ is the subspace of B-splines of order n (with $n \geq 1$) on the uniform grid $\mathcal{G}_j$ obtained by splitting γ into 2^j equal segments;
- $\tilde{V}_j \subset H^1(\gamma)$ is a subspace having the same dimension as V_j, which is *biorthogonal* to V_j in the following sense: denoting by $\{e_{j,k}, k = 0, \ldots, 2^j\}$ the B-splines basis function in V_j, the space $\tilde{V}_j$ has a basis $\{\tilde{e}_{j,k}, k = 0, \ldots, 2^j\}$ which satisfies

$$\int_\gamma e_{j,k} \tilde{e}_{j,k'} = \delta_{kk'}, \quad \forall k, k' = 0, \ldots, 2^j.$$

Moreover, $\{\tilde{e}_{j,k}\}$ is a Riesz's basis for $\tilde{V}_j$, that is the following norm equivalence holds uniformly in j:

$$\| \sum_{k=0}^{2^j} c_k \tilde{e}_{j,k} \|_{0,\gamma} \simeq \left(\sum_{k=0}^{2^j} |c_k|^2 \right)^{1/2}.$$

- the functions $\tilde{e}_{j,k}$ can be obtained as linear combination of the restriction to γ (identified through a suitable mapping with the interval $(0, 1)$), of the translates and dilates (with a dilation factor 2^j) of a compactly supported function $\tilde{e}$,

$$\tilde{e}_{j,k} = \sum_\ell c_{k,\ell}^j \tilde{e}(2^j x - \ell)|_{(0,1)},$$

which we assume to be itself *refinable*, i.e. to verify, for suitable values of the coefficients h_k,

$$\tilde{e}(s) = \sum_{k=0}^{K_2} h_k \tilde{e}(2s - k);$$

- all the scaling functions of V_j and $\tilde{V}_j$ vanish at the edges 0 and 1, except one function at each edge, that is

$$e_{j,0}(0) \neq 0 \quad \text{and} \quad e_{j,2^j}(1) \neq 0 \quad \forall j \geq j_0,$$

$$e_{j,k}(0) = 0 \quad \text{and} \quad e_{j,k}(1) = 0 \quad \forall j \geq j_0 \quad \text{and} \quad \forall k = 1, \cdots, 2^{j-1},$$

and

$$\tilde{e}_{j,0}(0) \neq 0 \quad \text{and} \quad \tilde{e}_{j,2^j}(1) \neq 0 \quad \forall j \geq j_0,$$

$$\tilde{e}_{j,k}(0) = 0 \quad \text{and} \quad \tilde{e}_{j,k}(1) = 0 \quad \forall j \geq j_0 \quad \text{and} \quad \forall k = 1, \cdots, 2^{j-1}.$$

- $\tilde{V}_j$ satisfies a Strang-Fix condition of order M, that is it contains polynomials up to degree $M - 1$, while, of course, V_j contains polynomials of order n. In particular, this implies that there exists coefficients a_k^l and b_k^l, $l = 0, \cdots, n$ and $\tilde{a}_k^l$ and $\tilde{b}_k^l$, $l = 0, \cdots, M - 1$ such that we will have for all $j \geq j_0$ and for $l = 0, \cdots, n$

$$2^{j/2}(2^j x)^l = \sum_{k=0}^{2^j} a_k^l e_{j,k}(x), \tag{16a}$$

$$2^{j/2}(2^j(1 - x))^l = \sum_{k=0}^{2^j} b_k^l e_{j,k}(x), \tag{16b}$$

and for $l = 0, \cdots, M - 1$

$$2^{j/2}(2^j x)^l = \sum_{k=0}^{2^j} \tilde{a}_k^l \tilde{e}_{j,k}(x), \tag{17a}$$

$$2^{j/2}(2^j(1 - x))^l = \sum_{k=0}^{2^j} \tilde{b}_k^l \tilde{e}_{j,k}(x). \tag{17b}$$

Let us now denote by $V_j^0 = V_j \cap H_0^1(\gamma)$ and $\tilde{V}_j^0 = \tilde{V}_j \cap H_0^1(\gamma)$ the space of functions of V_j and $\tilde{V}_j$ respectively, vanishing at the boundaries of the interval

$$V_j^0 = span < e_{j,k}, k = 1, \cdots, 2^j - 1 >$$
$$\tilde{V}_j^0 = span < \tilde{e}_{j,k}, k = 1, \cdots, 2^j - 1 > .$$

Following [3], it is possible to construct two subspaces $V_j^* \subset V_j$ and $\tilde{V}_j^* \subset \tilde{V}_j$ which will play the role of one of the two auxiliary spaces U_δ^+ and U_δ^- in the case of coupling wavelets and finite elements.

Let us briefly review the construction: let the vectors $(\alpha_k)_{k=0,\cdots,M-2}$ and $(\beta_k)_{k=2^j-M+2,\cdots,2^j}$ denote respectively the solutions of the two following linear systems:

$$\sum_{k=0}^{M-2} \tilde{a}_k^l \alpha_k = \tilde{a}_{M-1}^l \qquad \forall l = 0, \cdots, M - 2 \tag{18a}$$

$$\sum_{k=2^j-M+2}^{2^j} \tilde{b}_k^l \beta_k = \tilde{b}_{2^j-M+1}^l \qquad \forall l = 0, \cdots, M - 2. \tag{18b}$$

The solvability of the above linear system is discussed in ([3]). The space $\tilde{V}_j^*$ is obtained by defining

$$\tilde{V}_j^* = span < \tilde{e}_{j,k}^*, k = 1, \cdots, 2^j - 1 >$$

where

$$\tilde{e}_{j,k}^* = \tilde{e}_{j,k} + c_k \tilde{e}_{j,0}, \qquad \text{for } k = 1, \cdots, M - 1$$
$$\tilde{e}_{j,k}^* = \tilde{e}_{j,k}, \qquad \text{for } k = M, \cdots, 2^j - M$$
$$\tilde{e}_{j,k}^* = \tilde{e}_{j,k} + d_k \tilde{e}_{j,2^j}, \qquad \text{for } k = 2^j - M + 1, \cdots, 2^j - 1$$

with

$$c_k = -\frac{\alpha_k}{\alpha_0}, \quad k = 1, \cdots, M - 2 \qquad\qquad c_{M-1} = \frac{1}{\alpha_0}$$
$$d_k = -\frac{\beta_k}{\beta_{2^j}}, \quad k = 2^j - M + 2, \cdots, 2^j - 1 \qquad d_{2^j - M + 1} = \frac{1}{\beta_{2^j}}.$$

The same construction can be carried out for the space

$$V_j^* = span < \tilde{e}_{j,k}^*, k = 1, \cdots, 2^j - 1 >$$

with the obvious switching of the dual "ingredients".

The following theorem has been proven in [3]

Theorem 2. *The spaces V_j^* and $\tilde{V}_j^*$ have the same dimension as $\tilde{V}_j^0$, and V_j^0 respectively. They satisfy the* inf-sup *conditions*

$$\inf_{\eta \in V_j^0} \sup_{\zeta \in \tilde{V}_j^*} \frac{\int_\gamma \eta \zeta \, ds}{\|\eta\|_{H_{00}^{1/2}(\gamma)} \|\zeta\|_{-1/2,\gamma}} \geq \alpha_1$$

$$\inf_{\eta \in \tilde{V}_j^0} \sup_{\zeta \in V_j^*} \frac{\int_\gamma \eta \zeta \, ds}{\|\eta\|_{H_{00}^{1/2}(\gamma)} \|\zeta\|_{-1/2,\gamma}} \geq \alpha_2,$$

and a Strang-Fix condition with the same order as V_j and $\tilde{V}_j$ respectively. Moreover the two Riesz's bases $e_{j,k}^*$ and $\tilde{e}_{j,k}^*$ are constructed in such a way that the two following biorthogonality relations hold:

$$\int_\gamma e_{j,k} \tilde{e}_{j,k'}^* = \delta_{k,k'}, \qquad \int_\gamma \tilde{e}_{j,k} e_{j,k'}^* = \delta_{k,k'}, \qquad \forall k, k' = 1, \ldots, 2^j - 1. \quad (19)$$

Thanks to the refinability of the functions $\tilde{e}$ and φ, it is possible to compute integrals of the product of a wavelet type function ζ of the form (15) with any function in $\tilde{V}_j$ (and therefore in $\tilde{V}_j^0$ and in $\tilde{V}_j^*$) by reducing it to the solution of an eigenvector problem associated with the coefficients of the corresponding refinement equations ([7]). In fact, by linearity of the integral, the product of ζ times any function in $\tilde{V}_j$ reduces to computing several integrals of the form

$$\int_0^1 \tilde{e}(2^j x - l)\varphi(2^j x - k)(x)dx = \int_0^{2^j} \tilde{e}(y)\varphi(y + l - k)dy$$

$$= \frac{1}{2^j} \sum_{m=0}^{2^j} \int_m^{m+1} \tilde{e}(y)\varphi(y + l - k)dy.$$

Let us briefly recall how such integrals can be computed. Defining

$$I(m, k) = \int_m^{m+1} \tilde{e}(x)\varphi(x - k),$$

and recalling that the characteristic function of the interval $(0, 1)$ satisfies itself a refinement equation, namely

$$\chi_{(0,1)}(x) = \sum_{n=0}^1 \chi_{(0,1)}(2x - n),$$

we can write

$$I(m, k) = \sum_{n=0}^1 \sum_{p=0}^{K_2} \sum_{r=0}^{K_1} h_p g_r \int_{\mathbb{R}} \chi_{(0,1)}(2x - 2m - n)\tilde{e}(2x - p)\varphi(2x - 2k - r)dx$$

$$= \frac{1}{2} \sum_{n=0}^1 \sum_{p=0}^{K_2} \sum_{r=0}^{K_1} h_p g_r \int_{\mathbb{R}} \chi_{(0,1)}(y + p - n - 2m)\tilde{e}(y)\varphi(y + p - 2k - r)dy$$

$$= \frac{1}{2} \sum_{n=0}^1 \sum_{p=0}^{K_2} \sum_{r=0}^{K_1} h_p g_r I(2m + n - p, 2k + r - p)$$

$$= \frac{1}{2} \sum_{n=2m-p}^{2m-p+1} \sum_{p=0}^{K_2} \sum_{r=2k-p}^{2k-p+K_2} h_p g_{r+p-2k} I(n, r).$$

Since both φ and $\tilde{e}$ have compact support, $I(m, k)$ does not vanish only for a finite number of couples (m, k). It is not difficult to realize that the original problem has then been reduced to the solution of an eigenvector equation for the multi-indexed vector $(I(m, k))_{m,k}$, that has to be solved once for all. Under suitable mild assumptions on the functions $\tilde{e}$ and φ, we can assume that the function I satisfies a normalization hypothesis $\sum_m \sum_k I(m, k) = 1$, so that such eigenvector problem has a unique solution.

3.2 FEM Master / Wavelet Slave

For choosing the auxiliary spaces U_δ^+ and U_δ^- to use for coupling wavelets and finite elements in the mortar method, we distinguish two cases: *FEM* *master / Wavelet slave* and *Wavelet master / FEM slave*. In both cases, U_δ^+ and U_δ^- must be chosen in such a way that $P^-(u_h^+)$ and the integrals of the product $\lambda_h P^-(u_h^+)$ can be computed. Let us first consider the case of a finite element master side coupled with a wavelet type slave side. In this case we assume V_h^+ to be a finite element space of polynomials of degree $N-1$ on an unstructured, non-uniform grid, and we denote by h_+ the maximum size of the discretization length, while V_h^- and M_h are two wavelet type spaces at level j. Referring to the choice of M_h as proposed in [3], that is M_h is a suitable dual space of $V_h^-|_\gamma$, let us assume that M_h and V_h^- both contain polynomials of the same degree $L-1$. It is easy to check that the appropriate choice of the auxiliary spaces is obtained by setting $U_\delta^+ = V_j^*$, the space of B-splines of order n satisfying non-homogeneous boundary conditions, and $U_\delta^- = \tilde{V}_j^0$. In fact, this is the case where u_h^+ is a FEM type function, whose projection on U_δ^-, defined by (5), can be computed by taking advantage of the biorthogonality property (19) in the following way:

$$P^- u_h^+ = \sum_{k=1}^{2^j-1} \left(\int_\gamma u_h^+ e_{j,k}^* \, ds \right) \tilde{e}_{j,k}.$$

Recall now that $\tilde{e}_{j,k}$ is a linear combination of the translates and dilates of the refinable function $\tilde{e}$ and its integral with a wavelet type function can be computed using the previously mentioned technique, while e_{jk}^* is a B-spline whose integral against the elements of $V_h^+|_\gamma$ can be computed exactly. By applying Theorem 1 we can then estimate the effect of using the approximate integration technique proposed in Section 2. We have the following

Corollary 1. *If $u \in H^s(\Omega)$ with $2 \le s \le T$ it holds*

$$\|u - u_h\|_{1,*} \lesssim \left(2^{-j(s-1)} + h_+^{s-1} \right) \|u\|_{s,\Omega}, \tag{20}$$

where the limit T in the bound is $T = \min\{n + 5/2, M + 1/2, L, N\}$.

Proof. Using the abstract inequality (12), it is easy to see that in this case, setting $\delta = 2^{-j}$, two Jackson's type inequalities of the form (11a) and (11b) hold, with $R = M$ and $\tilde{R} = n + 1$. Then we can estimate

$$\|u - u_h\|_{1,*} \lesssim 2^{-j(s-1)}\|u\|_{s,\Omega} + 2^{-j(s-1)}\|\partial_{\nu_a} u\|_{s-3/2,\gamma}$$

$$+ h_+^{(s-1)}\|u\|_{s,\Omega_+} + 2^{-j(s-1)}\|u\|_{s,\Omega_-}$$

provided that $s \le N$ and $s \le L$. Furthermore, using a classical trace theorem we can write

$$\|u - u_h\|_{1,*} \lesssim 2^{-j(s-1)}\|u\|_{s,\Omega} + 2^{-j(s-1)}\|u\|_{s,\Omega_-}$$

$$+ h_+^{(s-1)}\|u\|_{s,\Omega_+} + 2^{-j(s-1)}\|u\|_{s,\Omega_-},$$

which easily implies (20).

3.3   Wavelet Master / FEM Slave

In this case we set V_h^+ to be a wavelet type space, while V_h^- and M_h are two finite element spaces defined on unstructured, non-uniform grids of mesh size equals to h_-, the grid for M_h being the trace on γ of the grid for V_h^-. In analogy with the *FEM master / Wavelet slave* case, the approximate integration is this time performed by setting $U_\delta^+ = \tilde{V}_j^*$ and $U_\delta^- = V_j^0$. The projection $P^-(u_h^+)$ in this case takes the form

$$P^- u_h^+ = \sum_{k=1}^{2^j-1} \left(\int_\gamma u_h^+ \tilde{e}_{j,k}^* \, ds \right) e_{j,k},$$

and weak continuity leads to the computation of FEM/FEM integrals. In this case it is easy to prove that Corollary 1 becomes

Corollary 2. *If $u \in H^s(\Omega)$ with $2 \leq s \leq T$ it holds*

$$\|u - u_h\|_{1,*} \lesssim \left(2^{-j(s-1)} + h_-^{s-1} \right) \|u\|_{s,\Omega}, \tag{21}$$

with $T = \min\{n + 3/2, M + 3/2, L, N\}$.

Remark 2. In both cases *FEM master / Wavelet slave* and *Wavelet master / FEM slave*, the two auxiliary spaces have been chosen in such a way that the two integrals defining the projectors are computable. Moreover, biorthogonality implies that *no linear system* has to be solved in order to compute the auxiliary projector.

References

1. C. Bernardi, Y. Maday, and A.T. Patera. (1994) A new nonconforming approach to domain decomposition: The mortar element method. In H. Brezis & J.-L. Lions eds., editor, *Nonlinear Partial Differential Equations and their Applications, Collège de France Seminar*, volume XI of *Notes Math. Ser. 299*, 13–51.
2. F. Brezzi and D. Marini. (1994) A three-field domain decomposition method. In A. Quarteroni, J. Periaux, Y.A. Kuznetsov, and O.B. Widlund, editors, *Domain Decomposition Methods in Science and Engineering*, volume 157 of *American Mathematical Society, Contemporary Mathematics*, 27–34.
3. S. Bertoluzza and V. Perrier. (2001) The mortar method in the wavelet context *Mathem. Mod. and Numer. Anal.* **35**, 647–674.
4. S. Bertoluzza and V. Perrier. (2000) Coupling wavelets and finite elements by the mortar method. Pubbl. I.A.N. **1191**. to appear in *C.R.A.S. Paris*.
5. L. Cazabeau, C. Lacour, and Y. Maday. (1997) Numerical quadratures and mortar methods. In Bristeau et al., editor, *Computational Sciences for the 21st Century*, 119–128. Wiley and Sons.
6. A. Cohen, I. Daubechies, and P. Vial. (1993) Wavelets on the interval and fast wavelet transforms. *ACHA* **1**, 54–81.

7. W. Dahmen and C.A. Michelli. (1993) Using the refinement equation for evaluating integrals of wavelets. *SIAM J. Numer. Anal.* **30**, 507–537.
8. R. Masson. (1996) Biorthogonal spline wavelets on the interval for the resolution of boundary problems. *M^3AS* **6**(6).
9. B. Wohlmuth. (2000) A mortar finite element method using dual spaces for the Lagrange multiplier. *SIAM Jour. Numer. Anal.* **38**(3), 989–1012.

Non-Conforming *hp* Finite Element Methods for Stokes Problems

Faker Ben Belgacem[1], Lawrence K. Chilton[2] and Padmanabhan Seshaiyer[3]

[1] Mathématiques pour l'Industrie et la Physique, Université Paul Sabatier,
    118 route de Narbonne, 31062 Toulouse Cedex 04, France
[2] Department of Mathematics and Statistics, Air Force Institute of Technology,
    2950 P Street, Bldg 640 WPAFB, OH 45433, USA
[3] Department of Mathematics and Statistics, Texas Tech University,
    Lubbock TX 79409-1042, USA

Abstract. In this paper, we present a non-conforming *hp* finite element formulation for the Stokes boundary value problem for viscous incompressible fluid flow in primal velocity-pressure variables. Within each subdomain the local approximation is designed using div-stable hp-mixed finite elements. We demonstrate via numerical experiments that the non-conforming method is optimal for various h, p and hp discretizations, including the case of exponential hp convergence over geometric meshes.

1   Introduction

Efficient numerical solution of problems in fluid mechanics is always of significant interest to the engineering community. However, there are several instances where the numerical method employed requires the assembly of many incompatible sub-discretizations during design. Often, these discretizations are independently modeled, or previously constructed meshes are already available. Therefore, to support a flexible meshing procedure, it is crucial that an efficient method be employed to join these independently modeled sub-meshes together. The use of non-conforming methods at the sub-domain level can help in this regard. It is also well known that, in order to obtain accurate solutions (exponential convergence) to such problems, one has to employ the *hp* version of the finite element method (see [16], [3], [2], [17] and the references therein). Therefore, one must combine non-conforming finite elements with the *hp* version to achieve the desired efficiency and accuracy in the numerical solution.

The mortar element method ([10], [4]) is an example of a non-conforming method which can be used to decompose and re-compose a domain into subdomains without requiring compatibility between the meshes on separate components. The *hp* version of this method was first developed for the Poisson problem in ([20], [18]) and the stability and convergence of this technique were also studied ([19], [21], [8], [22]). The mathematical theory of the mortar method for the incompressible Stokes equations for the *h*-version is presented

in [5], for spectral methods in [6] and for the *hp*-version in [7]. The purpose
of this paper is to numerically validate the non-conforming *hp* finite element
method for the Stokes problem in the presence of any mesh refinement and
any high polynomial degree.

The outline of the paper is as follows. In section 2, we describe the model
Stokes problem and its two-dimensional non-conforming finite element discretization. Section 3 discusses the convergence estimates for the velocity and
pressure. The implementation of the method is outlined in Section 4. Finally,
the numerical results are presented in Section 5.

2   The Stokes Problem and its Non-Conforming *hp* Discretization

2.1   The Model Problem

Let $\Omega \subset \mathbb{R}^2$ be a polygonal domain with boundary $\partial\Omega$. Consider the Stokes
boundary value problem for viscous incompressible fluid flow: Find a velocity
field $\mathbf{u} = (u_1, u_2)$ and a pressure p such that,

$$-\mu\, \Delta\mathbf{u} + \nabla p = \mathbf{f} \quad \text{in } \Omega \tag{1}$$

$$\nabla.\mathbf{u} = 0 \quad \text{in } \Omega \tag{2}$$

$$\mathbf{u} = 0 \quad \text{on } \partial\Omega \tag{3}$$

Here, $\mu > 0$ is the kinematic viscosity which is related to the Reynolds number
of the flow. The right hand side $\mathbf{f} = (f_1, f_2)$ is a given body force per unit
mass.

Let us now define $L_0^2(\Omega)$ to be the space of real-valued square-integrable
functions with vanishing mean value. Let $H_0^1(\Omega) = \{u \in H^1(\Omega) \mid u = 0 \text{ on } \partial\Omega\}$ and denote $\mathbf{H}_0^1(\Omega) = H_0^1(\Omega) \times H_0^1(\Omega)$. Here we have used standard
Sobolev space notation. Both spaces $L_0^2(\Omega)$ and $H_0^1(\Omega)$ are provided with the
usual norms and seminorms and $(\cdot\,,\cdot)$ is the usual $L^2(\Omega)$ inner product. We
can now rewrite (1)-(3) in a weak mixed formulation: Find a velocity field
$\mathbf{u} \in \mathbf{H}_0^1(\Omega)$ and a pressure $p \in L_0^2(\Omega)$ such that,

$$\mu(\nabla\mathbf{u}, \nabla\mathbf{v}) - (\nabla \cdot \mathbf{v}, p) = (\mathbf{f}, \mathbf{v}) \tag{4}$$

$$(\nabla \cdot \mathbf{u}, q) = 0 \tag{5}$$

for all $(\mathbf{v}, q) \in \mathbf{H}_0^1(\Omega) \times L_0^2(\Omega)$. Let us denote

$$b(\mathbf{v}, q) = -(\nabla \cdot \mathbf{v}, q)$$

For $\mathbf{f} \in \mathbf{L}^2(\Omega) = L^2(\Omega) \times L^2(\Omega)$, one can then show using Brezzi's saddle-
point theory (see [11], [14]) that the problem (4)-(5) is well posed and has

a unique solution $(\mathbf{u}, p) \in \mathbf{H}_0^1(\Omega) \times L_0^2(\Omega)$. This follows from the following continuous *inf-sup* condition,

$$\inf_{q \in L_0^2(\Omega)} \sup_{\mathbf{v} \in \mathbf{H}_0^1(\Omega)} \frac{b(\mathbf{v}, q)}{||\mathbf{v}||_{\mathbf{H}^1(\Omega)} \, ||q||_{L^2(\Omega)}} \geq \alpha > 0 \tag{6}$$

To discretize (4)-(5) by the finite element method, we choose finite dimensional spaces $\mathbf{V}_N \in \mathbf{H}_0^1(\Omega)$ and $M_N \in L_0^2(\Omega)$ of piecewise polynomials that approximate the velocity and pressure respectively. Our problem can then be stated as: Find a discrete velocity $\mathbf{u}_N \in \mathbf{V}_N$ and a discrete pressure $p_N \in M_N$ such that,

$$\mu(\nabla \mathbf{u}_N, \nabla \mathbf{v}_N) - (\nabla \cdot \mathbf{v}_N, p_N) = (\mathbf{f}, \mathbf{v}_N) \tag{7}$$
$$(\nabla \cdot \mathbf{u}_N, q_N) = 0 \tag{8}$$

for all $(\mathbf{v}_N, q_N) \in \mathbf{V}_N \times M_N$. Further if the finite element spaces $\mathbf{V}_N$ and M_N satisfy the following discrete *inf-sup* stability condition,

$$\inf_{q \in M_N} \sup_{\mathbf{v} \in \mathbf{V}_N} \frac{b(\mathbf{v}, q)}{||\mathbf{v}||_{\mathbf{H}^1(\Omega)} \, ||q||_{L^2(\Omega)}} \geq \alpha(N) > 0 \tag{9}$$

then the discrete problem (7)-(8) has a unique solution $(\mathbf{u}_N, p_N) \in (\mathbf{V}_N, M_N)$. This is the *standard conforming* finite element method.

2.2   A Non-Conforming *hp* Finite Element Discretization

Let us partition the domain Ω into S non-overlapping polygonal subdomains $\{\Omega_i\}_{i=1}^S$, which are geometrically conforming by which we mean that $\partial\Omega_i \cap \partial\Omega_j$ $(i < j)$ is either empty, a vertex, or a collection of entire edges of Ω_i and Ω_j. In the latter case, we denote this interface as Γ_{ij} $(i < j)$ and this will consist of individual common edges γ, $\gamma \subset \Gamma_{ij}$. Let us define the interface set Γ to be the union of the interface intersections $\partial\Omega_i \cap \partial\Omega_j$ $(i < j)$, which result in a non-empty Γ_{ij}. We further subdivide Ω_i into parallelograms by *regular* [12] families of meshes $\{\mathcal{T}_h^i\}$. Let the maximum size of triangulation of subdomain Ω_i be h_i. Note that the triangulations over different Ω_i are independent of each other, with no compatibility enforced across interfaces. Let us stress that only the velocity space and not the pressure space will be subjected to any particular continuity constraints.

For $K \subset \mathbb{R}^n$ and $k \geq 0$ integer, let $\mathcal{P}_k(K)$ denote the set of polynomials of total degree $\leq k$ on K while $\mathcal{Q}_k(K)$ denotes the set of polynomials of degree $\leq k$ in each variable. Denote $\mathbf{Q}_k(K) = \mathcal{Q}_k(K) \times \mathcal{Q}_k(K)$. Let $\mathbf{k}$ be a degree vector, $\mathbf{k} = \{k_1, k_2, \ldots, k_S\}$ which specifies the degree used over each subdomain and denote $k = \min_{1 \leq i \leq S} \{k_i\}$.

We assume then that the following families $\{\mathbf{V}_{h,k_i}^i\}$ of piecewise polynomial local velocity spaces are given on Ω_i:

$$\mathbf{V}_{h,k_i}^i = \{\mathbf{u} \in \mathbf{H}^1(\Omega_i) \mid \mathbf{u}|_K \in \mathbf{Q}_k(K) \ \text{ for } \ K \in \mathcal{T}_h^i, \ \mathbf{u} = 0 \ \text{ on } \ \partial\Omega_i \cap \partial\Omega\},$$

The local discrete pressure space is defined to be,

$$M^i_{h,k_i} = \{q \in L^2(\Omega_i) \mid q|_K \in \mathcal{P}_{k-1}(K)\}.$$

Note that the finite element combination $\mathbf{Q}_k/\mathcal{P}_{k-1}$ is uniformly divergence stable [9]. We now define a non-conforming space $\tilde{\mathbf{V}}_{h,\mathbf{k}}$ as follows,

$$\tilde{\mathbf{V}}_{h,\mathbf{k}} = \{\mathbf{u} \in \mathbf{L}^2(\Omega) \mid \mathbf{u}|_{\Omega_i} \in \mathbf{V}^i_{h,k_i}\}$$

Note that, $\tilde{\mathbf{V}}_{h,\mathbf{k}} \not\subset \mathbf{H}^1_0(\Omega)$ and hence cannot be used for finite element calculations. So, we use, instead, a subspace of $\tilde{\mathbf{V}}_{h,\mathbf{k}}$, denoted by $\mathbf{V}_{h,\mathbf{k}}$ (defined ahead), which enforces the inter-domain continuity in a *weak* sense. In addition to the meshes, the polynomial degrees may also be different across interfaces.

Since the meshes $\mathcal{T}^i_h$ are not assumed to conform across interfaces, two separate trace meshes can be defined on Γ_{ij}, one from Ω_i and the other from Ω_j. Given $\mathbf{u} \in \tilde{\mathbf{V}}_{h,\mathbf{k}}$, we denote the traces of $\mathbf{u}$ on Γ_{ij} from each of the domains Ω_i and Ω_j by $\mathbf{u}^i$ and $\mathbf{u}^j$, respectively. Then we can define the global non-conforming velocity space to be,

$$\mathbf{V}_{h,\mathbf{k}} = \left\{\mathbf{u} \in \tilde{\mathbf{V}}_{h,\mathbf{k}} \mid \int_\gamma (\mathbf{u}^i - \mathbf{u}^j)\,\chi\,ds = 0 \ \forall \chi \in \mathbf{S}^{\gamma,ij}_{h,\mathbf{k}} \ \forall \gamma \subset \Gamma_{ij} \subset \Gamma\right\}. \quad (10)$$

where $\mathbf{S}^{\gamma,ij}_{h,\mathbf{k}}$ is a space of Lagrange multipliers for each edge $\gamma \subset \Gamma_{ij}$. In the *mortar finite element method* (see [10], [4], [20], [8] and the references therein) the Lagrange multiplier space $\mathbf{S}^{\gamma,ij}_{h,\mathbf{k}}$ is defined in the following way. Let the mesh $\mathcal{T}^i_h$ induce a mesh $\mathcal{T}^i_h(\Gamma_{ij})$ on Γ_{ij}. Let $\gamma \subset \Gamma_{ij}$ and denote the subintervals of this mesh on γ by I_l, $0 \le l \le N$. Let,

$$S^{\gamma,ij}_{h,\mathbf{k}} = \{\chi \in C(\gamma)\mid \chi|_{I_l} \in \mathcal{P}_{k_i}(I_l), \ l = 1,\dots,N-1, \ \chi|_{I_l} \in \mathcal{P}_{k_i-1}(I_l) \ l = 0, N\}$$

Then we set, the global Lagrange multiplier space to be $\mathbf{S}^{\gamma,ij}_{h,\mathbf{k}} = S^{\gamma,ij}_{h,\mathbf{k}} \times S^{\gamma,ij}_{h,\mathbf{k}}$. Note that imposing the mesh and degree on $S^{\gamma,ij}_{h,\mathbf{k}}$ from the domain Ω_i as has been done here is quite arbitrary, and these can be taken from the domain Ω_j as well, without changing the results obtained. There are other choices for the Lagrange multiplier space that one can find in the literature [21].

The global pressure space is given by,

$$M_{h,\mathbf{k}} = \{q \in L^2_0(\Omega) \mid q|_{\Omega_i} \in M^i_{h,k_i}\} \quad (11)$$

The global pressure space $M_{h,\mathbf{k}}$ is provided with the $L^2(\Omega)$ - norm while the global velocity space is endowed with a discrete Hilbertian broken norm,

$$\|\mathbf{u}\|^2_* = \sum_{i=1}^S \|\mathbf{u}\|^2_{\mathbf{H}^1(\Omega_i)}$$

The non-conforming finite element discretization to (4)-(5) is then given as follows: Find $(\mathbf{u}_{h,\mathbf{k}}, p_{h,\mathbf{k}}) \in \mathbf{V}_{h,\mathbf{k}} \times M_{h,\mathbf{k}}$ satisfying,

$$a_S(\mathbf{u}_{h,\mathbf{k}}, \mathbf{v}) + b_S(\mathbf{v}, p_{h,\mathbf{k}}) = (\mathbf{f}, \mathbf{v}) \tag{12}$$

$$b_S(\mathbf{u}_{h,\mathbf{k}}, q) = 0 \tag{13}$$

where,

$$a_S(\mathbf{u}, \mathbf{v}) = \mu \sum_{i=1}^{S} (\nabla \mathbf{u}, \nabla \mathbf{v})_{\Omega_i} \qquad b_S(\mathbf{v}, q) = -\sum_{i=1}^{S} (\nabla \cdot \mathbf{v}, q)_{\Omega_i}$$

It can be shown (see [7] for more details) that there exists a constant α' that depends only on the shape of the subdomains $\{\Omega_i\}_{i=1}^{S}$, so that the following discrete *inf-sup* condition holds:

$$\inf_{q_{h,\mathbf{k}} \in M_{h,\mathbf{k}}} \sup_{\mathbf{v}_{h,\mathbf{k}} \in \mathbf{V}_{h,\mathbf{k}}} \frac{b_S(\mathbf{v}_{h,\mathbf{k}}, q_{h,\mathbf{k}})}{\|\mathbf{v}_{h,\mathbf{k}}\|_* \, \|q_{h,\mathbf{k}}\|_{L^2(\Omega)}} \geq \alpha' > 0 \tag{14}$$

As a consequence of (14) we then have the following theorem,

Theorem 2.1 *Problem (12)-(13) has a unique solution.*

3 Convergence Estimates

The space $M_{h,\mathbf{k}}$ satisfies the following error estimate (see [2]): For all $q \in L_0^2(\Omega)$ with $q_i = q|_{\Omega_i} \in H^l(\Omega_i)$, we have,

$$\inf_{q_{h,\mathbf{k}} \in M_{h,\mathbf{k}}} \|q - q_{h,\mathbf{k}}\|_{L^2(\Omega)} \leq C \sum_{i=1}^{S} \frac{h_i^{\nu}}{k_i^l} \|q_i\|_{H^l(\Omega_i)} \tag{15}$$

with $\nu = \min(l, k)$.

For the space $\mathbf{V}_{h,\mathbf{k}}$, the following estimate can be derived by following the proofs in [20] and [7]. Let $\mathbf{v} \in \mathbf{H}_0^1(\Omega)$ with $\mathbf{v}_i = \mathbf{v}|_{\Omega_i} \in \mathbf{H}^{l+1}(\Omega_i)$, $l > \frac{1}{2}$. Then we have,

$$\inf_{\mathbf{v}_{h,\mathbf{k}} \in \mathbf{V}_{h,\mathbf{k}}} \|\mathbf{v} - \mathbf{v}_{h,\mathbf{k}}\|_* \leq C \sum_{i=1}^{S} \frac{h_i^{\nu}}{k_i^l} \, |\log k_i|^{\frac{1}{2}} \, \|\mathbf{v}_i\|_{\mathbf{H}^{l+1}(\Omega_i)} \tag{16}$$

with $\nu = \min(l, k)$. For proving this result, however, we need to make a minor restriction on the spaces $\{\mathbf{V}_{h,\mathbf{k}}\}$ which is given by the following condition [13]: **Condition (M)** *There exist constants δ, C_0, κ, independent of the mesh parameter h and degree k, such that for any trace mesh on γ, given by $x_0 < x_1 < \ldots < x_{N+1}$, with $h_j = x_{j+1} - x_j$, we have,*

$$\frac{h_i}{h_j} \leq C_0 \delta^{|i-j|}$$

where δ satisfies $1 \leq \delta \leq \min\{(k+1)^2, \kappa\}$.

It must be pointed out that almost any mesh that is used in h, p or hp version will satisfy the above condition (see [18] for more details).

4   Mixed Method Implementation

It is somewhat cumbersome to implement the non-conforming method (12)-(13) due to the constraints,

$$\int_\gamma (\mathbf{u}^i - \mathbf{u}^j)\,\chi\,ds = 0 \quad \forall \chi \in S_{h,\mathbf{k}}^{\gamma,ij} \quad \forall \gamma \subset \Gamma_{ij} \subset \Gamma$$

imposed on $\mathbf{V}_{h,\mathbf{k}}$. In this section, we will rewrite the weak formulation in (12)-(13) as a convenient method for practical implementation.

Let us now introduce an auxiliary Lagrange multiplier unknown $\lambda_{h,\mathbf{k}}$ into the problem, belonging to the Lagrange multiplier space,

$$\mathbf{S}_{h,\mathbf{k}} = \mathbf{S}_{h,\mathbf{k}}(\Gamma) = \prod_{\Gamma_{ij} \subset \Gamma} \mathbf{S}_{h,\mathbf{k}}^{\gamma,ij}$$

Defining the bilinear form c_S on $\tilde{\mathbf{V}}_{h,\mathbf{k}} \times \mathbf{S}_{h,\mathbf{k}}$ by,

$$c_S(\mathbf{v},\chi) = \sum_{\gamma \subset \Gamma_{ij} \subset \Gamma} \int_\gamma (\mathbf{v}^i - \mathbf{v}^j)\,\chi\,ds,$$

our mixed problem then becomes: Find $(\mathbf{u}_{h,\mathbf{k}}, p_{h,\mathbf{k}}, \lambda_{h,\mathbf{k}}) \in \mathbf{V}_{h,\mathbf{k}} \times M_{h,\mathbf{k}} \times \mathbf{S}_{h,\mathbf{k}}$ satisfying,

$$a_S(\mathbf{u}_{h,\mathbf{k}},\mathbf{v}) + b_S(\mathbf{v},p_{h,\mathbf{k}}) + b_S(\mathbf{u}_{h,\mathbf{k}},q) + c_S(\mathbf{v},\lambda_{h,\mathbf{k}}) + c_S(\mathbf{u}_{h,\mathbf{k}},\chi) = (\mathbf{f},\mathbf{v})$$

5   Numerical Results

5.1   Rates of Convergence

It is well known that for non-smooth domains, quasiuniform h refinement will only give $O(h^{\alpha_0})$ convergence, where α_0 is the smallest singularity exponent. Hence the optimal $O(h^k)$ will not be realized when the polynomial degree $k \geq \alpha_0$. In such cases, *non-quasiuniform* mesh refinement can be used to improve the $O(h^{\alpha_0})$ convergence, and even recover the full $O(h^k)$ convergence. Here h now denotes $N^{-\frac{1}{d}}$, N being the number of degrees of freedom, and d being the dimension ($d = 2$ here).

For the singular function x^α on the interval $0 \leq x \leq 1$, it has been shown in [15] that the optimal 1-d mesh is the so-called *radical* mesh.

$$x_i = \left(\frac{i}{n}\right)^\beta \qquad i = 0, 1, \ldots, n \tag{17}$$

where the optimal exponent when the degree is k is

$$\beta = \frac{k + \frac{1}{2}}{\alpha - \frac{1}{2}}$$

With this mesh, the full $O(h^k)$ $(h = N^{-1})$ convergence is recovered in 1-d.

Let $\mathcal{A} = \{A_l\}$ be the set of points where the solution is singular. Then in 2-d, we use a radical mesh refinement with $O(N)$ elements analogous to (17), in a neighborhood of each $A_l \in \mathcal{A}$. We choose an exponent $\beta \geq 1$, let $\gamma = 1 - \frac{1}{\beta}$, and for each element K, denote diam(K) to be its diameter. Then if $d(K)$ $(D(K))$ is the minimum (maximum) distance of points in $\overline{K}$ from A_l, the radical meshes satisfy

$$C_1 \, h \, d^\gamma(K) \; \leq \; \text{diam}(K) \; \leq \; C_2 \, h \, D^\gamma(K), \quad A_l \notin K \tag{18}$$

$$C_1 \, h \, D^\gamma(K) \; \leq \; \text{diam}(K) \; \leq \; C_2 \, h \, D^\gamma(K), \quad A_l \in K \tag{19}$$

where $h = N^{-\frac{1}{2}}$. When the exponent β is properly optimized with respect to α and k, we can obtain $O(h^k)$ convergence, by combining radical meshes in the vicinity of appropriate A_l with adequate refinement for smooth components in the interior.

To ensure that the error using non-conforming methods is of the same order as that using conforming methods, we must ensure that Condition(M) is satisfied. For this, we note that for the mesh (18), (19), the trace on any Γ_{ij} containing A_l will be similar to (17). For (17), however, Condition(M) is easily verified to hold, taking $\alpha = e^{\beta-1}$ and $C_0 = 2^\beta - 1$ (see [18] for details).

As shown in [16], the conforming *hp* finite element method leads to *exponential* convergence when the mesh is refined *geometrically* in the vicinity of vertices. For notational convenience, we replace h by n here, where n is the number of layers of elements around each vertex.

Let $\{\mathcal{T}_n^i\}$ denote the family of meshes on Ω_i. These will be assumed to be *geometric* in the following sense. For each vertex N_l of Ω_i such that N_l is also a vertex of Ω, or a point where the boundary condition changes, we assume that in a neighborhood $\mathcal{N}_l$ of N_l, the elements of the mesh $\mathcal{T}_n^i$ are numbered by a double index $\tau_{i,j}^{(n)}$ with $i = 1, \ldots, \rho(j)$, $\rho(j) \leq \rho_0$ and $j = 1, 2, \ldots, n+1$. Let $h_{n,i,j} = \text{diam}(\tau_{i,j}^{(n)})$ and $d_{n,i,j}$ denote the distance between $\tau_{i,j}^{(n)}$ and N_l. Then if $N_l \notin \overline{\tau}_{i,j}^{(n)}$, for $i = 1, 2, \ldots, \rho(j)$, $j = 2, \ldots, n+1$,

$$C_1 q^{n+2-j} \leq d_{n,i,j} \leq C_2 q^{n+1-j}, \quad \kappa_1 d_{n,i,j} \leq h_{n,i,j} \leq \kappa_2 d_{n,i,j}$$

If $N_l \in \overline{\tau}_{i,j}^{(n)}$ then $j = 1$ and

$$\kappa_3 q^n \leq h_{n,i,j} \leq \kappa_4 q^n, \quad i = 1, \ldots, \rho(1),$$

where constants C_r and κ_r are uniform for all the meshes.

Outside the neighborhoods $\mathcal{N}_l$, we assume $\mathcal{T}_n^i$ consists of a conforming (quasiuniform) mesh. We consider continuous piecewise polynomials of degree k on the elements in $\mathcal{T}_n^i$. Note that the interface meshes will be traces of the geometrical meshes in $\mathcal{T}_n^i$. These are easily seen to satisfy Condition (M). An adaptive finite element methodology for implementing constrained *hp*-refinements on the highly graded meshes needed to achieve optimal exponential rates of convergence can be found in [1].

5.2   Computational Experiments

In this section, we perform calculations for the model problem (1)-(3) with viscosity $\mu = 1$ on the L-shaped domain Ω, shown in Figure 1. This domain is subdivided into two rectangular subdomains Ω_1 and Ω_2 by the interface AO. It is well known that this domain will result in a strong singularity at the reentrant corner O, and therefore in order to obtain good convergence results, the subdomain meshing must be suitably refined around O.

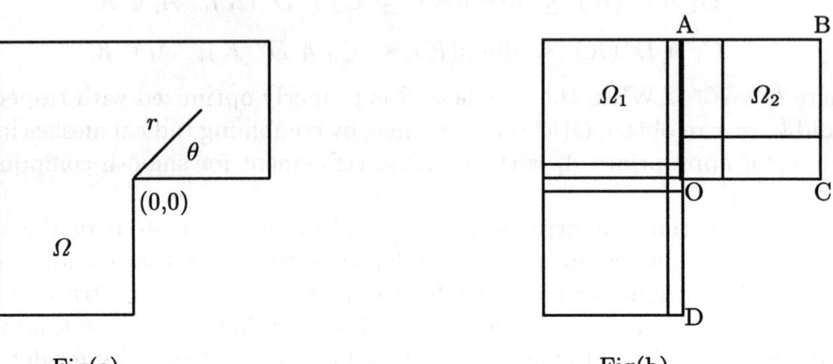

Fig. 1. (a) L-shaped domain (b) Tensor product mesh for $m = n = 2$

In our numerical calculations, we use an exact solution $(\mathbf{u}, p)$ which exhibits a corner singularity phenomena at the reentrant corner O. This solution reflects perfectly the typical (singular) behavior of solutions of the Stokes equations near reentrant corners. In polar coordinates (r, θ) at the origin, our exact solution is given by [23],

$$\mathbf{u}(r, \theta) = r^\lambda \begin{bmatrix} (1 + \lambda)\sin(\theta)\Psi(\theta) + \cos(\theta)\Psi'(\theta) \\ \sin(\theta)\Psi'(\theta) - (1 + \lambda)\cos(\theta)\Psi(\theta) \end{bmatrix}$$

$$p(r, \theta) = \frac{-r^{\lambda-1}[(1 + \lambda)^2\Psi'(\theta) + \Psi'''(\theta)]}{1 - \lambda}$$

with,

$$\Psi(\theta) = \frac{\sin((1 + \lambda)\theta)\cos(3\lambda\pi/2)}{(1 + \lambda) - \cos((1 + \lambda)\theta)} - \frac{\sin((1 - \lambda)\theta)\cos(3\lambda\pi/2)}{(1 - \lambda) + \cos((1 - \lambda)\theta)}$$

The exponent λ is the smallest positive solution of,

$$\sin(\lambda\omega) + \lambda\sin(\omega) = 0. \tag{20}$$

which gives $\lambda = 0.5444838205973307$.

Note that the solution satisfies, $\mathbf{u} = 0$ on the edges OC, OD and,

$$-\Delta\mathbf{u} + \nabla p = 0$$

We will now show that the non-conforming method described in this paper is stable and performs well for high polynomial degree and for strongly non-quasiuniform meshes.

We implement the non-conforming method using the *mixed* form described in section 4. We also consider *tensor product meshes*, where Ω_2 is divided into n^2 rectangles and Ω_1 is divided into $2m^2$ rectangles (see Figure 1). Since the mesh on Ω_1 is symmetric about $y = 0$, in the sequel we only describe the mesh on the top half.

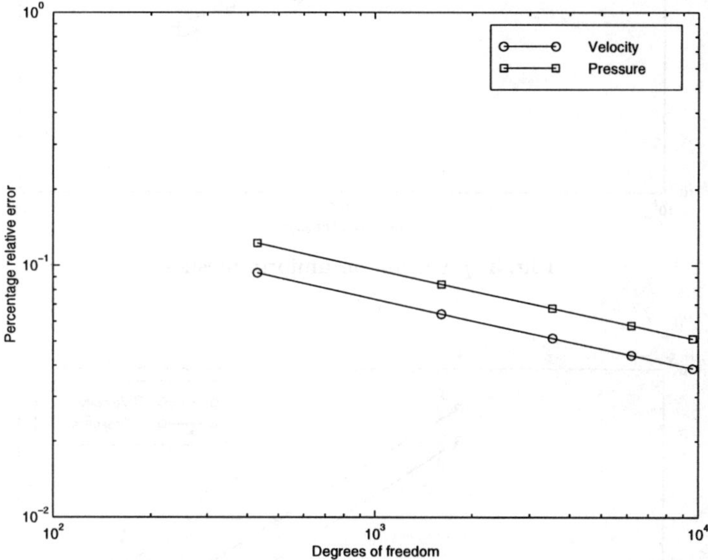

Fig. 2. *h*-version on uniform meshes

First, we consider the *h*-version on *uniform* meshes, by taking the n equally spaced grid points along both the x and y axes for Ω_2, and similarly for Ω_1, but with m points instead of n. We consider the combinations $(m, n) = \{(2, 3), (4, 6), \ldots, (10, 15)\}$ to get incompatible meshes. The approximation order for the velocity is chosen to be cubic and quadratic for the pressure. The percentage relative error in the discrete H^1-norm error for the velocity and L^2-norm error for the pressure are plotted against the number of degrees of freedom in Figure 2. We observe a rate of $O(h^\lambda)$ (where λ is the smallest positive solution of (20)), as expected for uniform meshes.

Next, we consider *p*-version on *uniform* meshes, by fixing the mesh with $(m, n) = (2, 3)$ and increasing the polynomial degree $k = 2, \ldots, 10$ to improve the accuracy. There results are illustrated in Figure 3.

In order to improve the $O(h^\lambda)$ convergence observed in Figure 2, we consider the *h*-version on *radical* meshes where we use equation (17) now for

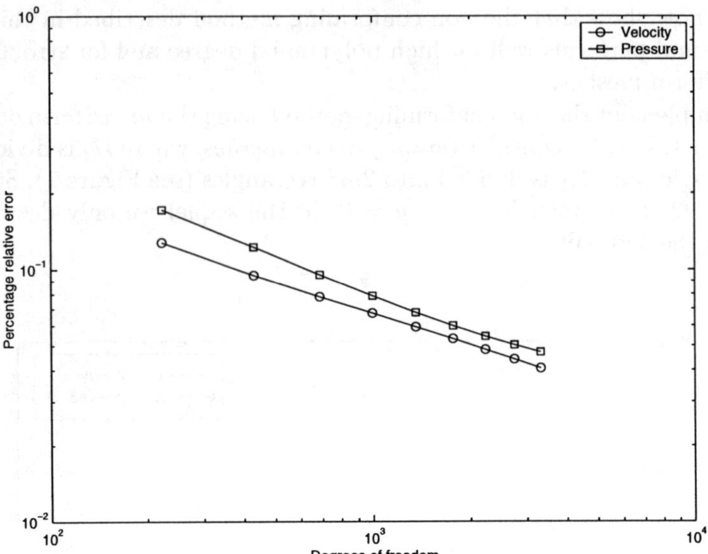

Fig. 3. p-version on uniform meshes

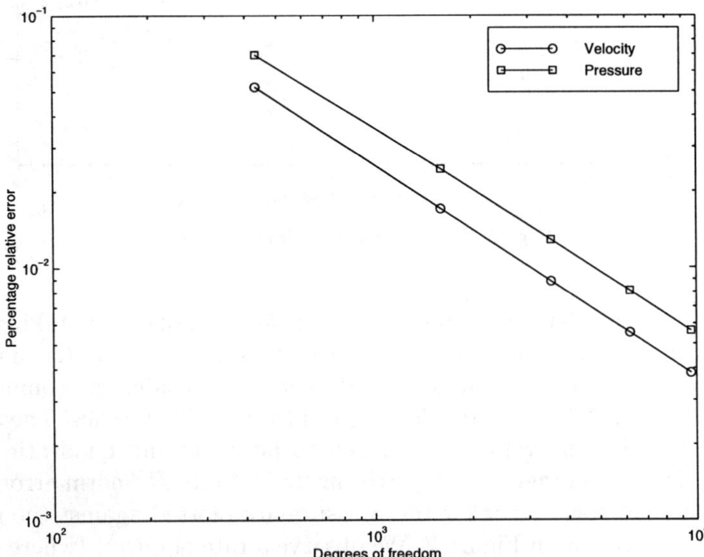

Fig. 4. h-version on radical meshes

choosing the grid points along both the x and y axes, with $\beta = 3$. We repeat
our computations as in the uniform mesh case and the results are illustrated

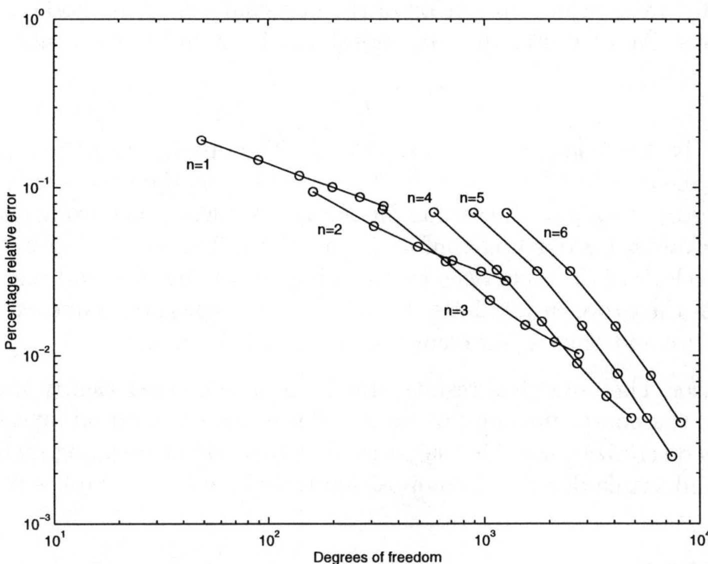

Fig. 5. *hp*-version for velocity on geometric meshes

in Figure 4. We clearly observe a rate much better than $O(h^\lambda)$, as expected for radical meshes.

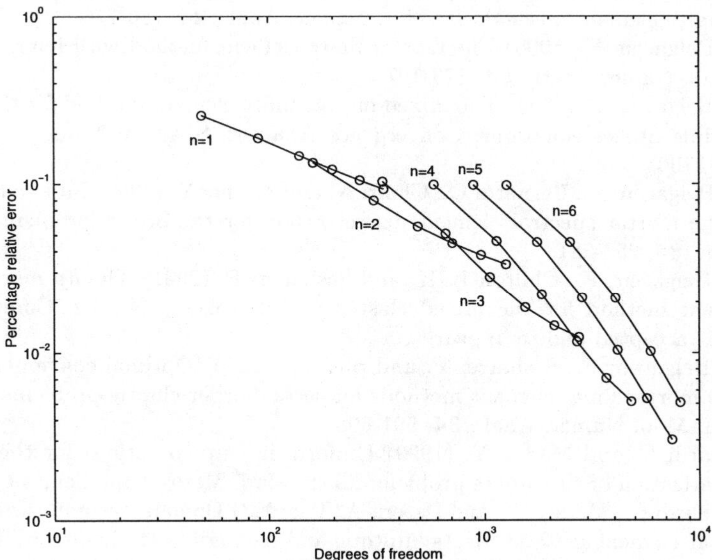

Fig. 6. *hp*-version for pressure on geometric meshes

Finally, we consider hp version of the non-conforming method on *geometric* meshes. We now take $m = n$, and along the x and y axes, take the grid points,

$$x_0 = 0, \quad x_j = \sigma_i^{n-j} \quad j = 1, \ldots, n$$

where σ_i is the geometric ratio used on Ω_i. The optimal value is 0.15 ([15]), but we take $\sigma_1 = 0.17$ and $\sigma_2 = 0.13$ to make the method non-conforming.

In Figure 5, we plot the results for the velocity when increasing the degree k for various n for our non-conforming method. The hp version is then the lower envelope of these curves – by changing both n and k simultaneously, we remain in the exponential phase. In Figure 6, we repeat the same experiment for pressure and once again recover exponential convergence.

Conclusion: The numerical results presented in this paper clearly show optimality of the non-conforming hp mortar finite element method for various h, p and hp discretizations. This suggests that this non-conforming method is a robust and viable domain decomposition technique for the Stokes problem.

References

1. Ainsworth M. and Senior B. (1999) hp-Finite Element Procedures on Non-Uniform Geometric Meshes: Adaptivity and Constrained Approximation. Grid Generation and Adaptive Algorithms. M. W. Bern and J. E. Flaherty and M. Luskin (eds.), IMA, Minnesota, **113**, 1-29.
2. Babuška I. and Suri M. (1994) The p and hp versions of the Finite Element Method: Basic Principles and Properties. SIAM Review, **36**, 578-632.
3. Babuška I. and Suri M. (1987) The optimal convergence rate of the p-version of the finite element method. SIAM J. Numer. Anal., **24**, 750-776.
4. Ben Belgacem F. (1999) The mortar finite element method with Lagrange Multipliers. Numer. Math. **84**, 173-197.
5. Ben Belgacem F. (2000) The mixed mortar finite element method for the incompressible Stokes equations: Convergence Analysis. SIAM J. Numer. Anal., **37**, 1085-1100.
6. Ben Belgacem F., Bernardi C., Chofri N. and Maday Y. (2000) Inf-sup condition for the mortar spectral element discretization for the Stokes problem. Numer. Math., **85**, 257-281.
7. Ben Belgacem F., Chilton L. K. and Seshaiyer P. (2001) The hp mortar finite element method for the mixed elasticity and stokes problems. Comp. Math. Appl., accepted and to appear.
8. Ben Belgacem F., Seshaiyer P. and Suri M. (2000) Optimal convergence rates of hp mortar finite element methods for second-order elliptic problems. RAIRO Math. Mod. Numer. Anal., **34**, 591-608.
9. Bernardi C. and Maday Y. (1999) Uniform inf-sup conditions for the spectral discretization of the Stokes problem. Math. Mod. Meth. Appl. Sci., **49**, 395-414.
10. Bernardi C., Maday Y., and Patera A. T. (1993) Domain decomposition by the mortar element method. In Asymptotic and Numerical Methods for PDEs with Critical Parameters. H. G. Kaper and M. Garbey, (eds.) NATO Adv. Sci. Inst. Ser. C Math. Phs. Sci. **384**, Kluwer, 269-286.

11. Brezzi F. and Fortin M. (1991). Mixed and Hybrid Finite Element Methods. Springer Verlag, New York, Springer Series in Computational Mathematics, **15**.

12. Ciarlet P. G. (1978) The Finite Element Method for Elliptic Problems. North Holland, Amsterdam.

13. Crouzeix M. and Thomée V. (1987) The stability in L^p and $W^{1,p}$ of the L^2 Projection on finite element function spaces. Math. Comp., **48**, 521-532.

14. Girault V. and Raviart P. A. (1980) Finite Element Methods for Navier-Stokes Equations, Springer Verlag.

15. Gui W. and Babuška I. (1986) The *hp* version of the finite element method in one dimension. Numer. Math., **3**, 577-657.

16. Guo B. and Babuška I. (1986) The *hp* version of the finite element method. Comput. Mech., **1**, 21–41 (Part I) 203-220 (Part II).

17. Schwab C. and Suri M. (1996) The *p* and *hp* versions of the finite element method for problems with boundary layers. Math. Comp., **65**, 1403-1429.

18. Seshaiyer P. (1998) Non-conforming *hp* finite element methods. Ph. D. Dissertation, University of Maryland, Baltimore County.

19. Seshaiyer P. and Suri M. (1998) Convergence results for non-conforming *hp* methods: The mortar finite element method. Contemp. Math. **218**, 467-473.

20. Seshaiyer P. and Suri M. (2000) Uniform *hp* convergence results for the mortar finite element method. Math. Comp. **69**, 521-546

21. Seshaiyer P. and Suri M. (2000) *hp* submeshing via non-conforming finite element methods. Comp. Meth. Appl. Mech. Engrg., **189**, 1011-1030.

22. Seshaiyer P. (2001) Stability and convergence of non-conforming *hp* finite element methods. Comp. Math. Appl., accepted and to appear.

23. Verfurth R. (1996) A review of a posteriori error estimation and adaptive mesh-refinement techniques. Wiley-Teubner, Chichester-Stuttgart.

A Defect Correction Method for Multi-Scale Problems in Computational Aeroacoustics

Georgi S. Djambazov, Choi-Hong Lai, Koulis A. Pericleous, and Zong-Kang Wang[1]

All at School of Computing and Mathematical Sciences, University of Greenwich, Greenwich, London SE10 9LS, UK

Abstract. Sound waves are propagating pressure fluctuations which are typically several orders of magnitude smaller than the pressure variations in the flow field that account for flow acceleration. On the other hand, these fluctuations travel at the speed of sound in the medium, not as a transported fluid quantity. Due to the above two properties, the Reynolds averaged Navier-Stokes (RANS) equations do not resolve the acoustic fluctuations. This paper discusses a defect correction method for this type of multi-scale problems in aeroacoustics.

1   Introduction

Many problems of fundamental and practical importance are of multi-scale nature. As a typical example, the velocity field in turbulent transport problems fluctuates randomly and contains many scales depending on the Reynolds number of the flow. In another typical example, which is the main concern of this paper, sound waves are several orders of magnitude smaller than the pressure variations in the flow field that account for flow acceleration. These sound waves are manifested as pressure fluctuations which propagate at the speed of sound in the medium, not as a transported fluid quantity. As a result, numerical solutions of the Navier-Stokes equations which describe fluid motion do not resolve the small scale pressure fluctuations. Computational scientists should be aware of the current electronic technology in floating point computation, which has implications on the precision of the data. Hence the finite size of data storage has obviously imposed limitations on the numerical accuracy achieved in solving a given mathematical model, even though it is perfectly correct in the description of the physics. On the other hand, direct numerical simulation to include the above multiple scales problem is still an expensive tool for sound analysis [1] based on the existing hardware technology. Therefore one may wish to seek for affordable alternative numerical algorithms.

One established method for the treatment of elliptic problems, where multi-scale phenomena do not exist, is to use the concept of a domain decomposition. Such technique is often motivated by parallel computing. One implementation of the method is to use it directly on the continuously partial differential equation and the results in various subdomains are then put together using certain techniques. Another implementation concept is to use it on the discretised system. In other words, a global grid is required before the partitioning of the domain. The authors take the first approach of the continuous problem and examine the corresponding coarse grid and fine grid problems taking into account of the multi-scale phenomena in the derivation of the respective models.

In essence, there are at least three different scales embedded in the flow variables, namely (i) the mean flow, (ii) flow perturbations or aerodynamic sources of sound, and (iii) the acoustic perturbation. While flow perturbation or aerodynamic sources of sound may be easier to recover, it is not true for the acoustic perturbation because of its comparatively small magnitude. From an engineering perspective, much of the larger scales behaviour may be resolved with the state-of-the-art CFD packages which implement various numerical methods of solving Navier-Stokes equations. This paper examines, in more detail, a defect correction method, first proposed in [2], and suitably adapted for the derivation of the coarse space mathematical model in order to recover smaller scales that have been left behind. The authors have demonstrated the accurate computation of the mean flow and flow perturbations in [3][4][5]. In the present study, a two-scale decomposition of flow variables is considered, i.e. the flow variable U is written as $\bar{u} + u$, where $\bar{u}$ denotes the mean flow and part of aerodynamic sources of sound and u denotes the remaining part of the aerodynamic sources of sound and the acoustic perturbation.

This paper follows the basic principle of the defect correction with suitable modification for time dependent problems and applies it to the recovery of the propagating acoustic perturbation. The method relies on the use of a lower order partial differential equation defined on the same computational domain where a residue exists such that the acoustic perturbation may be retrieved through a properly defined coarse mesh.

This paper is organised as follows. First, the derivation of a lower order partial differential equation resulting from the Navier-Stokes equations for the coarse space is given. Truncation errors due to the model reduction are examined. Second, accurate representation of residue on the coarse mesh is discussed. The coarse mesh is designed in such a way as to allow various frequencies of noise to be studied. Suitable interpolation operators are studied for the two different meshes. Third, numerical tests are performed for different mesh parameters to illustrate the concept. Finally, future work is discussed.

2   The Defect Correction Method

The aim here is to solve the non-linear equation

$$\mathcal{L}\{U\}U := \mathcal{L}\{\bar{u} + u\}(\bar{u} + u) = 0 \, , \tag{1}$$

where $\mathcal{L}\{U\}$ is a time-dependent non-linear operator depending on U. A concrete example of $\mathcal{L}\{U\}$ is given below. For simplicity, U is considered to have two different scales of magnitudes as $\bar{u} + u$. Note that $u \ll \bar{u}$ and that

$$\frac{1}{\delta t} \int_{t_0}^{t_0+\delta t} u \, dt \to 0 \, ,$$

with δt much larger than any significant period of the perturbation velocity. This integral essentially conveys the message that u is a certain fluctuation and will be damped out over the time interval δt. The problem here is thus purely related to the scales of magnitude of the dependent variables. In the case of sound generated by the motion of fluid, it is natural to imagine $\mathcal{L}\{U\}$ as the Navier-Stokes operator and, therefore, $\bar{u}$ as the mean flow and u as the acoustic perturbation as described in Section 1. For a 2-D problem,

$$\bar{u} = \begin{bmatrix} \bar{\rho} \\ \bar{v}_1 \\ \bar{v}_2 \end{bmatrix} \qquad u = \begin{bmatrix} \rho \\ v_1 \\ v_2 \end{bmatrix} \, ,$$

where ρ is the density of fluid and v_1 and v_2 are the velocity components along the two spatial axes. Using the summation notation of subscripts, the 2-D Navier-Stokes problem $\mathcal{L}\{u\}u = 0$ may be written as

$$\frac{\partial \rho}{\partial t} + \frac{\partial(\rho v_j)}{\partial x_j} = 0,$$

$$\frac{\partial v_i}{\partial t} + v_j \frac{\partial v_i}{\partial x_j} + \frac{1}{\rho}\frac{\partial P}{\partial x_i} - \frac{\mu}{\rho}\nabla^2 v_i = 0,$$

where P is the pressure and $(\mu/\rho)\nabla^2 v_i$ is the viscous force along i-th axis.

Suppose (1) may be split and re-written as

$$\mathcal{L}\{\bar{u} + u\}(\bar{u} + u) \equiv \mathcal{L}\{\bar{u}\}\bar{u} + E\{\bar{u}\}u + K[\bar{u}, u], \qquad (2)$$

where $\mathcal{L}\{\bar{u}\}$ and $E\{\bar{u}\}$ are operators depending on the knowledge of $\bar{u}$ and $K[\bar{u}, u]$ is a functional depending on the knowledge of both $\bar{u}$ and u. In order to obtain a solution to $\bar{u}$, one requires to solve a discretised form of $\mathcal{L}\{\bar{u}\}\bar{u} = 0$. Therefore one may use a CFD analysis package, which effectively solves a discretised form of $\mathcal{L}\{\bar{u}\}\bar{u} = 0$ instead of $\mathcal{L}\{\bar{u} + u\}(\bar{u} + u) = 0$. Following the concept of truncation error in a finite difference method, it is possible to define the truncation error due to the removal of the perturbation part of the flow variable, i.e.

$$\tau = \mathcal{L}\{\bar{u} + u\}(\bar{u} + u) - \mathcal{L}\{\bar{u}\}(\bar{u} + u). \qquad (3)$$

Using the relation $\mathcal{L}\{\bar{u}\}(\bar{u} + u) = \mathcal{L}\{\bar{u}\}\bar{u} + E\{\bar{u}\}u$, the truncation error due to the removal of the perturbation part is thus given by

$$\tau = K[\bar{u}, u]. \qquad (4)$$

Note that this truncation error is not related to the discretisation of continuous model but only related to the reduction of a more complex mathematical model to a less complex mathematical model. From the knowledge of physics of fluids, the acoustic perturbations ρ and v_j are of very small magnitude (this is not true for their derivatives), and therefore, K may be considered negligible due to the reason that any feedback from the propagating waves to the flow may be completely ignored, except in some cases of acoustic resonance, which we are not concerned with here. In other words the contribution due to the perturbation part has negligible effect on the main background flow of the fluid. The consequence of this is that one can apply a time-dependent discretisation, as in any CFD analysis packages, to obtain numerical approximations at every time step without considering the perturbation part at this stage. Such approximations due to a CFD analysis package may be denoted as $\bar{u}^*$ and their usage is discussed below. The small contribution to τ has made the concept of defect correction applicable to the present study.

Following the concept of defect correction, $\bar{u}$ may be considered as an approximate solution to (1). Hence one can evaluate the residue of (1) as

$$R \equiv \mathcal{L}\{\bar{u} + u\}(\bar{u} + u) - \mathcal{L}\{\bar{u}\}\bar{u} = -\mathcal{L}\{\bar{u}\}\bar{u} \; ,$$

which may then be substituted into (2) to give

$$E\{\bar{u}\}u + K[\bar{u}, u] = R \; , \tag{5}$$

As discuss above, $K[\bar{u}, u]$ is small and can then be neglected. Hence the problem in (5) is a linear problem and may be solved more easily to obtain the acoustics perturbation u. A non-linear iterative solver is required in order to obtain u for cases when $K[\bar{u}, u]$ is not negligible.

Expanding $\mathcal{L}\{\bar{u} + u\}(\bar{u} + u) = 0$ for $\mathcal{L}$ being the Navier-Stokes operator and re-arranging the terms we obtain

$$\frac{\partial \rho}{\partial t} + \bar{v}_j \frac{\partial \rho}{\partial x_j} + \bar{\rho}\frac{\partial v_j}{\partial x_j} + [v_j \frac{\partial(\bar{\rho} + \rho)}{\partial x_j} + \rho\frac{\partial(\bar{v}_j + v_j)}{\partial x_j}] = -[\frac{\partial \bar{\rho}}{\partial t} + \bar{v}_j\frac{\partial \bar{\rho}}{\partial x_j} + \bar{\rho}\frac{\partial \bar{v}_j}{\partial x_j}],$$

and

$$\frac{\partial v_i}{\partial t} + \bar{v}_j \frac{\partial v_i}{\partial x_j} + \frac{1}{\bar{\rho}}\frac{\partial P}{\partial x_i} - \frac{\mu}{\bar{\rho}}\nabla^2 v_i \tag{6}$$

$$+[\frac{\rho}{\bar{\rho}}\frac{\partial(\bar{v}_i + v_i)}{\partial t} + (v_j + \frac{\rho}{\bar{\rho}}(\bar{v}_j + v_j))\frac{\partial(\bar{v}_i + v_i)}{\partial x_j}] = -[\frac{\partial \bar{v}_i}{\partial t} + \bar{v}_j\frac{\partial \bar{v}_i}{\partial x_j} + \frac{1}{\bar{\rho}}\frac{\partial \bar{P}}{\partial x_i} - \frac{\mu\nabla^2\bar{v}_i}{\bar{\rho}}].$$

It can be seen that (6) may be written in the form of (5) where

$$E\{\bar{u}\}u = \begin{bmatrix} \frac{\partial \rho}{\partial t} + \bar{v}_j\frac{\partial \rho}{\partial x_j} + \bar{\rho}\frac{\partial v_j}{\partial x_j} \\ \frac{\partial v_i}{\partial t} + \bar{v}_j\frac{\partial v_i}{\partial x_j} + \frac{1}{\bar{\rho}}\frac{\partial P}{\partial x_i} - \frac{\mu}{\bar{\rho}}\nabla^2 v_i \end{bmatrix}, \tag{7}$$

$$K[\bar{u}, u] = \begin{bmatrix} v_j\frac{\partial(\bar{\rho}+\rho)}{\partial x_j} + \rho\frac{\partial(\bar{v}_j+v_j)}{\partial x_j} \\ \frac{\rho}{\bar{\rho}}\frac{\partial(\bar{v}_i+v_i)}{\partial t} + (v_j + \frac{\rho}{\bar{\rho}}(\bar{v}_j + v_j))\frac{\partial(\bar{v}_i+v_i)}{\partial x_j} \end{bmatrix}, \tag{8}$$

$$R = \begin{bmatrix} -[\frac{\partial \bar{\rho}}{\partial t} + \bar{v}_j \frac{\partial \bar{\rho}}{\partial x_j} + \bar{\rho} \frac{\partial \bar{v}_j}{\partial x_j}] \\ -[\frac{\partial \bar{v}_i}{\partial t} + \bar{v}_j \frac{\partial \bar{v}_i}{\partial x_j} + \frac{1}{\bar{\rho}} \frac{\partial \bar{P}}{\partial x_i} - \frac{\mu}{\bar{\rho}} \nabla^2 \bar{v}_i] \end{bmatrix} \equiv -\mathcal{L}\{\bar{u}\}\bar{u}. \tag{9}$$

Hence the equation $E\{\bar{u}\}u = R$, with E given by (7), which is known as the linearised Euler equation, can be solved with the knowledge of $\bar{u}$. The numerics and the techniques involved here are often referred to as Computational AeroAcoustics (CAA) methods.

The remaining question is to obtain the approximate solution $\bar{u}$ to the original problem (2). It is well known that CFD analysis packages provide excellent methods for the solution of $\mathcal{L}\{\bar{u}\}\bar{u} = 0$. Therefore one requires to use a Reynolds averaged Navier-Stokes package supplemented with turbulence models such as [7,8] to provide a solution of $\bar{u}$. One requires $\bar{u}$ to be as accurate as possible to capture all the physics of interest, such as flow turbulence and the presence of vortices. Finally, the approximate solution $\bar{u}^*$ obtained from the CFD package may be used to compute the residue as $-\mathcal{L}\{\bar{u}^*\}\bar{u}^*$.

3   Coarse Grid Sound Source Retrieval

In order to simulate accurately the approximate solution, $\bar{u}$, to the original problem, $\mathcal{L}\{U\}U = 0$, the QUICK differencing scheme [9] is used which produces sufficiently accurate results of $\bar{u}$ for the purpose of evaluating the residue as defined in (9). A sufficiently fine mesh has to be used in order to preserve vorticity motion. However, much coarser mesh may be used for the numerical solutions of linearised Euler equations [3–5]. It certainly has to obey the Courant limit and also to account for the fact that the acoustic wavelength may be larger than a typical flow feature which needs to be resolved, e.g. a travelling vortex [10]. The present defect correction method requires to calculate the residue on the CFD mesh and to transfer these residuals onto the acoustic mesh. Physically, the residue is effectively the sound source that would have disappeared without using the present retrieval technique.

Let h denote the mesh to be used in the Reynolds averaged Navier-Stokes solver. Instead of evaluating $\bar{u}$, one would solve the discretised approximation $\mathcal{L}_h\{\bar{u}_h\}\bar{u}_h = 0$ to obtain $\bar{u}_h$. The residue on the fine mesh h can be computed as $\mathcal{L}\{\bar{u}_h\}\bar{u}_h$ by means of a higher order approximation [5]. Let H denote the mesh for the linearised Euler equations solver. Again instead of evaluating u, one would solve the discretised approximation $E_H\{\bar{u}_H\}u_H = R_H$ to obtain u_H. Here R_H is the projection of R onto the mesh H. Let $I_{\{h,H\}}$ be a restriction operator to restrict the residue computed on the fine mesh h to the coarser mesh H. The restricted residue can then be used in the numerical solutions of linearised Euler equations. Therefore the two-level numerical scheme is (for non-resonance problems) :

$$\text{Solve } \mathcal{L}_h\{\bar{u}_h\}\bar{u}_h = 0$$
$$R_H := -I_{\{h,H\}}\mathcal{L}\{\bar{u}_h\}\bar{u}_h$$
$$\bar{u}_H := I_{\{h,H\}}\bar{u}_h$$

$$\text{Solve } E_H\{\bar{u}_H\}u_H = R_H$$
$$U_H := \bar{u}_H + u_H$$

Here U_H denotes the discretised approximation of the resultant solution on mesh H. Note that R_H cannot be computed as $\mathcal{L}\{\bar{u}_h\}I_{\{h,H\}}\bar{u}_h$ because $\mathcal{L}$ is a non-linear operator.

In the actual implementation, a pressure-density relation which also defines the speed of sound c in air is used:

$$\frac{\partial P}{\partial \rho} = c^2 \approx 1.4\frac{\bar{P}}{\bar{\rho}}, \tag{10}$$

and the first component of the linearised Euler equations in (7) becomes

$$\frac{\partial P}{\partial t} + \bar{v}_j\frac{\partial P}{\partial x_j} + \bar{\rho}c^2\frac{\partial v_j}{\partial x_j} = -c^2[\frac{\partial \bar{\rho}}{\partial t} + \bar{v}_j\frac{\partial \bar{\rho}}{\partial x_j} + \bar{\rho}\frac{\partial \bar{v}_j}{\partial x_j}]. \tag{11}$$

The purpose of this substitution is to make sure that the new fluctuations P and v_i do not contain a hydrodynamic component, and hence can be resolved on regular Cartesian meshes [4] which is essential for the accurate representation of the acoustic waves or the fluctuation quantity u. On the other hand, an unstructured mesh may be used to obtain $\bar{u}_h$. The two different meshes overlap one another on the computational domain. The computational domain for the linearised Euler equations is not necessarily the same as the one for the CFD solutions. It must be large enough to contain at least the longest wavelength of a particular problem under consideration or a number of wavelengths where propagation is of interest. The numerical example as shown in Section 4 does not contain any complicating solid objects, the restriction operator $I_{\{h,H\}}$ may then be chosen as an arithmetic averaging process [10].

4   Numerical Experiments

The propagation of the following one-dimensional pulse is considered: an initial pressure distribution with a peak in the origin generates two opposite acoustic waves in both directions. The exact solution of this problem (12) can be verified by substitution in the linearised Euler equations.

$$\begin{aligned} P &= f(x - ct) + f(x + ct), \\ \bar{\rho}cv_1 &= f(x - ct) - f(x + ct), \\ f(x) &= \begin{cases} \frac{A}{2}(1 + \cos 2\pi\frac{x}{\lambda}), |x| < \frac{\lambda}{2} \\ 0, |x| \geq \frac{\lambda}{2} \end{cases} \end{aligned} \tag{12}$$

Here A is the amplitude and λ is the wavelength of the two sound waves that start from the origin ($x = 0$) at $t = 0$. The example was reported in [2]. This paper provides a detailed numerical study on various aspects of the grid parameters being used in the two-level method. The CFD domain is of 12 wavelengths and the CAA domain is of 14 wavelengths.

The effects of the following parameters on the solution accuracy are studied. (a) the ratio H:h, (b) number of points per wavelength, and (c) the restriction operator for residual transfer from fine grid to coarse grid. In all cases, the norm $\|P_H - P\|_\infty$ is compared. Here P_H is the approximation obtained on the coarse mesh (CAA) after correction and P is the exact solution of the pressure variable.

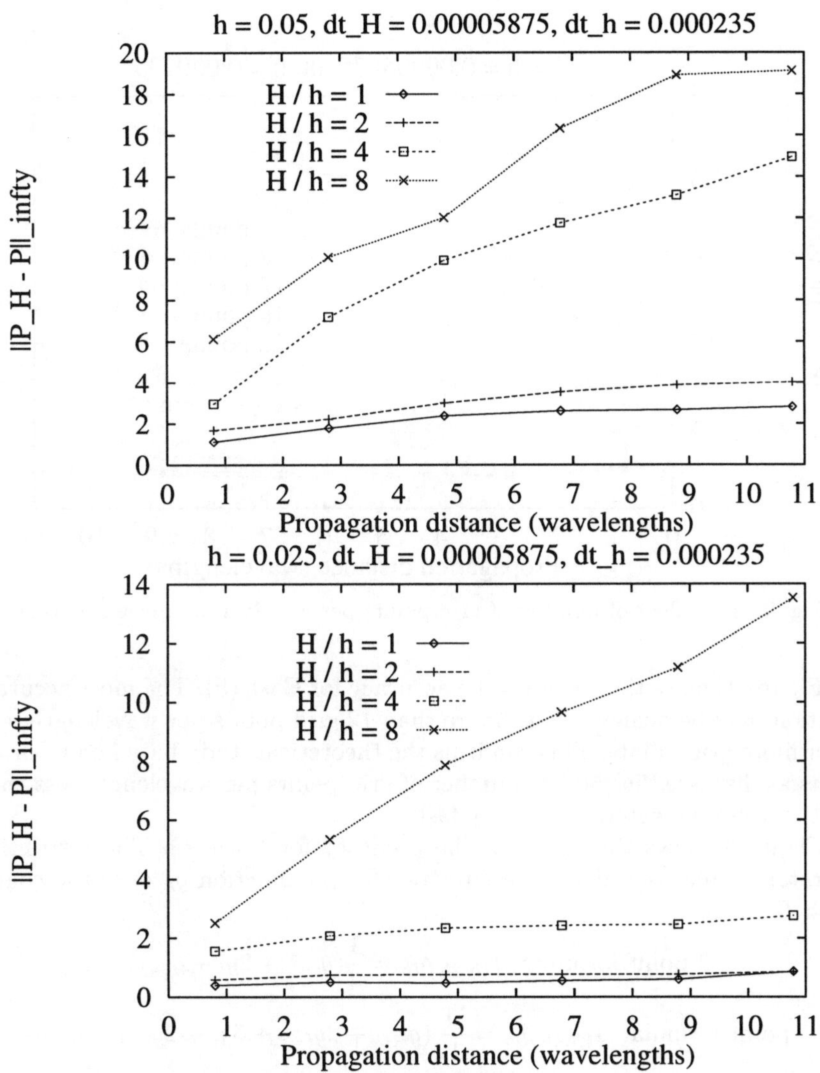

Fig. 1. The effect of mesh ratio $H{:}h$ on the accuracy.

Let δt_h and δt_H be the step lengths in the temporal axis for the CFD mesh and the CAA mesh respectively. Figure 1 shows the effect on the accuracy for Case (a). Here δt_h and δt_H are chosen to be 0.000235 and 0.00005875 respectively. Two different mesh sizes for the CFD are chosen and they are 0.05 and 0.025. It can be seen that when h is not fine enough, say $h = 0.05$, to resolve some of the physics, it is still possible to use the mesh $H = 2h$ or $H = h$ to recover the small scale signal. If a finer mesh was used, say $h = 0.025$, it is possible to use $H \leq 4h$. This property essentially links with the Courant number of the coarse mesh for CAA [5], i.e. H, and is also confirmed in the test performed for Case (b).

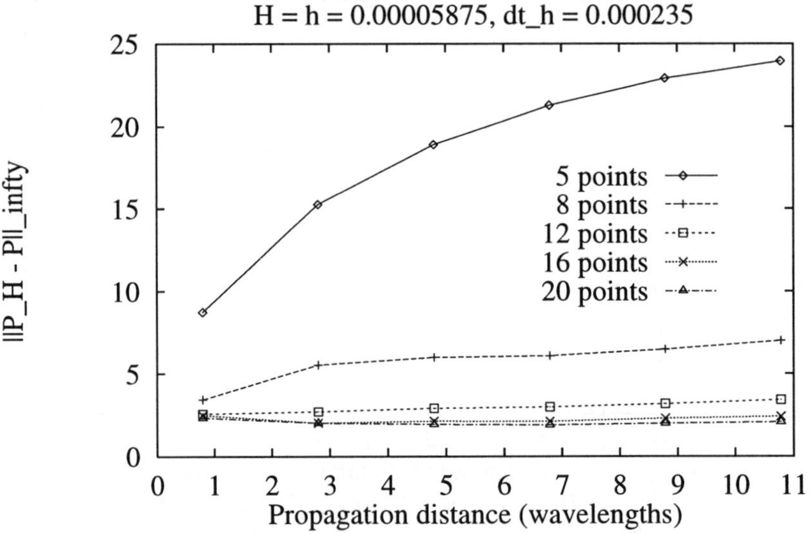

Fig. 2. The effect of number of grid points per wavelength on the accuracy.

Figure 2 shows the effect on the accuracy for Case (b). The most accurate solution may be achieved with more than 12 grid points per wavelength, e.g. 16 or more grid points. This confirms the theoretical study based on Courant limits as discussed in [5]. For number of grid points per wavelength less than 12, the accuracy deteriorates very fast.

Figure 3 shows the effect on the accuracy for Case (c). The restriction operators being used in this test to transfer the function g_h onto the coarse mesh H includes

$$3 \text{ point formula: } I_{\{h,2h\}}g_h = \frac{1}{4}(g_{i-1} + 2g_i + g_{i+1})$$

$$5 \text{ point formula: } I_{\{h,4h\}}g_h = \frac{1}{12}(g_{i-2} + 2g_{i-2} + 6g_i + 2g_{i+1} + g_{i+2})$$

$$7 \text{ point formula: } I_{\{h,6h\}}g_h = \frac{1}{16}(g_{i-3}+2g_{i-2}+3g_{i-1}+4g_i+3g_{i+1}+2g_{i+2}+g_{i+3})$$

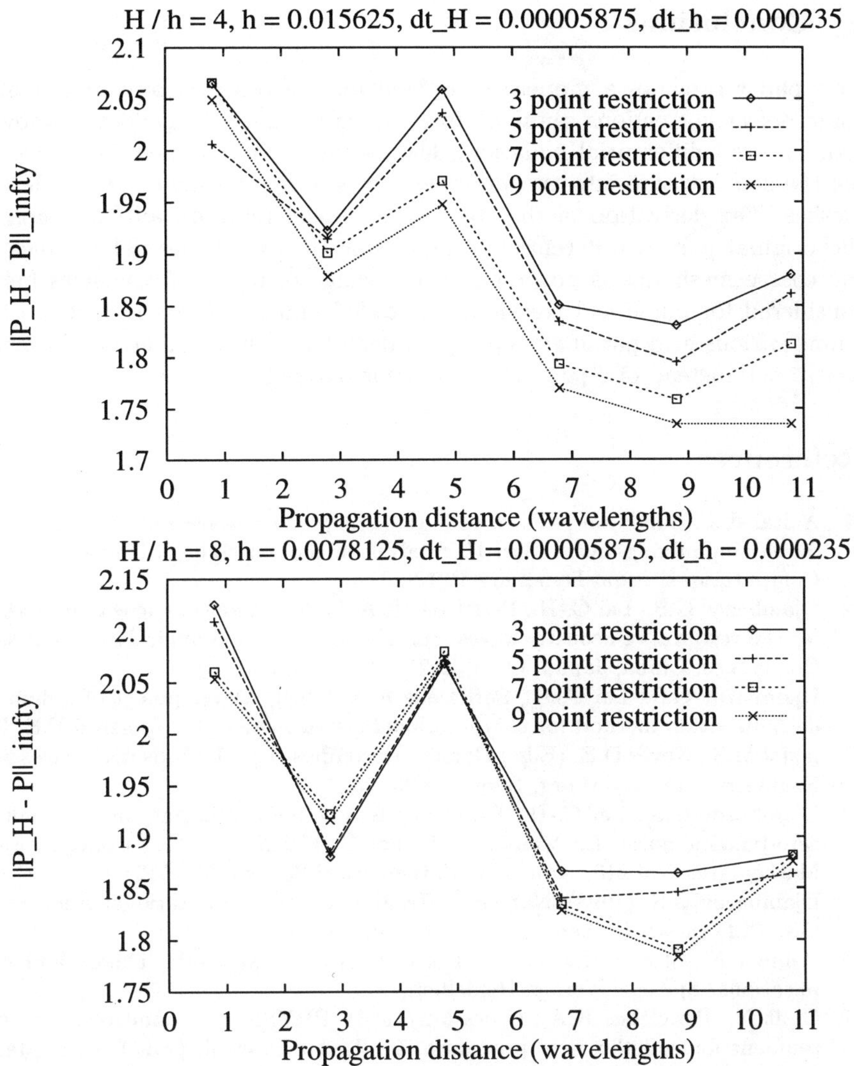

Fig. 3. The effect of restriction operators on the accuracy.

$$9 \text{ point formula: } I_{\{h,8h\}} g_h = \frac{1}{48}(g_{i-4} + 2g_{i-3}$$

$$+ 6g_{i-2} + 8g_{i-1} + 14g_i + 8g_{i+1} + 6g_{i+2} + 2g_{i+3} + g_{i+4})$$

For very fine CFD mesh, one can retrieve the small scale signal even on a relatively coarse mesh. In the present study, with $h = 0.0078125$ one can use $H \leq 8h$ while still maintaining the accuracy. The accuracy exhibited by using the coarse mesh $H = 8h = 0.0625$ is compatible with the result for Case (a) as depicted in Figure 1.

5   Conclusions

This paper provides a numerical method for the retrieval of sound signals using defect corrections obtained from a coarse space defined with a lower order partial differential equation. The essential concept here is to decouple the computation into two different scales of magnitude on two different meshes. The derivation of the coarse grid model relies on an expansion of the original partial differential equation for the two scales. The choice of the coarse mesh size is governed by the range of noise's frequencies being considered for numerical treatment. Detailed numerical experiments to examine various grid parameters are provided. The truncation error of solving $\mathcal{L}\{\bar{u}\}\bar{u} = 0$ instead of $\mathcal{L}\{\bar{u} + u\}\bar{u} + u = 0$ is derived.

References

1. Avital E.J., Sandham N.D., Luo K.H. (1998) Mach wave radiation by time-developing mixing layers. Part II: Analysis of the source field. Theoretical and Computational Fluid Dynamics 12:73-90
2. Djambazov G.S., Lai C.-H., Pericleous K.A. (1999) A defect correction method for the retrieval of acoustic waves. In: Abstract - 12th Domain Decomposition Conference, Chiba, Japan, October 25 - 29, 1999, 93
3. Djambazov G.S., Lai C.-H., Pericleous K.A. (1999) Development of a domain decomposition method for computational aeroacoustics. In: Bjorstad P.E, Espedal M.S., Keyes D.E. (Eds.) Domain Decomposition Methods in Sciences and Engineering IX. DDM.org, Bergen, 719-725
4. Djambazov G.S., Lai C.-H., Pericleous K.A. (1998) Efficient computation of aerodynamic noise. In: Mandel J., Farhat C., Cai X.-C. (Eds.) Contemporary Mathematics Vol 218, American Mathematical Society, 506-512
5. Djambazov G.S. (1998) Numerical Techniques for Computational Aeroacoustics. PhD thesis, University of Greenwich, London
6. Böhmer K., Stetter H.J. (1984) Defect Correction Methods: Theory and Applications, Springer Verlag, Heidelberg
7. Croft N., Pericleous K.A., Cross M. (1995) PHYSICA: A multiphysics environment for complex flow processes. In: Taylor C. et al. (Eds.) Num. Meth. Laminar & Turbulent Flow IX/2, Pineridge Press, U.K., 1296
8. CHAM Ltd, Wimbledon, UK (1995) PHOENICS, Version 2.1.3
9. Leonard B.P. (1979) A stable and accurate convective modelling procedure based on quadratic upstream interpolation. Computer Methods in Applied Mechanics and Engineering 19:59-98
10. Djambazov G.S., Lai C.-H., Pericleous K.A. (2000) On the coupling of Navier-Stokes and linearised Euler equations for aeroacoustic simulation. Comput Visual Sci 3:9-12

Domain Decomposition Methods for Time-Harmonic Maxwell Equations: Numerical Results

Ana Alonso Rodríguez[1] and Alberto Valli[2]

[1] Dipartimento di Matematica, Università degli Studi di Milano, via Saldini 50, 20133 Milano, Italy
[2] Dipartimento di Matematica, Università degli Studi di Trento, 38050 Povo (Trento), Italy

Abstract. We present a series of numerical results illustrating the performance of some non-overlapping domain decomposition algorithms for time-harmonic Maxwell equations in different physical situations. For the full-Maxwell equations with damping we consider the well-known Dirichlet/Neumann and Neumann/Neumann methods. Numerical evidence will show that both schemes are convergent with a rate independent of the mesh size. For the low-frequency model in a conductor, we consider again the Dirichlet/Neumann and the Neumann/Neumann algorithms. Both methods turn out to be efficient and robust. Finally, for the eddy-current problem, we implement an iterative procedure coupling a scalar problem in the insulator and a vector problem in the conductor.

1 The Time-Harmonic Maxwell Equations in a Conductor

The time-harmonic Maxwell equations are derived from the complete Maxwell equations assuming that both the electric field $\mathcal{E}$ and the magnetic field $\mathcal{H}$ are of the form $\mathcal{E}(t, \mathbf{x}) = \mathrm{Re}[\mathbf{E}(\mathbf{x}) \exp(i\omega t)]$, $\mathcal{H}(t, \mathbf{x}) = \mathrm{Re}[\mathbf{H}(\mathbf{x}) \exp(i\omega t)]$, where $\omega \neq 0$ is a given angular frequency. Let $\Omega \subset \mathbb{R}^3$ be a bounded Lipschitz polyhedron with unit outward normal $\mathbf{n}$. Let $\varepsilon(\mathbf{x})$, $\mu(\mathbf{x})$ and $\sigma(\mathbf{x})$ denote respectively the dielectric constant, the magnetic permeability and the conductivity of the medium. The time-harmonic Maxwell equations read:

$$\begin{cases} \operatorname{curl}\mathbf{H} - (i\,\omega\varepsilon + \sigma)\mathbf{E} = \mathbf{J} & \text{in } \ \Omega \\ \operatorname{curl}\mathbf{E} + i\,\omega\mu\mathbf{H} = \mathbf{0} & \text{in } \ \Omega, \end{cases} \tag{1}$$

where $\mathbf{J} = \mathbf{J}(\mathbf{x})$ is a known function specifying the applied current density. (See [6] for a complete presentation of time-harmonic Maxwell equations.)

We shall assume that the tangential trace of $\mathbf{E}$ is given on the boundary of Ω (for instance, equal to $\mathbf{0}$ for a perfectly conducting boundary), namely,

$$\mathbf{E} \times \mathbf{n} = \boldsymbol{\Upsilon} \text{ on } \partial\Omega. \tag{2}$$

In the general case of anisotropic inhomogeneous media the coefficients ε, μ and σ are 3×3 symmetric real matrices with entries in $L^\infty(\Omega)$. The matrices

ε and μ are assumed to be uniformly positive definite in Ω. The conductivity σ is an uniformly positive definite matrix in a conductor (instead, it is equal to 0 in an insulator).

As μ is non-singular, we may eliminate the magnetic field $\mathbf{H}$ in (1) and consider the following boundary value (*full-Maxwell problem*):

$$\begin{cases} \operatorname{curl}(\mu^{-1}\operatorname{curl}\mathbf{E}) - \omega^2\varepsilon\mathbf{E} + i\omega\sigma\mathbf{E} = -i\omega\mathbf{J} & \text{in } \Omega \\ \mathbf{E} \times \mathbf{n} = \Upsilon & \text{on } \partial\Omega. \end{cases} \tag{3}$$

When the frequency ω is small, by checking the effective value of the dielectric constant, the magnetic permeability and the conductivity in a metallic conductor, the term $\omega^2\varepsilon\mathbf{E}$ can be dropped out and one is left with the *low-frequency problem*:

$$\begin{cases} \operatorname{curl}(\mu^{-1}\operatorname{curl}\mathbf{E}) + i\omega\sigma\mathbf{E} = -i\omega\mathbf{J} & \text{in } \Omega \\ \mathbf{E} \times \mathbf{n} = \Upsilon & \text{on } \partial\Omega. \end{cases} \tag{4}$$

Being σ uniformly positive definite in Ω (i.e., Ω is a conductor), problem (3) and problem (4) are well posed. In fact, the bilinear form associated with both problems is

$$a(\mathbf{u},\mathbf{v}) := \int_\Omega [\mu^{-1}\operatorname{curl}\mathbf{u} \cdot \operatorname{curl}\overline{\mathbf{v}} - \omega^2\tilde{\varepsilon}\mathbf{u} \cdot \overline{\mathbf{v}} + i\omega\sigma\mathbf{u} \cdot \overline{\mathbf{v}}]$$

with $\tilde{\varepsilon} = \varepsilon$ for the full-Maxwell problem with damping and $\tilde{\varepsilon} = 0$ for the low-frequency problem, and, since σ is uniformly positive definite in Ω, it is coercive in $H(\operatorname{curl};\Omega)$, the space of complex vector functions $\mathbf{u}$ in $(L^2(\Omega))^3$ with $\operatorname{curl}\mathbf{u}$ in $(L^2(\Omega))^3$.

2   Domain Decomposition Algorithms for the Time-Harmonic Maxwell Equations in a Conductor

Let the bounded domain Ω be decomposed in two subdomains Ω_1 and Ω_2 such that $\overline{\Omega} = \overline{\Omega_1} \cup \overline{\Omega_2}$ and $\Omega_1 \cap \Omega_2 = \emptyset$. We will set $\Gamma := \overline{\Omega_1} \cap \overline{\Omega_2}$ and on Γ we consider $\mathbf{n}_\Gamma$, the unit outward normal vector to Ω_1.

In each subdomain we want to solve

$$\begin{cases} \operatorname{curl}(\mu^{-1}\operatorname{curl}\mathbf{E}_j) - \omega^2\tilde{\varepsilon}\mathbf{E}_j + i\omega\sigma\mathbf{E}_j = -i\omega\mathbf{J} & \text{in } \Omega_j \\ \mathbf{E}_j \times \mathbf{n} = \Upsilon & \text{on } \partial\Omega_j \cap \partial\Omega, \end{cases} \tag{5}$$

$j = 1, 2$, with the interface conditions

$$(\mathbf{E}_1 \times \mathbf{n}_\Gamma)_{|\Gamma} = (\mathbf{E}_2 \times \mathbf{n}_\Gamma)_{|\Gamma}$$
$$(\mu^{-1}\operatorname{curl}\mathbf{E}_1 \times \mathbf{n}_\Gamma)_{|\Gamma} = (\mu^{-1}\operatorname{curl}\mathbf{E}_2 \times \mathbf{n}_\Gamma)_{|\Gamma}.$$

We consider two families of domain decomposition methods: the γ-Dirichlet/Robin methods and the γ-Robin/Robin methods. For each value of the (complex) parameter γ we have a different algorithm.

Fixed a relaxation parameter θ and given a initial guess $\boldsymbol{\lambda}^0$, the γ-Dirichlet/Robin iterative algorithm reads:

$$\begin{cases} \operatorname{curl}(\mu^{-1}\operatorname{curl}\mathbf{E}_1^{k+1}) - \omega^2\tilde{\varepsilon}\mathbf{E}_1^{k+1} + i\omega\sigma\mathbf{E}_1^{k+1} = -i\omega\mathbf{J} & \text{in } \Omega_1 \\[2ex] \mathbf{E}_1^{k+1} \times \mathbf{n}_\Gamma = \boldsymbol{\Upsilon} & \text{on } \partial\Omega \cap \partial\Omega_1 \\[2ex] \mathbf{E}_1^{k+1} \times \mathbf{n}_\Gamma = \boldsymbol{\lambda}^k & \text{on } \Gamma \end{cases}$$

$$\begin{cases} \operatorname{curl}(\mu^{-1}\operatorname{curl}\mathbf{E}_2^{k+1}) - \omega^2\tilde{\varepsilon}\mathbf{E}_2^{k+1} + i\omega\sigma\mathbf{E}_2^{k+1} = -i\omega\mathbf{J} & \text{in } \Omega_2 \\[2ex] \mathbf{E}_2^{k+1} \times \mathbf{n}_\Gamma = \boldsymbol{\Upsilon} & \text{on } \partial\Omega \cap \partial\Omega_2 \\[2ex] (\mu^{-1}\operatorname{curl}\mathbf{E}_2^{k+1}) \times \mathbf{n}_\Gamma - \gamma\,\mathbf{n}_\Gamma \times (\mathbf{E}_2^{k+1} \times \mathbf{n}_\Gamma) \\ \quad = (\mu^{-1}\operatorname{curl}\mathbf{E}_1^{k+1}) \times \mathbf{n}_\Gamma - \gamma\,\mathbf{n}_\Gamma \times (\mathbf{E}_1^{k+1} \times \mathbf{n}_\Gamma) & \text{on } \Gamma \end{cases}$$

$$\boldsymbol{\lambda}^{k+1} = (1-\theta)\boldsymbol{\lambda}^k + \theta\,(\mathbf{E}_2^{k+1} \times \mathbf{n}_\Gamma)_{|\Gamma} \quad \text{on } \Gamma.$$

The γ-Robin/Robin method is given by (for $j = 1, 2$):

$$\begin{cases} \operatorname{curl}(\mu^{-1}\operatorname{curl}\mathbf{E}_j^{k+1}) - \omega^2\tilde{\varepsilon}\mathbf{E}_j^{k+1} + i\omega\sigma\mathbf{E}_j^{k+1} = -i\omega\mathbf{J} & \text{in } \Omega_j \\[2ex] \mathbf{E}_j^{k+1} \times \mathbf{n}_\Gamma = \boldsymbol{\Upsilon} & \text{on } \partial\Omega \cap \partial\Omega_j \\[2ex] \mathbf{E}_j^{k+1} \times \mathbf{n}_\Gamma = \boldsymbol{\lambda}^k & \text{on } \Gamma \end{cases}$$

$$\begin{cases} \operatorname{curl}(\mu^{-1}\operatorname{curl}\boldsymbol{\Phi}_1^{k+1}) - \omega^2\tilde{\varepsilon}\boldsymbol{\Phi}_1^{k+1} + i\omega\sigma\boldsymbol{\Phi}_1^{k+1} = 0 & \text{in } \Omega_1 \\[2ex] \boldsymbol{\Phi}_1^{k+1} \times \mathbf{n}_\Gamma = 0 & \text{on } \partial\Omega \cap \partial\Omega_1 \\[2ex] (\mu^{-1}\operatorname{curl}\boldsymbol{\Phi}_1^{k+1}) \times \mathbf{n}_\Gamma + \gamma\,\mathbf{n}_\Gamma \times (\boldsymbol{\Phi}_1^{k+1} \times \mathbf{n}_\Gamma) \\ \quad = -(\mu^{-1}\operatorname{curl}\mathbf{E}_1^{k+1}) \times \mathbf{n}_\Gamma + (\mu^{-1}\operatorname{curl}\mathbf{E}_2^{k+1}) \times \mathbf{n}_\Gamma & \text{on } \Gamma \end{cases}$$

$$\begin{cases} \operatorname{curl}(\mu^{-1}\operatorname{curl}\boldsymbol{\Phi}_2^{k+1}) - \omega^2\tilde{\varepsilon}\boldsymbol{\Phi}_2^{k+1} + i\omega\sigma\boldsymbol{\Phi}_2^{k+1} = 0 & \text{in } \Omega_2 \\[2ex] \boldsymbol{\Phi}_2^{k+1} \times \mathbf{n}_\Gamma = 0 & \text{on } \partial\Omega \cap \partial\Omega_2 \\[2ex] (\mu^{-1}\operatorname{curl}\boldsymbol{\Phi}_2^{k+1}) \times \mathbf{n}_\Gamma - \gamma\,\mathbf{n}_\Gamma \times (\boldsymbol{\Phi}_2^{k+1} \times \mathbf{n}_\Gamma) \\ \quad = (\mu^{-1}\operatorname{curl}\mathbf{E}_1^{k+1}) \times \mathbf{n}_\Gamma - (\mu^{-1}\operatorname{curl}\mathbf{E}_2^{k+1}) \times \mathbf{n}_\Gamma & \text{on } \Gamma \end{cases}$$

$$\boldsymbol{\lambda}^{k+1} = \boldsymbol{\lambda}^k + \theta\,(\boldsymbol{\Phi}_1^{k+1} \times \mathbf{n}_\Gamma + \boldsymbol{\Phi}_2^{k+1} \times \mathbf{n}_\Gamma)_{|\Gamma} \quad \text{on } \Gamma.$$

We note that setting γ equal zero we have the well-known Dirichlet/Neumann and Neumann/Neumann algorithms, respectively.

For the discretization of problem (5) the curl-conforming finite elements introduced by Nédélec (see [7], [8]) can be used.

Remark 1. Concerning the convergence of these iteration-by-subdomain procedures for the discrete problem we have the following results:

- for the low-frequency model it can be proven that using the first family of curl-conforming finite elements (see [7]), both the the Dirichlet/Neumann method and the Neumann/Neumann method converge with a rate independent of the mesh size, provided that the relaxation parameter θ is properly chosen (see [3] and [9]);

- when $\tilde{\varepsilon}$ is uniformly positive definite, we need to consider $\gamma \neq 0$ in order to prove convergence. More precisely, let $Z = 1 - iB$ be a complex number such that the bilinear form $b(\mathbf{u}, \mathbf{v}) := Z\,a(\mathbf{u}, \mathbf{v})$ is real coercive (i.e., there exists $\alpha > 0$ such that $\mathrm{Re}(b(\mathbf{u}, \mathbf{u})) \geq \alpha \|\mathbf{u}\|^2_{H(\mathrm{curl};\Omega)}$ for each $\mathbf{u} \in H(\mathrm{curl};\Omega)$) and consider the discrete problem corresponding to the first family of curl-conforming finite element. Then, for both the γ-Dirichlet/Robin and the γ-Robin/Robin method it can be proven that, choosing $\gamma = \eta Z$ with $\eta \in \mathbb{R}$ large enough, there exists an interval $\mathcal{I} = (0, \theta^*)$ such that choosing θ in $\mathcal{I}$ the iteration-by-subdomain procedure converges uniformly in h (see [4]).

However, numerical experiments show that the 0-Dirichlet/Robin and the 0-Robin/Robin algorithms (namely, the Dirichlet/Neumann and the Neumann/Neumann algorithms, respectively) are indeed convergent also for the full-Maxwell equations with damping (i.e. assuming $\tilde{\varepsilon}$ and σ uniformly positive definite in Ω. The rate of convergence seems to be independent of the mesh size and the number of iterations is in general quite small (provided that the acceleration parameter θ is properly chosen). Consequently, it is apparent that choices of the parameter γ different from 0 are not necessary in numerical computations.

3   Numerical Results for the Time-Harmonic Maxwell Equations in a Conductor

We present some numerical tests illustrating the performances of the Dirichlet/Neumann and the Neumann/Neumann algorithms described in the previous section. We consider model problems with scalar constant parameters μ, ε, σ, and very simple geometric situations: the computational domain is always the parallelepiped $\Omega = (0, 2) \times (0, 1) \times (0, 1)$, which will be decomposed into two subdomains $\Omega_1 = (0, x_\Gamma) \times (0, 1) \times (0, 1)$ and $\Omega_2 = (x_\Gamma, 2) \times (0, 1) \times (0, 1)$. The numerical mesh is uniform, and each element of the grid is a cube of side h. We employ the first family of edge elements introduced by Nédélec (see [7]), with 12 degrees of freedom for each element, one for each edge. In Table 1 we indicate the total number of degrees of freedom for different values of h.

Table 1. Number of degrees of freedom for different values of h

h	1/4	1/6	1/8	1/9	1/10	1/12
DOF	240	960	2464	3600	5040	8976

For the numerical computations we have used the standard tool-box sparfun of MATLAB^{TM} 5.2. In particular, for solving the linear system we adopt the `gmres` function with TOL $= 10^{-6}$.

In the iteration-by-subdomain procedure we have used the following stopping test

$$\sum_{i=1}^{2} \frac{\|\mathbf{E}_{i,h}^{k+1} - \mathbf{E}_{i,h}^{k}\|_{H(\mathrm{curl};\Omega_i)}^{2}}{\|\mathbf{E}_{i,h}^{k+1}\|_{H(\mathrm{curl};\Omega_i)}^{2}} \leq 10^{-6} \, ,$$

and in all the tests we start the iterations setting $\boldsymbol{\lambda}^0 = \mathbf{0}$.

3.1   Full-Maxwell Equations with Damping

In the model problem for the full-Maxwell problem with damping we have a frequency of order of the gigahertz, $\omega = c\,10^9$ Hz; we set $\mu = 10^{-6}$ H/m, $\varepsilon = 10^{-10}$ F/m and we consider a domain measured in decimeters. Hence our model problem reads:

$$\begin{cases} \mathrm{curl\ curl}\ \mathbf{E} + (i\,c\,k - c^2)\mathbf{E} = \mathbf{F} & \text{in } \Omega \\ \mathbf{E} \times \mathbf{n} = \boldsymbol{\Upsilon} & \text{on } \partial\Omega \end{cases} \tag{6}$$

where $k = 10\,\sigma \in \mathbb{R}$.

The first numerical test concerns the data $\mathbf{F}$ and $\boldsymbol{\Upsilon}$ obtained, according to (6), choosing $\mathbf{E}(x,y,z) = (e^z \sin(xy),\ e^x(y+z),\ \cos(xz))$ and $c = k = 1$. We set $x_\Gamma = 3/2$. In Table 2 we show the number of iterations required to achieve convergence with different values of h. We note that, though this situation is not covered by the theoretical results quoted in Remark 1 , the number of iterations is constant with respect to h and quite small. We have

Table 2. $\mathbf{E}(x,y,z) = (e^z \sin(xy),\ e^x(y+z),\ \cos(xz))$, $x_\Gamma = 3/2$, $c = k = 1$

algorithm \ h	1/4	1/6	1/8	1/10	1/12
$D/N\ \theta = 0.5$	5	5	5	5	5
$N/N\ \theta = 0.25$	4	4	4	4	4

repeated the computations for three other sets of data ($\mathbf{E}(x,y,z) = (z^2 +$

$i \sin z,\ zx^2 + i\,\cos x,\ xy^2 + i\,e^y)$, $\mathbf{E}(x,y,z) = (z^2,\,x^2,\,y^2)$ and $\mathbf{E}(x,y,z) = (\sin z,\,\cos x,\,e^y)$ with $c = k = 1$) and $x_\Gamma = 1/2$, and the convergence results have been quite similar. For instance, in Figures 1 and 2 one can see the convergence history for the data corresponding to function $\mathbf{E}(x,y,z) = (z^2 + i\,\sin z,\ zx^2 + i\,\cos x,\ xy^2 + i\,e^y)$ for three different values of h.

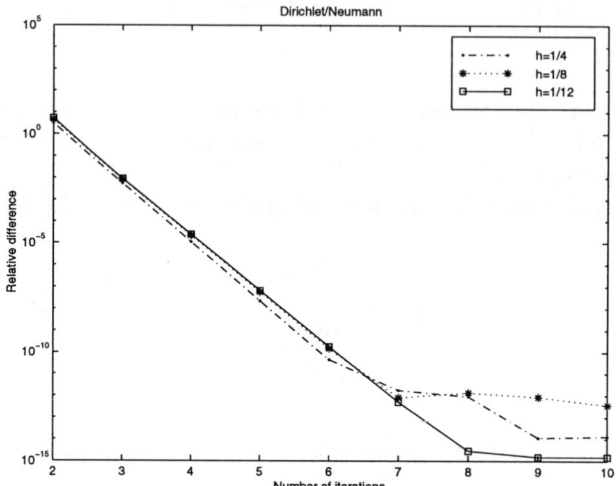

Fig. 1. Convergence history of the Dirichlet/Neumann method for different values of h

We plot the relative difference

$$\sum_{i=1}^{2} \frac{\|\mathbf{E}_{i,h}^{k+1} - \mathbf{E}_{i,h}^{k}\|_{H(\mathrm{curl};\Omega_i)}^2}{\|\mathbf{E}_{i,h}^{k+1}\|_{H(\mathrm{curl};\Omega_i)}^2}$$

as a function of the number of iterations.

From Table 3 we deduce that the number of iterations depends only slightly on the position of the interface. The dependence seems to be weaker for the Neumann/Neumann method with $\theta = 0.25$.

Table 3. $\mathbf{E}(x,y,z) = (\sin z,\,\cos x,\,e^y)$, $h = 1/9$, $c = k = 1$.

algorithm \\ x_Γ	1/3	2/3	1	4/3	5/3
$D/N\ \theta = 0.5$	7	4	3	4	6
$N/N\ \theta = 0.25$	4	3	3	3	4

In Table 4 it is shown that the choice $\theta = 0.5$ for the Dirichlet/Neumann scheme (with $c = k = 1$) is nearly optimal. We have considered different values of x_Γ.

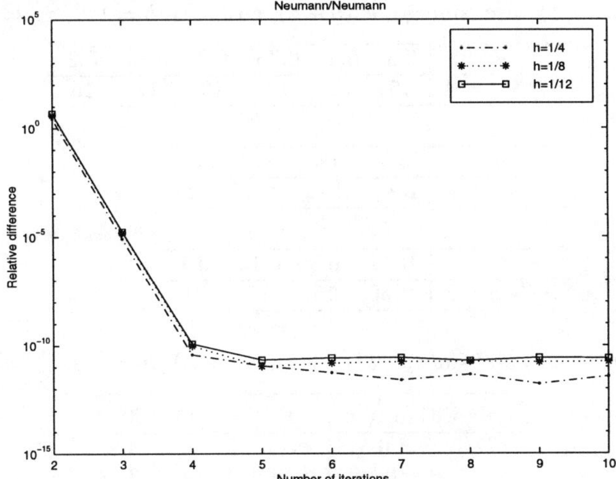

Fig. 2. Convergence history of the Neumann/Neumann method for different values of h

Table 4. Number of iterations for D/N algorithm. $\mathbf{E}(x, y, z) = (\sin z, \cos x, e^y)$, $h = 1/9$, $c = k = 1$.

$\theta \ \backslash \ x_\Gamma$	1/3	4/3	5/3
0.4	7	7	8
0.45	6	6	7
0.5	7	4	6
0.55	9	6	6
0.6	11	7	7

Still taking $k = 1$, for higher values of the frequency ($c > 1$) the number of iterations increases, and the choice of the relaxation parameter needs to be more accurate (see Tables 5-6).

Table 5. $\mathbf{E}(x, y, z) = (e^z \sin(xy), e^x(y + z), \cos(xz))$, $h = 1/9$, $x_\Gamma = 4/3$, $k = 1$.

algorithm $\backslash$ c	0.25	0.5	1	2	4	8
$D/N \ \theta = 0.5$	4	4	4	5	46	div.
$N/N \ \theta = 0.25$	3	3	3	4	9	> 50

Finally, in Table 7 we have fixed $c = 8$ and we observe that the number of iterations decreases if the damping increases.

In conclusion, even if the theoretical results reported in Remark 1 ensure convergence only for particular values of $\gamma \in \mathbb{C}$, numerical experiments show that the 0-Dirichlet/Robin and the 0-Robin/Robin algorithms (namely the

Table 6. $\mathbf{E}(x,y,z) = (e^z \sin(xy), e^x(y+z), \cos(xz))$, $h = 1/9$, $x_\Gamma = 4/3$, $k = 1$.

θ	0.5	0.4	0.35	0.3	0.25	0.2
D/N $c = 4$	46	15	11	10	12	15

θ	0.5	0.4	0.3	0.2
D/N $c = 8$	div.	div.	> 50	24

θ	0.25	0.2	0.15	0.1
N/N $c = 8$	> 50	16	15	21

Table 7. $\mathbf{E}(x,y,z) = (e^z \sin(xy), e^x(y+z), \cos(xz))$, $h = 1/9$, $x_\Gamma = 4/3$, $c = 8$.

algorithm \ k	1	5	10	50
$D/N\ \theta = 0.5$	div.	5	4	3
$N/N\ \theta = 0.25$	> 50	4	3	3

Dirichlet/Neumann and the Neumann/Neumann algorithms, respectively) are convergent with a rate of convergence independent of the mesh size, and that the two methods have a good performance.

3.2   The Low-Frequency Model

We remember that in this case theoretical results ensure convergence of both Dirichlet/Neumann and the Neumann/Neumann algorithms applied to the discrete problem, with a rate independent of the mesh size h (see [3] and [9]). Now we present some numerical results that illustrate the performance of the algorithms.

Our model problem for the low-frequency case reads:

$$\begin{cases} \operatorname{curl} \operatorname{curl} \mathbf{E} + iK\mathbf{E} = \mathbf{F} & \text{in } \Omega \\ \mathbf{E} \times \mathbf{n} = \boldsymbol{\Upsilon} & \text{on } \partial\Omega \end{cases} \tag{7}$$

with $K := \omega\mu\sigma \in \mathbb{R}$.

As for the full-Maxwell problem, we consider the data $\mathbf{F}$ and $\boldsymbol{\Upsilon}$ obtained, according to (7), choosing $\mathbf{E}(x,y,z) = (e^z \sin(xy), e^x(y+z), \cos(xz))$ and $K = 1$. We set $x_\Gamma = 3/2$ and we consider different values of h. Table 8 shows the number of iterations required to achieve convergence. We note that the number of iterations is constant with respect to h and quite small. We have repeated the computations for other data and the convergence results have been similar.

We have also considered different positions of the interface (see Table 9). The results are analogous to those for the full-Maxwell problem: the number of iterations depends slightly on the position of the interface and this dependence seems to be weaker for the Neumann/Neumann algorithm.

Table 8. $\mathbf{E}(x, y, z) = (e^z \sin(xy), e^x(y + z), \cos(xz))$, $x_\Gamma = 3/2$, $K = 1$.

algorithm \ h	1/4	1/6	1/8	1/10	1/12
$D/N\ \theta = 0.5$	5	5	5	5	5
$N/N\ \theta = 0.25$	4	4	4	4	4

Table 9. $\mathbf{E}(x, y, z) = (yz^2 + i\sin(z), zx^2 + i\cos(x), xy^2 + ie^y)$, $h = 1/9$, $K = 1$.

algorithm \ x_Γ	1/3	2/3	1	4/3	5/3
$D/N\ \theta = 0.5$	6	4	3	4	6
$N/N\ \theta = 0.25$	4	3	3	3	4

Finally, we take other values of K (see Table 10) and we see that the number of iterations is almost constant with respect to K.

Table 10. $\mathbf{E}(x, y, z) = (yz^2 + i\sin(z), zx^2 + i\cos(x), xy^2 + ie^y)$, $h = 1/9$, $x_\Gamma = 4/3$.

algorithm \ K	10^{-2}	1	10^2	10^4
$D/N\ \theta = 0.5$	4	4	3	3
$N/N\ \theta = 0.25$	3	3	3	3

4 The Eddy-Current Problem and a Domain Decomposition Algorithm for its Solution

Now we consider the *eddy current problem*, in which the displacement current term $i\omega\varepsilon\mathbf{E}$ is neglected and the conductivity σ is equal to zero in a non-empty open subset $\Omega_I \subset \Omega$, i.e., a part of the domain is an insulator (see, e.g., [6]). We denote $\Omega_C = \Omega \setminus \overline{\Omega_I}$, $\Gamma := \overline{\Omega_I} \cap \overline{\Omega_C}$ and $\mathbf{n}_l$ the unit outward normal vector to Ω_l for $l = I, C$. We assume that $\operatorname{div}\mathbf{J}_I = 0$ where $\mathbf{J}_I := \mathbf{J}_{|\Omega_I}$. For simplicity in the presentation, we also assume that Ω_I is simply connected and $\partial\Omega_I$ is connected. Hence the eddy-current problem (i.e., the low-frequency model for heterogeneous media) reads

$$
\begin{cases}
\operatorname{curl}\mathbf{H}_C - \sigma\mathbf{E}_C = \mathbf{J}_C & \text{in } \Omega_C \\
\operatorname{curl}\mathbf{E}_C + i\omega\mu\mathbf{H}_C = 0 & \text{in } \Omega_C \\
\mathbf{E}_C \times \mathbf{n}_C = \Upsilon & \text{on } \partial\Omega \cap \partial\Omega_C \\
\operatorname{curl}\mathbf{H}_I = \mathbf{J}_I & \text{in } \Omega_I \\
\operatorname{div}(\varepsilon\mathbf{E}_I) = 0 & \text{in } \Omega_I \\
\operatorname{curl}\mathbf{E}_I + i\omega\mu\mathbf{H}_I = 0 & \text{in } \Omega_I \\
\mathbf{E}_I \times \mathbf{n}_I = \Upsilon & \text{on } \partial\Omega \cap \partial\Omega_I,
\end{cases}
$$

with the interface conditions

$$
\begin{aligned}
(\mathbf{E}_C \times \mathbf{n}_C)_{|\Gamma} &= -(\mathbf{E}_I \times \mathbf{n}_I)_{|\Gamma} \\
(\mathbf{H}_C \times \mathbf{n}_C)_{|\Gamma} &= -(\mathbf{H}_I \times \mathbf{n}_I)_{|\Gamma}.
\end{aligned}
$$

(For a mathematical justification of this model see [1] and [5]).

In this situation it is possible to eliminate $\mathbf{E}_I$ and $\mathbf{H}_C$. Noting that $\operatorname{div}_\tau(\mathbf{E}_I \times \mathbf{n}_I) = \operatorname{curl}\mathbf{E}_I \cdot \mathbf{n}_I$, where div_τ denotes the tangential divergence operator, we can consider the following problem

$$\begin{cases} \operatorname{curl}(\mu^{-1}\operatorname{curl}\mathbf{E}_C) + i\omega\sigma\mathbf{E}_C = -i\omega\mathbf{J}_C & \text{in } \Omega_C \\ \mathbf{E}_C \times \mathbf{n}_C = \Upsilon & \text{on } \partial\Omega \cap \partial\Omega_C \\ \operatorname{curl}\mathbf{H}_I = \mathbf{J}_I & \text{in } \Omega_I \\ \operatorname{div}(\mu\mathbf{H}_I) = 0 & \text{in } \Omega_I \\ \mu\mathbf{H}_I \cdot \mathbf{n}_I = -(i\omega)^{-1}\operatorname{div}_\tau\Upsilon & \text{on } \partial\Omega \cap \partial\Omega_I, \end{cases}$$

with the interface conditions

$$(i\omega)^{-1}(\operatorname{curl}\mathbf{E}_C \cdot \mathbf{n}_C)_{|\Gamma} = (\mu\mathbf{H}_I \cdot \mathbf{n}_I)_{|\Gamma}$$
$$(i\omega)^{-1}(\mu^{-1}\operatorname{curl}\mathbf{E}_C \times \mathbf{n}_C)_{|\Gamma} = (\mathbf{H}_I \times \mathbf{n}_I)_{|\Gamma}.$$

We can decompose $\mathbf{H}_I$ in the following way:

$$\mathbf{H}_I = \mu^{-1}\operatorname{curl}\mathbf{q}_I + (i\omega)^{-1}\nabla\varphi_I,$$

where $\mathbf{q}_I$ is the solution of

$$\begin{cases} \operatorname{curl}(\mu^{-1}\operatorname{curl}\mathbf{q}_I) = \mathbf{J}_I & \text{in } \Omega_I \\ \operatorname{div}\mathbf{q}_I = 0 & \text{in } \Omega_I \\ \mathbf{q}_I \times \mathbf{n}_I = 0 & \text{on } \partial\Omega_I \end{cases}$$

and φ_I satisfies

$$\begin{cases} \operatorname{div}(\mu\nabla\varphi_I) = 0 & \text{in } \Omega_I \\ \int_{\Omega_I}\varphi_I = 0 \\ \mu\nabla\varphi_I \cdot \mathbf{n}_I = -\operatorname{div}_\tau\Upsilon & \text{on } \partial\Omega \cap \partial\Omega_I \\ \mu\nabla\varphi_I \cdot \mathbf{n}_I = \operatorname{curl}\mathbf{E}_C \cdot \mathbf{n}_C & \text{on } \Gamma. \end{cases}$$

We note that $\mathbf{q}_I$ can be computed independently of $\mathbf{E}_C$, hence we are led to the coupled problem:

$$\begin{cases} \operatorname{curl}(\mu^{-1}\operatorname{curl}\mathbf{E}_C) + i\omega\sigma\mathbf{E}_C = -i\omega\mathbf{J}_C & \text{in } \Omega_C \\ \mathbf{E}_C \times \mathbf{n}_C = \Upsilon & \text{on } \partial\Omega \cap \partial\Omega_C \\ \mu^{-1}\operatorname{curl}\mathbf{E}_C \times \mathbf{n}_C = i\omega\mu^{-1}\operatorname{curl}\mathbf{q}_I \times \mathbf{n}_I + \nabla\varphi_I \times \mathbf{n}_I & \text{on } \Gamma \\ \operatorname{div}(\mu\nabla\varphi_I) = 0 & \text{in } \Omega_I \\ \int_{\Omega_I}\varphi_I = 0 \\ \mu\nabla\varphi_I \cdot \mathbf{n}_I = -\operatorname{div}_\tau\Upsilon & \text{on } \partial\Omega \cap \partial\Omega_I \\ \mu\nabla\varphi_I \cdot \mathbf{n}_I = \operatorname{curl}\mathbf{E}_C \cdot \mathbf{n}_C & \text{on } \Gamma \end{cases} \qquad (8)$$

Note that $\operatorname{curl}\mathbf{q}_I \cdot \mathbf{n}_I = \operatorname{div}_\tau(\mathbf{q}_I \times \mathbf{n}_I) = 0$ on $\partial\Omega_I$. Moreover, $\operatorname{curl}\mathbf{E}_C \cdot \mathbf{n}_C = \operatorname{div}_\tau(\mathbf{E}_C \times \mathbf{n}_C)$ on Γ. Hence to solve problem (8) we propose the following iterative algorithm:

$$\begin{cases} \text{div}\,(\mu\nabla\varphi_I^{k+1}) = 0 & \text{in}\ \ \Omega_I \\[2mm] \int_{\Omega_I}\varphi_I^{k+1} = 0 \\[2mm] \mu\nabla\varphi_I^{k+1}\cdot\mathbf{n}_I = -\text{div}\,_\tau\Upsilon & \text{on}\ \partial\Omega\cap\partial\Omega_I \\[2mm] \mu\nabla\varphi_I^{k+1}\cdot\mathbf{n}_I = \text{div}\,_\tau\boldsymbol{\lambda}^k & \text{on}\ \Gamma \end{cases} \tag{9}$$

$$\begin{cases} \text{curl}\,(\mu^{-1}\text{curl}\,\mathbf{E}_C^{k+1}) + i\omega\sigma\mathbf{E}_C^{k+1} = -i\omega\mathbf{J}_C\ \ \text{in}\ \ \Omega_C \\[2mm] \mathbf{E}_C^{k+1}\times\mathbf{n}_C = \Upsilon & \text{on}\ \partial\Omega\cap\partial\Omega_C \\[2mm] \mu^{-1}\text{curl}\,\mathbf{E}_C^{k+1}\times\mathbf{n}_C = i\omega\mu^{-1}\text{curl}\,\mathbf{q}_I\times\mathbf{n}_I \\[1mm] \qquad\qquad\qquad\qquad +\nabla\varphi_I^{k+1}\times\mathbf{n}_I & \text{on}\ \Gamma \end{cases} \tag{10}$$

$$\boldsymbol{\lambda}^{k+1} = (1-\theta)\boldsymbol{\lambda}^k + \theta\,(\mathbf{E}_C^{k+1}\times\mathbf{n}_C)_{|\Gamma}. \tag{11}$$

Remark 2. The finite element approximation of (9)-(11) is easily achieved, as both problem (9) and (10) are classical, and we can use standard piecewise polynomials for the approximation in $H^1(\Omega_I)$ and the Nédélec finite elements for the approximation in $H(\text{curl};\Omega_C)$. In [2] it has been proven that there exists an interval $\mathcal{I} = (0,\theta^*)$ such that choosing θ in $\mathcal{I}$ the iterative procedure converges uniformly in h.

5   Numerical Results for the Eddy Current Problem

For testing the iterative algorithm presented in the previous section we consider a model problem with scalar constant parameters μ and σ. The computational domain is again the parallelepiped $\Omega = (0,2)\times(0,1)\times(0,1)$ and we take $\Omega_I := (0,x_\Gamma)\times(0,1)\times(0,1)$ and $\Omega_C := (x_\Gamma,2)\times(0,1)\times(0,1)$. The numerical mesh is uniform, and each element of the grid is a cube of side h.

The aim of these numerical experiments is to verify the effectiveness of the iteration-by-subdomain procedure. Firstly, we consider a model problem with $\mathbf{J} = \mathbf{0}$ (hence $\mathbf{J}_C = \mathbf{0}$ and $\mathbf{q}_I = \mathbf{0}$) and $\Upsilon = \mathbf{0}$, and, starting with a non-zero initial datum $\boldsymbol{\lambda}^0 = (\mathbf{E}_C^0\times\mathbf{n}_C)_{|\Gamma}$, we study the convergence of the solution to 0. Since in particular μ is constant, setting $\psi_I^k = \mu\varphi_I^k$, the iterative procedure can be rewritten as:

$$\begin{cases} \Delta \psi_I^{k+1} = 0 & \text{in } \Omega_I \\[2mm] \displaystyle\int_{\Omega_I} \psi_I^{k+1} = 0 \\[2mm] \dfrac{\partial \psi_I^{k+1}}{\partial n_I} = 0 & \text{on } \partial\Omega \cap \partial\Omega_I \\[2mm] \dfrac{\partial \psi_I^{k+1}}{\partial n_I} = \operatorname{div}_\tau \boldsymbol{\lambda}^k & \text{on } \Gamma \end{cases}$$

$$\begin{cases} \operatorname{curl}\operatorname{curl}\mathbf{E}_C^{k+1} + i\kappa\mathbf{E}_C^{k+1} = \mathbf{0} & \text{in } \Omega_C \\[2mm] \mathbf{E}_C^{k+1} \times \mathbf{n}_C = \mathbf{0} & \text{on } \partial\Omega \cap \partial\Omega_C \\[2mm] \operatorname{curl}\mathbf{E}_C^{k+1} \times \mathbf{n}_C = \nabla\psi_I^{k+1} \times \mathbf{n}_I & \text{on } \Gamma \end{cases}$$

$$\boldsymbol{\lambda}^{k+1} = (1-\theta)\boldsymbol{\lambda}^k + \theta\,(\mathbf{E}_C^{k+1} \times \mathbf{n}_C)_{|\Gamma},$$

where $\kappa := \omega\sigma\mu$.

We use the following stopping test:

$$\|\nabla\psi_{I,h}^{k+1} - \nabla\psi_{I,h}^{k}\|_{L^2(\Omega_I)}^2 + \|\mathbf{E}_{C,h}^{k+1} - \mathbf{E}_{C,h}^{k}\|_{H(\operatorname{curl};\Omega_C)}^2 \le 10^{-8} \,.$$

In the test case 1 we take a real function $\mathbf{E}_C^0$, given by

$$\mathbf{E}_{C,\,test\,1}^0 = (e^z \sin(xy),\, e^x(y+z),\, \cos(xz))$$

whereas in the test case 2 the initial value is complex:

$$\mathbf{E}_{C,\,test\,2}^0 = (e^z + i\sin(xy), yz + ie^x, i\cos(xz)).$$

In Table 11 and Table 12 we show the number of iterations for the test case 1 and for the test case 2, respectively, with different values of h and x_Γ. The coefficient κ is set equal to 50 and the relaxation parameter θ is equal to 1. It can be seen that the number of iterations required to achieve convergence is almost constant with respect to h (in fact, it decreases slightly as the mesh size decreases).

From Table 13 we can see that the choice of the relaxation parameter $\theta = 1$ seems to be optimal.

Finally, we consider different values of the coefficient κ and we observe (see Table 14) that the number of iterations required to achieve convergence does not change with κ.

Table 11. Number of iterations. Test case 1, $\theta = 1$

$x_\Gamma \backslash h$	1/4	1/6	1/8	1/10	1/12
1/2	5	4	4	4	3
1	4	4	4	4	3
3/2	4	4	4	4	3

Table 12. Number of iterations. Test case 2, $\theta = 1$

$x_\Gamma \backslash h$	1/4	1/6	1/8	1/10	1/12
1/2	5	4	4	4	4
1	4	4	3	3	3
3/2	4	4	4	4	4

Table 13. Number of iterations. Test case 1, $h = 1/8$

$x_\Gamma \backslash \theta$	0.9	0.95	1	1.05	1.1
1/2	5	4	4	5	5
1	5	5	4	5	6
3/2	5	5	4	5	6

Table 14. Number of iterations. Test case 2, $\theta = 1$, $h = 1/8$

$x_\Gamma \backslash \kappa$	10^{-1}	1	10	10^2	10^3
1/2	4	4	4	4	4
1	3	3	3	3	3
3/2	4	4	4	4	4

In the second set of numerical experiments we construct the data in the following way: we take a function $\mathbf{E}_C \in H(\mathrm{curl}\,; \Omega_C)$, we compute its interpolant $\mathbf{E}_{C,h}$ in the finite element space, and using this function we determine $\varphi_{I,h}$ solving the discrete analogous of problem $(8)_4$-$(8)_7$ (with $\mathbf{\Upsilon} = \mathbf{0}$). Then we calculate the data of the discrete problem corresponding to $(8)_1$-$(8)_3$, in such a way that $\mathbf{E}_{C,h}$ turns out to be the discrete solution. In particular, we consider $\mathbf{E}_C(x,y,z) = (i\,y(1-y)z(1-z)\sin(xy), z(1-z)e^x, y(1-y)(1+i\cos z))$ (so that $\mathbf{E}_C \times \mathbf{n}_C = \mathbf{\Upsilon} = \mathbf{0}$ on $y = 0$, $y = 1$, $z = 0$ and $z = 1$), and constant coefficients $\omega = 50$, $\mu = 10^{-6}$, $\sigma_{|\Omega_C} = 10^6$. In this set of experiments we initialize the iterations with $\boldsymbol{\lambda}^0 = \mathbf{0}$ and we study the convergence to 0 of the relative difference between two subsequent iterates:

$$\frac{\|\nabla\varphi_{I,h}^{k+1} - \nabla\varphi_{I,h}^k\|_{L^2(\Omega_I)}^2}{\|\nabla\varphi_{I,h}^{k+1}\|_{L^2(\Omega_I)}^2} + \frac{\|\mathbf{E}_{C,h}^{k+1} - \mathbf{E}_{C,h}^k\|_{H(\mathrm{curl};\Omega_C)}^2}{\|\mathbf{E}_{C,h}^{k+1}\|_{H(\mathrm{curl};\Omega_C)}^2}.$$

In Figure 3 we see the convergence histories for $x_\Gamma = 1$ and different values of h, whereas in Figure 4 and in Figure 5 we set $h = 1/10$ and consider different values of x_Γ.

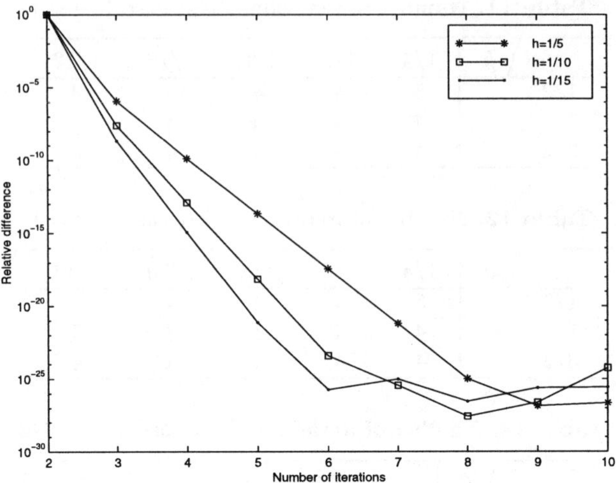

Fig. 3. Convergence history for $x_\Gamma = 1$ and different values of h

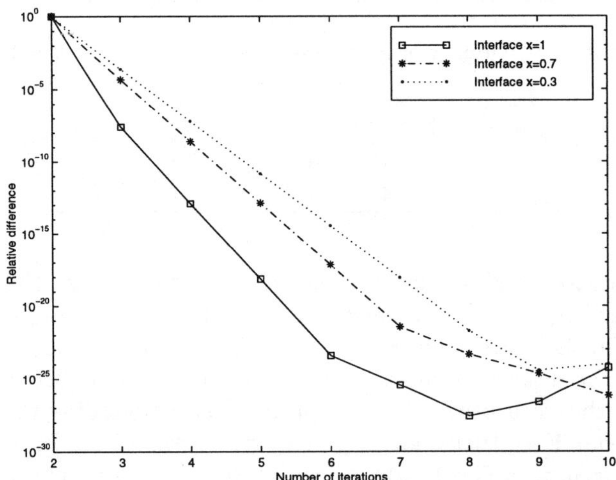

Fig. 4. Convergence history for $h = 1/10$ and different values of x_Γ

We can thus conclude that the algorithm we have proposed for solving the eddy-current problem is efficient: its convergence is fast and it seems to converge better for smaller values of the mesh size. The iteration-by-subdomain procedure is quite insensitive to the position of the interface; it converges faster when the two subdomains are of the same size but the performance of the method is good also otherwise. Finally, we observe that (at least in the case of constant coefficients) the algorithm is rather insensitive to the coefficients of the problem.

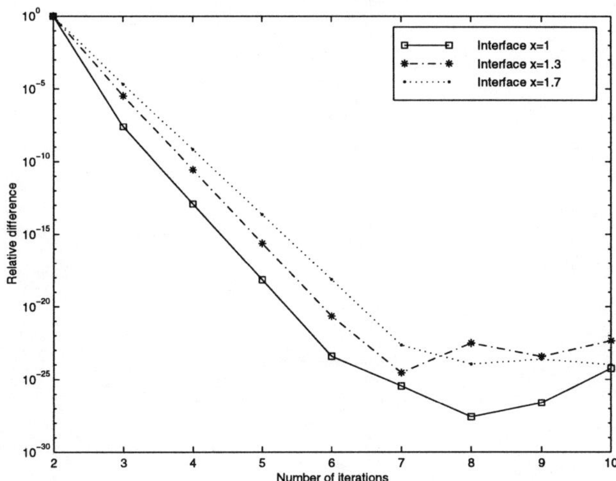

Fig. 5. Convergence history for $h = 1/10$ and different values of x_Γ (continued)

References

1. Alonso, (1999) A mathematical justification of the low-frequency heterogeneous time-harmonic Maxwell equations. Math. Models Methods Appl. Sci. **9**, 475–489
2. Alonso, A. and Valli, A. (1997) A domain decomposition approach for heterogeneous time-harmonic Maxwell equation. Comput. Meth. Appl. Mech. Engrg. **143**, 97–112
3. Alonso, A. and Valli, A. (1999) An optimal domain decomposition preconditioner for low-frequency time-harmonic Maxwell equations. Math. Comput. **68**, 607–631
4. Alonso, A. and Valli, A. (to appear) Domain decomposition algorithms for time-harmonic Maxwell equations with damping. Math. Model. Num. Anal.
5. Ammari, H., Buffa, A. and Nédélec, J.C. (2000) A justification of eddy currents model for the Maxwell equations. SIAM J. Appl. Math. **60**, 1805–1823
6. Bossavit, A. (1998) Computational Electromagnetism. Academic Press, San Diego 1998
7. Nédélec, J.C. (1980) Mixed finite elements in $\mathbb{R}^3$, Numer. Math. **35**, 315–341
8. Nédélec, J.C. (1986) A new family of mixed finite elements in $\mathbb{R}^3$, Numer. Math. **50**, 57–81
9. Quarteroni, A. and Valli, A. (1999) Domain Decomposition Methods for Partial Differential Equations. Oxford University Press, Oxford

Iterated Frequency Filtering Preconditioners

Yves Achdou[1] and Frédéric Nataf[2]

[1] UFR Mathématiques, Université Paris 7, Case 7012, 75251 PARIS Cedex 05,
   France and Laboratoire d'Analyse Numérique, Université Paris 6
[2] CMAP, CNRS UMR 7641, École Polytechnique, 91128 Palaiseau cedex, France

Abstract. For a symmetric and positive definite (SPD) matrix arising from the discretization of a partial differential equation with finite differences or finite elements on a structured grid in dimension d ($d \geq 3$), we propose a SPD block ILU preconditioner whose factorized form requires a smaller amount of memory than the original matrix. Moreover, the computing time for the preconditioner solves is linear with respect to the number of unknowns. The preconditioner is built in d stages: in a first stage, we use the tangential filtering decomposition of Wittum et al [15,16,12,13], and obtain a preconditioner which remains rather difficult to factorize. Then, in a second stage, we apply again the tangential filtering decomposition to the diagonal blocks of this first preconditioner. The final stage consists of factorizing exactly the blocks corresponding to one dimensional problems. Such preconditioners can also be computed adaptively and combined in a multiplicative way. Numerical tests are presented, in particular for problems with highly heterogeneous media.

1   Introduction

For many reasons, developing good preconditioners for elliptic partial differential equations discretized on Cartesian grids is still an important task:

- For applications involving non-linear partial differential equations in heterogeneous media, the simulation procedure uses as a key ingredient preconditioned iterative methods for large systems of linear equations. Such solvers are used within fixed point or Newton methods, and the linear systems change at each step of the non-linear solver. Therefore, it is very important that the CPU time for constructing the preconditioner should be small. Also, since the systems are large, the preconditioner should not be too demanding in term of memory. On the other hand, the preconditioner must be robust enough.
- In a different context, for very large linear problems, many domain decomposition methods do not need exact solves of the local problems, see [9], and it is possible to use preconditioned iterative methods in the subdomains. In particular, non-conforming domain decomposition methods like the mortar method [4] enable the use of non-matching grids, and therefore allow Cartesian grids in the subdomains, even with rather complex geometries.
- Such preconditioners are also a key tool in fictitious domain methods [7].

The aim of this work is to construct a block incomplete Choleski factorization of a sparse symmetric positive definite block-tridiagonal matrix arising from the discretization of a partial differential equation by finite elements or finite difference with a Cartesian grid in three dimensions or more. The aim is that both the amount of memory needed for the preconditioner and the complexity of a preconditioner solve are linear with respect to the number of unknowns. Also the preconditioner should be robust even for very heterogeneous problems.

Many research works have already addressed this topic, see [10], [11], [3] for descriptions and comparisons.

This work is an extension of previous articles of G. Wittum and his coworkers [15],[16],[12],[13], [14],[6],[5]: these papers exploit the fact that the orginal matrix admits an exact block LU or Choleski factorization, which is computed by an induction formula. The approximate sparse block factorization is constructed by a modified induction formula in such a way that it coincides with the original matrix when applied to a given test-vector. For this reason, these preconditioners are called frequency filtering decomposition. The choice of the test-vector affects the convergence.

Also, it is possible to combine in a multiplicative way preconditioners constructed with different test-vectors, (see [12], [13] for a description and an analysis of this method). The test-vectors can even be chosen in an adaptive way (see [14]).

The linear solve with such preconditioners has a linear complexity only for bidimensional problems. In this work, we modify the tangential filtering method in [12] in order to obtain linear complexity: for a matrix arising from a d dimensional partial differential equation, the idea is to construct the preconditioner in d steps by repeating the tangential filtering method in lower and lower dimensions. This results in a preconditioner which is symmetric and positive definite (SPD) if the original matrix is itself SPD.

Our method resembles the nested factorization proposed by Appleyard and Cheshire [2] (see also [10] for a review), because in both cases, the preconditioner solves consist of nested backward-forward substitutes. However the construction of our preconditioners is different for two reasons:

- our construction is not done in a nested way.
- we can use very general test-vectors whereas in [2], the test-vector is $(1 \ldots 1)^T$. Thus, the proposed preconditioner can be used in an adaptive procedure.

The paper is organized as follows: we begin by a review on frequency filtering factorizations and adaptive procedures. Then, we describe in § 4 and § 5 the iterated filtering procedure and give some properties of the preconditioner. Finally in § 6 and § x7, we present numerical tests with and without adaption of the test-vectors.

2 Filtering Factorizations

Given a second order linear elliptic operator $\mathcal{L}$, we are interested in solving the elliptic partial differential equation

$$\mathcal{L}(v) = f \tag{1}$$

in a given domain $\Omega \subset \mathbb{R}^3$ with appropriate boundary conditions. In this paper, we limit ourselves to symmetric and positive definite operators, although the techniques described below can be extended to non-symmetric operators. Discretizing the elliptic operator with a finite element or finite difference method on a structured mesh, we obtain a large system of linear equations

$$\boldsymbol{Kv} = \boldsymbol{F} \, . \tag{2}$$

We assume that the grid is obtained as a deformation of a Cartesian grid on the unit cube, so for three integers n_x, n_y, n_z, $v \in \mathbb{R}^{n_x \times n_y \times n_z}$ and $\boldsymbol{K}$ is a symmetric and positive definite (SPD) matrix of order $n_x \times n_y \times n_z$. We do not require $\boldsymbol{K}$ to be diagonally dominant. We number the unknowns lexicographically, so the vector $\boldsymbol{v}$ is a collection of n_x sub-vectors $v_i \in \mathbb{R}^{n_y \times n_z}$.

$$v = \left(v_1^T, \ldots, v_{n_x}^T\right)^T \, . \tag{3}$$

The sub-vector v_i contains the unknowns located in a crossection of the grid. In turn, the vector v_i is itself a collection of n_y sub-vectors in $\mathbb{R}^{n_z}$.

$$v_i = \left(v_{i,1}^T, \ldots, v_{i,n_y}^T\right)^T \, . \tag{4}$$

According to (3), the matrix $\boldsymbol{K}$ has a block-tridiagonal structure

$$\boldsymbol{K} = \begin{pmatrix} D_1 & L_1^T & & \\ L_1 & D_2 & \ddots & \\ & \ddots & \ddots & L_{n_x-1}^T \\ & & L_{n_x-1} & D_{n_x} \end{pmatrix}, \tag{5}$$

and each block is a matrix of order $n_y \times n_z$. The diagonal blocks D_i are SPD. According to (4), the blocks D_i and L_i have also a block-tridiagonal structure, and each block of D_i (respectively L_i) is a matrix of order n_z. The diagonal blocks of D_i are SPD, which is not necessarily the case of the diagonal blocks of L_i.

An exact block factorization of the matrix $\boldsymbol{K}$ is given by

$$\boldsymbol{K} = \begin{pmatrix} T_1 & & & \\ L_1 & T_2 & & \\ & \ddots & \ddots & \\ & & L_{n_x-1} & T_{n_x} \end{pmatrix} \begin{pmatrix} T_1^{-1} & & & \\ & T_2^{-1} & & \\ & & \ddots & \\ & & & T_{n_x}^{-1} \end{pmatrix} \begin{pmatrix} T_1 & L_1^T & & \\ & T_2 & \ddots & \\ & & \ddots & L_{n_x-1}^T \\ & & & T_{n_x} \end{pmatrix}, \tag{6}$$

where T_i, $i = 1 \ldots n_x$ are SPD matrices of order $n_y \times n_z$. The matrices T_i satisfy the induction formula

$$T_i = \begin{cases} D_1 & i = 1, \\ D_i - L_{i-1} T_{i-1}^{-1} L_{i-1}^T & 1 < i \leq n_x . \end{cases} \tag{7}$$

Defining $T = \text{Block-Diag}(T_1, \ldots, T_{n_x})$ and L by

$$L = \begin{pmatrix} 0 & & & \\ L_1 & \ddots & & \\ & \ddots & \ddots & \\ & & L_{n_x-1} & 0 \end{pmatrix} ,$$

the factorization (6) can be written compactly

$$K = (L + T) T^{-1} (L^T + T) . \tag{8}$$

Unlike the matrices D_i, the matrices T_i are not sparse, so this factorization cannot be used in practice for large problems.

In order to build an incomplete block-factorization of K, the idea is to replace the blocks T_i, $i = 1, \ldots, n_x$ by suitable sparse or block-sparse approximations $\bar{T}_i$:

$$M = \begin{pmatrix} \bar{T}_1 & & & \\ L_1 & \bar{T}_2 & & \\ & \ddots & \ddots & \\ & & L_{n_x-1} & \bar{T}_{n_x} \end{pmatrix} \begin{pmatrix} \bar{T}_1^{-1} & & & \\ & \bar{T}_2^{-1} & & \\ & & \ddots & \\ & & & \bar{T}_{n_x}^{-1} \end{pmatrix} \begin{pmatrix} \bar{T}_1 & L_1^T & & \\ & \bar{T}_2 & \ddots & \\ & & \ddots & L_{n_x-1}^T \\ & & & \bar{T}_{n_x} \end{pmatrix} , \tag{9}$$

or in a compact form with obvious notations

$$M = (L + \bar{T}) \bar{T}^{-1} (L^T + \bar{T}) . \tag{10}$$

The matrix M may be used as a preconditioner.

Remark 1. The systems $MX = F$ are solved in two sweeps

$$Y = F - \bar{T}^{-1} LY, \quad X = \bar{T}^{-1}(Y - L^T X) . \tag{11}$$

so one needs only solving linear systems of the form $\bar{T}_i V_i = G_i$.

Remark 2. Note that

$$M = L + L^T + \text{Block-Diag}(\bar{T}_1, \ \bar{T}_2 + L_1 \bar{T}_1^{-1} L_1^T, \ \ldots, \ \bar{T}_{n_x} + L_{n_x-1} \bar{T}_{n_x}^{-1} L_{n_x-1}^T) , \tag{12}$$

so only the block-diagonal part of M differs from that of K.

Remark 3. From the two previous remarks, since we just need to solve systems with M (and not for example multiplying a vector by M), it is enough to compute and store (when it is possible) the factorized form of the matrices $\bar{T}_i$ instead of $\bar{T}_i$. The other data, namely L, are already available as a part of K.

The main tool for the constructions of the matrices $\bar{T}$ proposed in this work is described in the next section.

2.1 Pointwise Tangential Frequency Filtering

C. Wagner [12], pursuing the works of G. Wittum [16] [15] has proposed a tangential frequency filtering decomposition of the matrix K.
The general form of the preconditioner M is given by (9) and the induction formula

$$\bar{T}_i = \begin{cases} D_1 & i = 1 \, , \\ D_i + \Theta_i \bar{T}_{i-1} \Theta_i - \Theta_i L_{i-1}^T - L_{i-1} \Theta_i & 1 < i \le n_x \, , \end{cases} \tag{13}$$

where Θ_i, $i = 2, \ldots, n_x$ is a matrix of order $n_y \times n_z$, not defined yet.

Lemma 1 (Wagner [12]). *The blocks $\bar{T}_i$ given by (13) satisfy*

$$\bar{T}_i \ge T_i \, , \tag{14}$$

and $\bar{T}$ is SPD.
Moreover, $M - K = Block\text{-}Diag(N_1, \ldots, N_{n_x})$, with

$$N_i = (\Theta_i \bar{T}_{i-1} - L_{i-1}) \bar{T}_{i-1}^{-1} (\bar{T}_{i-1} \Theta_i - L_{i-1}^T) \, . \tag{15}$$

The matrix $M - K$ is symmetric and positive semi-definite.

Proof. The inequality (14) is clearly true for $i = 1$. Assuming that $\bar{T}_{i-1} \ge T_{i-1}$, we have

$$\begin{aligned} \bar{T}_i &= D_i - L_{i-1} \Theta_i - \Theta_i L_{i-1}^T + \Theta_i \bar{T}_{i-1} \Theta_i \\ &\ge D_i - L_{i-1} \Theta_i - \Theta_i L_{i-1}^T + \Theta_i T_{i-1} \Theta_i \, , \end{aligned}$$

$$T_i = D_i - L_{i-1} T_{i-1}^{-1} L_{i-1}^T \, ,$$

This implies that

$$\begin{aligned} \bar{T}_i - T_i &\ge L_{i-1} T_{i-1}^{-1} L_{i-1}^T - L_{i-1} \Theta_i - \Theta_i L_{i-1}^T + \Theta_i T_{i-1} \Theta_i \\ &= (-L_{i-1} + \Theta_i T_{i-1}) T_{i-1}^{-1} (-L_{i-1}^T + T_{i-1} \Theta_i) \\ &\ge 0 \, . \end{aligned}$$

From (13) and (12), the ith block of $(M - K)$ is given by

$$N_i = L_{i-1} \bar{T}_{i-1}^{-1} L_{i-1}^T - L_{i-1} \Theta_i - \Theta_i L_{i-1}^T + \Theta_i \bar{T}_{i-1} \Theta_i \, ,$$

or, in a factorized form by (15).

The idea is to construct the approximate factorization M, by (9) and the induction formula (13) by requiring that $Mt = Kt$ for a given test-vector $t = \left(t_1^T, \ldots, t_{n_x}^T\right)^T \in \mathbb{R}^{n_x \times n_y \times n_z}$.

Definition 1 (Pointwise Tangential Filtering). Let $t \in \mathbb{R}^{n_x \times n_y \times n_z}$ be a vector with no zero component. The pointwise tangential filtering factorization of the block-tridiagonal matrix K for the test-vector $t \in \mathbb{R}^{n_x \times n_y \times n_z}$ is given by (9) where the blocks $\bar{T}_i$ are given by the induction formula (13) where Θ_i, $i = 2, \ldots, n_x$ is a diagonal matrix of order $n_y \times n_z$ which satisfies

$$\Theta_i t_i = \bar{T}_{i-1}^{-1} L_{i-1}^T t_i \ . \tag{16}$$

Remark 4. The diagonal matrix Θ_i is uniquely defined by (16), because the components of t_i are non-zero.

Remark 5. If M is of the form (9) with (13), then from (15), $Mt = Kt$ yields that for all $i = 2, \ldots, n_x$,

$$(-L_{i-1} + \Theta_i \widetilde{T}_{i-1})\bar{T}_{i-1}^{-1}(-L_{i-1}^T + \bar{T}_{i-1}\Theta_i)t_i = 0 \ .$$

Equation (16) follows from the fact that $\bar{T}_{i-1}$ is SPD.

Remark 6. The term *tangential* in the name of the method can be explained by the fact that the function $z \mapsto ((M - K)z, z)$ is a positive semi-definite quadratic form in the variable $z - t$.

Lemma 1 ensures that the spectral radius of $I - M^{-1}K$ is smaller than one, so the preconditioned Richardson method will converge. In [12], in the case where K is of the form

$$K = \begin{pmatrix} D & L & & \\ L & D & \ddots & \\ & \ddots & \ddots & L \\ & & L & D \end{pmatrix} \ , \qquad L = L^T \ .$$

C. Wagner gives a sharp spectral analysis when the vectors t_i are all equal and chosen as an eigenvector of $D^{-1}L$.

The preconditioner M is a function of the matrix K and of a test-vector t. It will be denoted

$$M = \mathcal{M}(K, t) \ .$$

For a semi-discrete version of the operator $-\Delta + \eta$, it is possible to tune the test-vector t in order to minimize the condition number of the preconditioned operator, see [1] and the optimal value of the condition number varies like $h^{-\frac{2}{3}}$. With this choice, the number of iterations of the preconditioned conjugate gradient method varies like $h^{-\frac{1}{3}}$.

Remark 7. Note that there are other filtering strategies : the Tangential Filtering in the Average was proposed by A. Buzdin[5], pursuing the works of G. Wittum [16] [15] and C. Wagner [12]. More recently, A. Buzdin and G.Wittum [6] have proposed the Two Vectors Filtering in the Average.

3   Adaptive Filtering

A key issue in the use of the preconditioners of § 2 is the choice of the test-vector(s) t. In the case of the operator $\eta - \Delta$ discretized on a uniform mesh for instance, it is possible to find out the optimal choice of the test-vector, see [1]. In most other cases (see §6), the numerical experiments show that a suitable choice of the test-vector ensures a fast convergence. However, in some difficult and rather rare cases, stagnation of the preconditioned conjugate gradient may happen for a given choice of the test-vector. This phenomena is also present in the fixed point algorithm considered in [14]:

$$v^{(i+1)} = (\mathrm{Id} - \mathcal{M}(K, t^1)^{-1} K)(v^{(i)}) - \mathcal{M}(K, t^1)^{-1} f$$

for some test-vector t^1. A remedy is proposed which consists in using a sequence of test-vectors instead of a single test-vector. This sequence is not given in advance but is adaptively computed. Its construction is based on the convergence behavior of the above iterative method.

The initial error $v - v^{(0)}$ is written as a linear combination of the eigenvectors w_C^k (with eigenvalue μ_k) of the iteration matrix $C(K, t^1) = \mathrm{Id} - \mathcal{M}(K, t^1)^{-1} K$

$$v - v^{(0)} = \sum_k c_k \, w_C^k \, .$$

The error after i iteration steps is given by

$$v - v^{(i)} = C^i(K, t^1)(v - v^{(0)}) = \sum_k c_k \, \mu_k^i \, w_C^k \, .$$

The convergence rate is determined by the large eigenvalues of $C(K, t^1)$ and the error is dominated by the corresponding eigenvectors. Hence a preconditioner which damps the error corresponding to the dominating error components is desired. The idea is thus to iterate a few steps with the initial test-vector t^1, estimate a dominating eigenvector by $t^2 = v^{(i+1)} - v^{(i)}$ and then to construct a new approximate inverse $\mathcal{M}(K, t^2)$: let $C(K, t^2) = \mathrm{Id} - \mathcal{M}(K, t^2)^{-1} K$, the algorithm reads

$$v^{(i+1/2)} = C(K, t^1)(v^{(i)}) - \mathcal{M}(K, t^1)^{-1} f \, ,$$
$$v^{(i+1)} = C(K, t^2)(v^{(i+1/2)}) - \mathcal{M}(K, t^2)^{-1} f \, .$$

The process of combining in a multiplicative way the preconditioners when the algorithm stagnates can be repeated until convergence, so a sequence of preconditioners is constructed. Combined with these preconditioners, a Jacobi or Gauss-Seidel smoother may be employed.

4   The Dimension-wise Iterated Frequency Filtering Preconditioner

Applying one of the preconditioners defined in § 2 requires the solution of $2 \times n_x$ problems corresponding to cross-sections. Using direct methods for these intermediate systems would result in a preconditioner whose CPU time and memory requirements would not be linear w.r.t. the number of unknowns. Another possibility is to solve these intermediate problems with an iterative preconditioned procedure. Then the memory requirement and the CPU time would depend on the choice of the preconditioner. The drawback of this approach is that the overall preconditioning strategy becomes nonlinear, because it involves nested iterative methods.

In order to keep the overall cost of the preconditioner linear, we prefer a third option, i.e. to modify the frequency filtering decomposition preconditioners in (8) by replacing the matrices $\bar{T}_i$ in (9) by incomplete LU factorizations $\tilde{T}_i$, themselves built with the same filtering techniques described in § 2 (so the filtering is applied another time for two dimensional problems).

For a matrix K as in (5), the preconditioner is built in two steps:

- A test-vector t is chosen and an incomplete LU factorization M is computed:

$$M = \mathcal{M}(K, t) . \tag{17}$$

- A new preconditioner $\widetilde{M}$ is computed by

$$
\widetilde{M} =
\begin{pmatrix}
\tilde{T}_1 & & & \\
L_1 & \tilde{T}_2 & & \\
& \ddots & \ddots & \\
& & L_{n_x-1} & \tilde{T}_{n_x}
\end{pmatrix}
\begin{pmatrix}
\tilde{T}_1^{-1} & & & \\
& \tilde{T}_2^{-1} & & \\
& & \ddots & \\
& & & \tilde{T}_{n_x}^{-1}
\end{pmatrix}
\begin{pmatrix}
\tilde{T}_1 & L_1^T & & \\
& \tilde{T}_2 & \ddots & \\
& & \ddots & L_{n_x-1}^T \\
& & & \tilde{T}_{n_x}
\end{pmatrix},
\tag{18}
$$

where $\tilde{T}_i = \mathcal{M}(\bar{T}_i, t_i)$, $t = (t_i)_{1 \le i \le n_x}$. This process is denoted in a compact form by $\widetilde{M} = \widetilde{\mathcal{M}}(K, t)$.

With this choice, the matrices $\tilde{T}_i$ are of the form (6), but now the diagonal blocks in the factorization (6) are tridiagonal matrices, so it is possible to factorize them exactly with a linear memory cost. Finally, the factorized form of $\tilde{T}_i$ is linear in terms of memory requirements, and from Remark 3, the same is true for the factorized form of $\widetilde{M}$.

The cost of $\tilde{M}$ is clearly linear in terms of memory. It is linear in CPU time as well, since the matrices $\tilde{T}_i$ are defined by their block-LU factorization, their inversion has a linear cost w.r.t. the number of unknowns $n_x \times n_y \times n_z$. This is obtained at the expense of a loss of the filtering property: we have now in general, $Kt \neq \widetilde{M}t$. The remainder term in the factorization $\widetilde{N} = \widetilde{M} - K$

is block diagonal, $\widetilde{N} = \text{Block-Diag}(\tilde{N}_i)_{1 \le i \le n_x}$:

$$\tilde{N}_i = -D_i + \tilde{T}_i + L_{i-1}\tilde{T}_{i-1}^{-1}L_{i-1}^T$$
$$= -D_i + \tilde{T}_i + L_{i-1}\bar{T}_{i-1}^{-1}L_{i-1}^T - \bar{T}_i - L_{i-1}\bar{T}_{i-1}^{-1}L_{i-1}^T + \tilde{T}_i + L_{i-1}\tilde{T}_{i-1}^{-1}L_{i-1}^T$$
$$= N_i + (\tilde{T}_i - \bar{T}_i) + L_{i-1}(\tilde{T}_{i-1}^{-1} - \bar{T}_{i-1}^{-1})L_{i-1}^T,$$

where N_i is defined in (15). Therefore, even if pointwise tangential filtering is used at each step in (17) and (18), the filtering property is lost:

$$(\widetilde{M} - K)(t) = N(t) + ((\tilde{T}_i - \bar{T}_i)(t_i))_{1 \le i \le n_x}$$
$$+ (L_{i-1}(\tilde{T}_{i-1}^{-1} - \bar{T}_{i-1}^{-1})L_{i-1}^T(t_i))_{1 \le i \le n_x}$$
$$= (L_{i-1}\tilde{T}_{i-1}^{-1}(\bar{T}_{i-1} - \tilde{T}_{i-1})\bar{T}_{i-1}^{-1}L_{i-1}^T(t_i))_{1 \le i \le n_x}.$$

Recall $N(t) = 0$ by construction of M (see Lemma 1) as well as $(\tilde{T}_i - \bar{T}_i)(t_i) = 0$ by construction of $\tilde{T}_i$. Nevertheless if $\tilde{T}_i$ is obtained by using a pointwise tangential filtering, $\widetilde{M}$ keeps the essential property of being symmetric definite positive since we have by Lemma 1 that $\tilde{T}_i \ge \bar{T}_i > 0$. We have

Lemma 2. *Under the assumption that the matrix K is SPD, the preconditioner $\widetilde{M} = \widetilde{\mathcal{M}}(K, t)$ and the blocks $\tilde{T}_i$ defined above are SPD.*

Remark 8. So far, we have presented the construction of the preconditioner in dimension $d = 3$ but the above ideas can work in arbitrary dimensions. Also the preconditioner has a highly recursive structure, and generic object-oriented programming can be used for a convenient implementation in arbitrary dimension d (see [1]).

5   The Complete Algorithm

To summarize, the proposed preconditioner (tested below) is a combination of the ideas of recursion and adaptivity.

First, an Iterated Tangential Frequency Filtering Decomposition $\tilde{M}$ is built from the matrix of the problem K and from a given test-vector t. It is used as a preconditioner in a preconditioned conjugate gradient algorithm (PCG) with a maximum number of iterations MaxIt and a stopping criterion TOL. If the PCG has not converged in MaxIt iterations, the adaptive algorithm (see § 3) is used. It is a preconditioned Ridchardson method (with the same stopping criterion TOL) that uses various Iterated Frequency Filtering Decompositions and smoothers combined in a multiplicative way. Since memory is limited, the preconditioners are stored in a first in-first out (FIFO) list with maximal size MaxSize (if the size of the list is MaxSize and if a new preconditioner needs to be added, then the first one in the list is removed). A new Iterated Tangential Frequency Filtering Decomposition is added to the list, when one iteration of the Ridchardson method has not decreased the residual by a factor smaller than $\varrho_{\max}$.

6   Numerical Experiments Without Adaptivity

We present bidimensional results based on a finite difference implementation and tridimensional results based on a Q1 finite element discretization of an elliptic partial differential equation. These results have been obtained on a Pentium III - 800Mhz with Linux OS. For all the tests below, the stopping criteria is to reduce the two-norm of the residual by a factor 10^{-10}. Unless explicitly mentioned, a diagonal scaling is performed before entering the iterative method.

6.1   Bidimensional Results

In [8], the two vectors filtering in the average approach has been compared to standard ILU factorizations, using matlab: ILU('0') (no fill-in) and ILU(ε) with a drop tolerance $\varepsilon = 10^{-4}$. The two frequencies filtering approach outperformed the standard ILU factorizations by a large factor. For example, for a Poisson problem in the unit square with 400^2 nodes, the construction of the preconditioner needs 1.44 megaflops for ILU('0'), 2668 megaflops for ILU(10^{-4}) and 1.93 megaflops for the filtering preconditioner. The solution process takes 2734 megaflops with ILU('0'), 949 megaflops with ILU(10^{-4}) and 368 megaflops with the filtering preconditioner.

The new results in dimension two presented here correspond to the tangential filtering preconditioner. The test-vector is chosen in agreement with the analysis performed in [1] in the case of the Poisson problem in a unit square, with homogeneous Dirichlet conditions. Table 1 contains the results for the Poisson problem: the overall times (third column) includes the setup times and the times spent for the iterative procedure; the setup times are given in the fourth column. Note that the setup time does not depend on the PDE. The next series of tests concerns a problem with highly discontinuous scalar

Table 1. Results for the Poisson problem in two dimensions

h	Iteration counts	overall time (s)	setup time (s)
1/128	27	0.98	0.26
1/256	34	4.80	1.00
1/512	47	23.14	3.90
1/1024	62	135.00	14.30

coefficients

$$-\mathrm{div}(\varepsilon(\mathrm{x})\nabla\mathrm{u}) = \mathrm{f} \quad \text{in } \Omega\,, \quad \mathrm{u} = 0 \quad \text{on } \partial\Omega\,,$$

where $\varepsilon(x) = 10^{-6}$ if $x \in [\frac{1}{4}, \frac{3}{4}]^2$ and $\varepsilon(x) = 1$ otherwise.

Table 2. The results for highly discontinuous coefficients in dimension two

h	Iteration counts	overall time (s)	setup time (s)
1/128	32	1.10	0.25
1/256	46	6.10	1.00
1/512	51	24.86	3.86
1/1024	84	173.80	14.30

6.2 Tridimensional Results

In all the test-cases below, we use a Q1 finite element method for discretizing the different partial differential equations. It yields a matrix with a 27 points stencil about four times larger than the one resulting from a finite difference or finite volume scheme.

The Poisson Problem with a Uniform Grid

Table 3 contains the results for the Poisson problem with a uniform grid.

Table 3. Results for the Poisson problem in dimension three

h	Iteration counts	overall time (s)	setup time (s)
1/32	16	2.7	0.8
1/62	22	23.4	6.2
1/100	29	64.5	32.6

Non Uniform and Very Anisotropic Grids

The Laplace operator is discretized on a highly non-uniform and anisotropic grid, i.e. the grid is refined with a geometric progression in each direction. The results are shown in Table 4. The parameter α_x (resp. α_y, α_z) is the ratio of the geometric progression in the x direction, $\alpha_x = 1$ corresponds to a uniform grid. Note also that no diagonal scaling of the matrix is performed in this case.

Smoothly Varying Coefficients

We consider a variable coefficient operator on a uniform grid

$$-\nabla \cdot \Phi(x,y,z)\nabla u = f \quad \text{in } \Omega, \quad u = 0 \quad \text{on } \partial\Omega , \tag{19}$$

with $\Phi(x,y,z) = a + \sin(b(x + y + 2z))$. The results are presented in Table 5. For example, the second line corresponds to the case where φ oscillates smoothly between $5.\,10^{-3}$ and 2.

Table 4. Non uniform and very anisotropic grids in tridimensional geometrically refined grids. The ratio ϱ is the average error reduction.

N	$\alpha_x = 1.5$ $\alpha_y = 1$ $\alpha_z = 1$	$\alpha_x = 1$ $\alpha_y = 1.5$ $\alpha_z = 1$	$\alpha_x = 1$ $\alpha_y = 1$ $\alpha_z = 1.5$	$\alpha_x = 1.5$ $\alpha_y = 1.5$ $\alpha_z = 1.5$
33^3	$h_{x,\min} = 7.6e-5$ $h_y = 0.029$ $h_z = 0.029$	$h_x = 0.029$ $h_{y,\min} = 7.6e-5$ $h_z = 0.029$	$h_x = 0.029$ $h_y = 0.029$ $h_{z,\min} = 7.6e-5$	$h_{x,\min} = 7.6e-5$ $h_{y,\min} = 7.6e-5$ $h_{z,\min} = 7.6e-5$
	iter 12 ϱ 0.135	iter 25 ϱ 0.384	iter 26 ϱ 0.41	iter 23 ϱ 0.363
65^3	$h_{x,\min} = 9.93e-8$ $h_y = 0.015$ $h_z = 0.015$	$h_x = 0.015$ $h_{y,\min} = 9.93e-8$ $h_z = 0.015$	$h_x = 0.015$ $h_y = 0.015$ $h_{z,\min} = 9.93e-8$	$h_{x,\min} = 9.93e-8$ $h_{y,\min} = 9.93e-8$ $h_{z,\min} = 9.93e-8$
	iter 13 ϱ 0.164	iter 63 ϱ 0.693	iter 47 ϱ 0.61	iter 51 ϱ 0.63

Table 5. Iteration counts for smoothly varying coefficients in dimension three

N		17^3	33^3	65^3
$a = 1.05$	$b = 5$	13	19	29
$a = 1.005$	$b = 5$	15	22	33
$a = 1.05$	$b = 20$	14	18	26

Problems with Jumps in the Coefficients

We consider again problem (19), but now the coefficient Φ is a very discontinuous function of (x, y, z) that takes the values 1 or $\varepsilon = 10^{-6}$. The test cases are named after Table 6 and the results are given in Table 7.

Table 6. Different test cases : the coefficient φ equals ε in the given region and 1 elsewhere

step1	step2	bar
$(0,1)^2 \times (0.3, 0.7)$	$(0,1) \times (0.3, 0.7) \times (0,1)$	$(0,1) \times (0.3, 0.7)^2$
cube	diamond	checkerboard
$(0.3, 0.7)^3$	$\lvert x - \tfrac{1}{2} \rvert + \lvert y - \tfrac{1}{2} \rvert + \lvert z - \tfrac{1}{2} \rvert < 0.3$	see Figure 1

7   Numerical Experiments with the Adaptive Method

Here, we present tests for similar cases as in § 6.2, but now we use the adaptive method presented in § 3 and § 5. Here, no diagonal scaling is performed on the

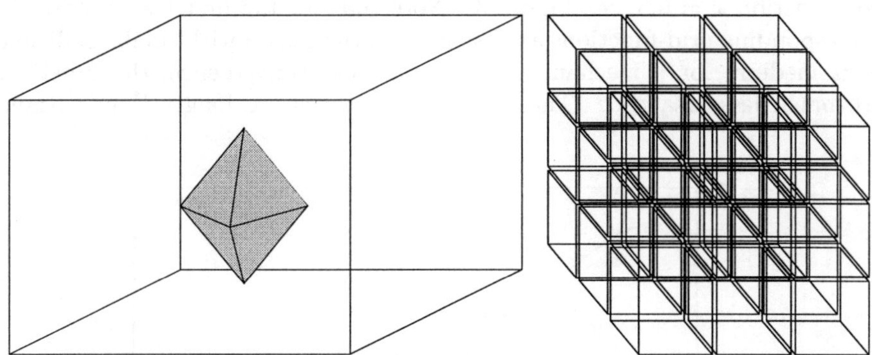

Fig. 1. Description of two test-cases.**(a)** the diamond test-case: $\Phi = \varepsilon$ in the grey region. **(b)** the checkerboard test-case: the domain is divided into smaller cubes and Φ is constant in the cubes and takes alternatively the values one and ε.

Table 7. Results for highly heterogeneous problems with uniform grids

N		31^3	61^3	81^3	101^3
step1	#iter	34	27	39	50
($\varepsilon = 10^{-6}$)	CPU time (sec.)	6.2	51.1	104.6	267.4
step2	#iter	23	23	58	36
($\varepsilon = 10^{-6}$)	CPU time (sec.)	3.6	27.1	143.5	211.2
bar	#iter	22	23	75	57
($\varepsilon = 10^{-6}$)	CPU time (sec.)	3.5	28.9	170.5	283.3
cube	#iter	24	32	70	62
($\varepsilon = 10^{-6}$)	CPU time (sec.)	3.7	36.26	171.6	329.9
diamond	#iter	20	30	35	42
($\varepsilon = 10^{-6}$)	CPU time (sec.)	3.0	31.5	88.3	215.8
checkerboard	#iter	24	32	60	45
($\varepsilon = 10^{-6}$)	CPU time (sec.)	3.4	33.3	140.0	229.1

matrix of the problem. We focus here on the representative test-case called cube in Table 6, but we actually obtained very similar behaviors for all the test-cases in Table 6. The mesh is uniform with 60^3 nodes. The convergence history is displayed on Figure 2 . If the residual has not been decreased by a factor $\varrho_{\max}$ set to 0.3 in this test, a new iterated tangential frequency filtering decomposition is computed and combined in a multiplicative way with the preceding ones. Also we combine these preconditioners with a Jacobi smoother. In this test, four tangential frequency filtering decompositions have been used : the updates of the global preconditioner can be seen by slope variations on the convergence history, see Figure 2 . The test-vectors $(t_i)_{i=1\ldots4}$ stand for tridimensional functions, and we display these functions on the cross-section $x = 0.5$, see Figure 3 . Note that for the first test-vectors the corresponding grid-functions are localized in the region with small coefficient (soft medium for a mechanical analogy). Near convergence, the functions are much less localized. The overall CPU times are longer than without

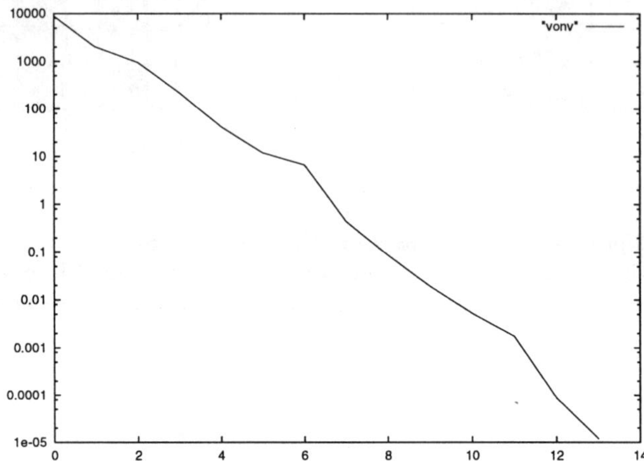

Fig. 2. Convergence history with the adaptive method for the cube test-case

adaptivity, because the preconditioner is more complex : the CPU-time is 89 s for 61^3 nodes, and 280 s for 81 nodes (respectively 36 s and 171 s without adaptivity, see Table 7). For the checkerboard test-case, the results are similar : we display on Figure 4 the successive four test-vectors needed for convergence. Here also, the functions are localized in the region where Φ is small.

References

1. Achdou Y. and Nataf F. (submitted, 2001) Dimension-wise Iterated Frequency filtering Decomposition.
2. Appleyard J.R and Cheshire I.M. (1983) Nested factorization. In Seventh SPE Symposium on Reservoir Simulation paper number 12264. 315–324,
3. Benzi M. and Tuma M. (1999) A comparative study of sparse approximate inverse preconditioners. Appl. Num. Math. **30**, 305–340
4. Bernardi C., Maday Y. and Patera A. (1989) A new nonconforming approach to domain decomposition: the mortar element method. Nonlinear Partial Differential Equations and their Applications, eds H. Brezis and J.L. Lions, Pitman.
5. Buzdin A. (1998) Tangential decomposition. Computing **61**, 257–276
6. Buzdin A. and Wittum G. (1999) Zwei-Frequenzenzerlegung. Technical report, IWR, Universität Heidelberg, **INF 368**

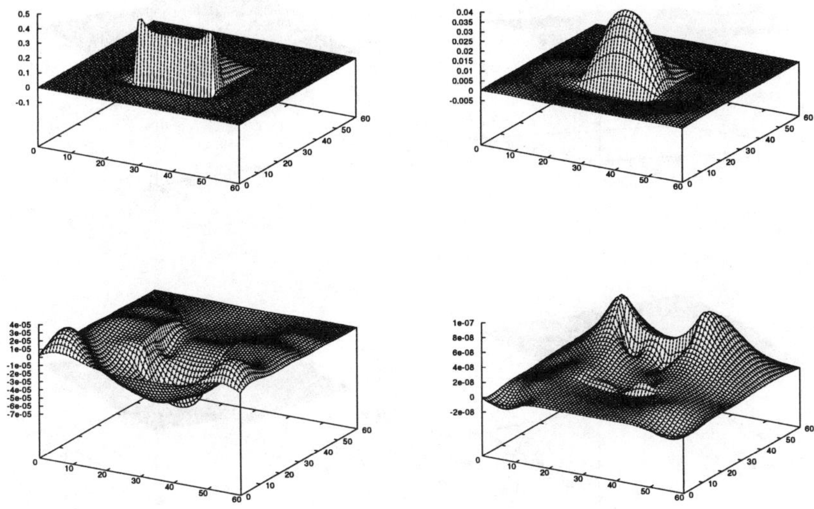

Fig. 3. The grid functions corresponding to the successive four test-vectors (cross-section $x = 0.5$) for the cube test-case

7. Finogenov, S. A. and Kuznetsov, Yu. A. (1988) Two-stage fictitious components method for solving the Dirichlet boundary value problem. Soviet Journal of Numerical Analysis and Mathematical Modelling, **3** 4, 301–323

8. Gander M. and Nataf F. (2000) AILU: A Preconditioner Based on the Analytic Factorization of the Elliptic Operator. Numerical Linear Algebra **7**, 505–526

9. Kuznetsov Y.A. (1995) Efficient iterative solvers for elliptic finite element problems on nonmatching grids. Russian Journal of Numerical Analysis and Mathematical Modelling **10** 3, 187–211

10. Meurant G. (1999) Computer solution of large linear systems. North-Holland Publishing Co., Amsterdam.

11. Saad Y. (1996) Iterative Methods for Sparse Linear Systems. PWS Publishing Company

12. Wagner C. (1997) Tangential frequency filtering decompositions for symmetric matrices. Numer. Math. **78** (1), 119–142

13. Wagner C. (1997) Tangential frequency filtering decompositions for unsymmetric matrices. Numer. Math. **78** (1), 143–163

14. Wagner C. and Wittum C. (1997) Adaptive filtering. Numer. Math. **78** (2), 305–328.

15. Wittum G. (1991) An ILU-based smoothing correction scheme. In Parallel algorithms for partial differential equations, Proc. 6th GAMM-Semin., Kiel/Ger., Notes Numer. Fluid Mech., **31**, 228–240

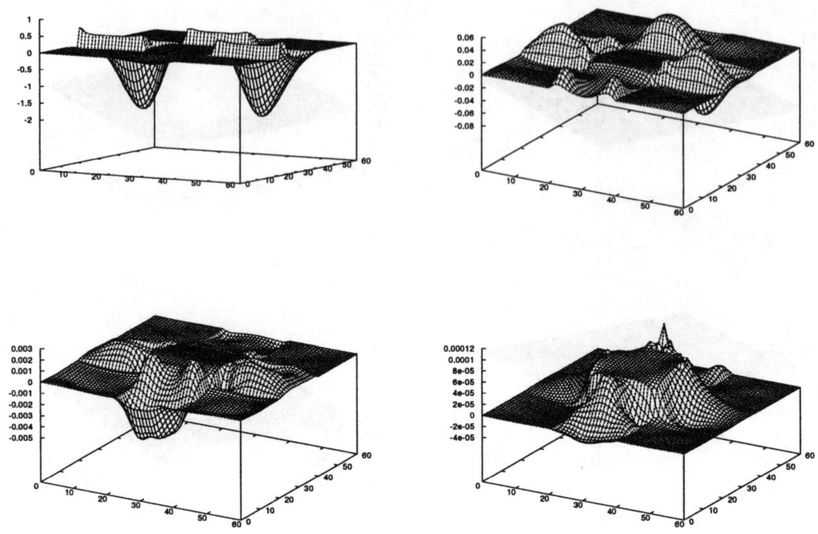

Fig. 4. The grid functions corresponding to the successive four test-vectors (cross-section $x = 0.5$) for the checkerboard test-case

16. Wittum G. (1992) Filternde Zerlegungen. Schnelle Löser fuer grosse Gleichungssysteme. Teubner Skripten zur Numerik, Stuttgart.

A "Parareal" Time Discretization for Non-Linear PDE's with Application to the Pricing of an American Put

Guillaume Bal[1] and Yvon Maday[2]

[1] Department of Applied Physics & Applied Mathematics, Columbia University, New York, NY 10027, USA
[2] Laboratoire d'Analyse Numérique, Université Pierre et Marie Curie, Boîte courrier 187, 75252 Paris Cedex 05, France

Abstract. In this paper, we introduce a new implementation of the "parareal" time discretization aimed at solving unsteady nonlinear problems more efficiently, in particular those involving non-differentiable partial differential equations. As in the former implementation [3], the main goal of this scheme is to parallelize the time discretization to obtain an important speed up. As an application in financial mathematics, we consider the Black-Scholes equations for an American put. Numerical evidence of the important savings in computational time is also presented.

1 Introduction

In a recent note [3], the concept of "parareal" time discretization has been introduced to speed up the simulation of time-dependent partial differential equations through the use of parallel architectures. This discretization is based on

- the decomposition of the time interval $(0, T)$ into a sequence of sub-intervals (T^n, T^{n+1}) of size ΔT ($\Delta T = T/N$ for some integer N)
- the iterative use of two time discretization schemes, the first one with time step ΔT used over $(0, T)$, the second one with a smaller time step δt ($\delta t = \Delta T/M$ for some integer M) used over each sub-interval (T^n, T^{n+1}) independently, thus possibly in parallel.

As shown in [3], the global accuracy of the iterative process after a few iterations is comparable to that of the fine discretization of time step δt used over the whole interval $(0, T)$. Assuming that a large number of processors are at hand, this new discretization may allow us to get real time solutions when the fine discretization fails to do so. We refer to [1] and [2] for other approaches of parallelization in time that seem limited to differential systems.

In [3], the method has been proposed and implemented on linear and nonlinear parabolic problems. Here we introduce a slightly modified implementation, which turns out to be equivalent for linear problems, but gives better answers for nonlinear problems and more importantly allows us to tackle non-differentiable problems (whereas the former implementation was based on the linearization of the PDE). An example of such non-differentiable problems is provided by the Black-Scholes equation for an American put in financial mathematics to which we dedicate our last section.

2 Problem Formulation

We consider the following first-order time-dependent partial differential equation

$$F(\partial_t u, u) = 0, \tag{1}$$

where F is a functional of $\partial_t u$, u, and possibly derivatives of u with respect to other variables than time. To solve (1) on $(0, T)$ with initial condition $u(t = 0) = u_0$ approximately, assuming that a unique solution exists, we have at our disposal a one step "coarse" solver

$$E_{\Delta T}(T^n, U^n; U^{n+1}) = 0, \tag{2}$$

which to an approximation U^n of $u(T^n)$ yields U^{n+1}, an approximation of $u(T^{n+1})$. It is standard that using this iterative scheme allows us to define an approximation of the solution $u(t)$ for each $t = T^n$, $n = 1, .., N$. We assume in addition that we have a fine, one step solver

$$\mathcal{F}_{\delta t}(s, \sigma, \varphi; \psi) = 0,$$

which, to any $s \in (0, T), \sigma \in (0, T - s)$ and φ associates an approximation ψ of the solution $u(s + \sigma)$ to (1) with initial condition $u(t = s) = \varphi$.

Our objective is to devise a method that combines the use of the coarse solver on the whole interval $(0, T)$ and the fine solver in parallel on each small intervals (T^n, T^{n+1}) and allows us to provide an approximation of u over the whole interval $(0, T)$ with the same accuracy as a global use of $\mathcal{F}$ (with $s = 0$ and $\sigma = T$). To simplify notation, we assume that the fine solver is exact in the remainder of this section .

The iterative scheme is defined as follows. For $k = 1$, we compute a coarse approximation of each $u(T^n)$:

$$\begin{aligned} E_{\Delta T}(T^n, U_1^n; U_1^{n+1}) &= 0 \\ U_1^0 &= u_0. \end{aligned} \tag{3}$$

Then for $k \geq 1$, we define the solutions

$$\begin{aligned} F(\partial_t y_k^n, y_k^n) &= 0 \\ y_k^n(n\Delta T) &= U_k^n, \end{aligned} \tag{4}$$

for $0 \leq n \leq N - 1$; next the jumps

$$S_k^n = y_k^{n-1}(n\Delta T) - U_k^n, \tag{5}$$

for $1 \leq n \leq N$; and finally the corrected coarse solutions

$$E_{\Delta T}(T^n, U_{k+1}^n; U_{k+1}^{n+1} - \sum_{\ell=1}^{k} S_\ell^{n+1}) = 0, \quad 1 \leq n \leq N \tag{6}$$

$$U_{k+1}^0 = u_0.$$

Another definition of the scheme can be done by introducing a predictor and a corrector. We can say that we first apply the coarse solver,

$$E_{\Delta T}(T^n, U_{k+1}^n; \tilde{U}_{k+1}^{n+1}) = 0$$

and then add the successive corrective jumps, $U_{k+1}^{n+1} = \tilde{U}_{k+1}^{n+1} + \sum_{\ell=1}^{k} S_\ell^{n+1}$.

Remark 1. The iterative scheme is optimal in the sense that if U_{k-1}^m is exact for $m \leq n$, then so is U_k^{n+1}.

Remark 2. It is also interesting to understand what has to be done if a good approximation U_0^n of $u(T^n)$ is available for $n = 1, .., N$. Before defining the fine solutions (4), we have to initialize the corrector by introducing $S_0^n = \tilde{U}_0^n - U_0^n$ with $\tilde{U}_0^n$ defined by

$$E_{\Delta T}(T^{n-1}, U_0^{n-1}; \tilde{U}_0^n) = 0,$$

and then proceed as before by replacing the sum in (6) by $\sum_{\ell=0}^{k} S_\ell^{n+1}$.

3   Analysis of the Algorithm for Linear PDE's

Let us consider the following linear partial differential equation

$$\frac{\partial u}{\partial t} + A(u) = 0,$$

where A is a linear operator over a Hilbert space X and u is searched in $C^1(0, T; X)$. This problem can be put in the framework of (1) by setting $F(a, b) = a + A(b)$.

Let us first mention that, for this simple situation, the current formulation is actually equivalent to that in [3]. We can indeed show that

$$U_{k+1}^{n+1} = U_k^{n+1} + S_k^{n+1} + \text{prop}\{S_{k+1}^m\}_{m=1}^n$$

where $\text{prop}\{S_{k+1}^m\}_{m=1}^n$ is the propagated jump at step $k + 1$ and time n by the coarse solver (see [3] for the notations and definition of the propagated "jump"). To avoid confusion between the different formulations of the "parareal" scheme, we refer to [3] for additional details.

In the remainder of this section we prove that the accuracy of the current formulation of the "parareal" scheme increases with the number of iterations, first for a linear scalar differential equation, and then for more general linear parabolic equations.

3.1   Analysis of the Algorithm on a Differential Equation

Let us first consider the iterative scheme for the simple linear equation $F(a, b) = a + \lambda b$, where $\lambda \geq 0$, or equivalently

$$\begin{aligned} \partial_t u + \lambda u &= 0 \\ u(t = 0) &= u_0. \end{aligned} \tag{7}$$

Since the equation is linear, we assume that $u_0 = 1$ to simplify. First-order approximations of this equation are obtained for instance by

$$E_{\Delta T}(t, b; a) = \frac{a - b}{\Delta T} + \lambda \begin{cases} a & \text{implicit scheme} \\ b & \text{explicit scheme.} \end{cases} \tag{8}$$

For the implicit scheme, introducing $\delta = \lambda \Delta T$, we easily check that $U_0^n = (1 + \delta)^{-n} = e^{-\delta n} + O(n\delta^2)$, where $e^{-\delta n} = u(T^n)$ is the exact solution. We then observe that

$$S_k^n = e^{-\delta} U_{k-1}^{n-1} - U_{k-1}^n,$$

and therefore

$$U_k^{n+1} = \frac{1}{1 + \delta} U_k^n + \sum_{l=1}^{k} (e^{-\delta} U_{l-1}^n - U_{l-1}^{n+1}), \tag{9}$$

hence, subtracting the same relation with k replaced by $k - 1$,

$$U_k^{n+1} = \frac{1}{1 + \delta}(U_k^n - U_{k-1}^n) + e^{-\delta} U_{k-1}^n. \tag{10}$$

Let us define $V_k^n = e^{n\delta} U_k^n$, which is supposed to converge to 1 as $k \to \infty$. Upon multiplying the above expression by $e^{n\delta}$, we have

$$V_k^{n+1} = V_{k-1}^n + f(\delta)(V_k^n - V_{k-1}^n), \tag{11}$$

where

$$f(\delta) = \begin{cases} e^{\delta}(1 + \delta)^{-1} & \text{implicit scheme} \\ e^{\delta}(1 - \delta) & \text{explicit scheme.} \end{cases} \tag{12}$$

We say that the scheme is of *order m* if

$$|1 - f(\delta)| \leq C_f \delta^{m+1}. \tag{13}$$

Both the explicit and implicit schemes in (12) are of order 1. We can however construct higher order schemes, for instance by following the Richardson rule. For U^n given, we apply the implicit scheme of time step $\Delta T/2$ twice to obtain V and the implicit scheme of time step ΔT once to obtain W. We then define

$U^{n+1} = 2V - W$. We easily check that this scheme is of order 2 and that $U^N - u(T) = O((\Delta T)^2)$. The corresponding function f is

$$f(\delta) = e^\delta \left(\frac{2}{\left(1 + \dfrac{\delta}{2}\right)^2} - \frac{1}{1+\delta} \right) = 1 + O(\delta^3).$$

Let us now define the error terms ε_n^k via

$$V_k^n = e^{\delta n} U_k^n = 1 + \varepsilon_k^n. \tag{14}$$

We have the following

Proposition 1. *Let $u(t)$ be the solution of (7) and U_k^n the solution of the iterative scheme (3)—(6), where $E_{\Delta T}$ is a coarse scheme of order $m \geq 1$ according to the definition (13). Then the error terms ε_k^n satisfy the following estimate*

$$|\varepsilon_k^n| \leq C_k n^{k+1} \delta^{(m+1)(k+1)}. \tag{15}$$

In particular, we have

$$U_k^N = u(T) + O(\delta^{m(k+1)})$$

and

$$y_k^n(t) = u(t) + O(n^k \delta^{(m+1)k}).$$

Proof. Let us denote $f = f(\delta)$. From (11) and (14), we have

$$\varepsilon_k^{n+1} = f(\varepsilon_k^n - \varepsilon_{k-1}^n) + \varepsilon_{k-1}^n,$$

or equivalently

$$\varepsilon_k^{n+1} = f\varepsilon_k^n + (1 - f)\varepsilon_{k-1}^n. \tag{16}$$

Upon solving the above equations successively, we verify that

$$\varepsilon_k^{n+1} = (1 - f) \sum_{p=1}^{n} f^{n-p} \varepsilon_{k-1}^p \tag{17}$$

because $\varepsilon_{k-1}^0 = 0$ for all $k \geq 1$. Let us now prove the assertion by induction. We verify that

$$\varepsilon_0^n = 1 - f^n = (1 - f) \sum_{p=0}^{n-1} f^p.$$

Notice that f^n is bounded independent of n so $|\varepsilon_0^n| \leq C_0 n \delta^{m+1}$. Let us now assume that $|\varepsilon_{k-1}^n| \leq C_{k-1} n^k \delta^{(m+1)k}$. We then deduce from (17) that

$$|\varepsilon_k^{n+1}| \leq C\delta^{m+1} \sum_{p=1}^{n} p^k \delta^{(m+1)k} \leq C_k (n + 1)^{k+1} \delta^{(m+1)(k+1)},$$

since $\sum_{p=1}^{n} p^k \leq n^{k+1}$. This concludes the proof of the proposition.

3.2   Analysis of the Algorithm on a Parabolic Equation

We now generalize the above analysis to the following type of parabolic problems

$$\partial_t u + P(u) = 0$$
$$u(t = 0, x) = u_0(x), \tag{18}$$

where $P(u)$ is a pseudo-differential operator

$$P(u) = \widehat{P(\xi)\hat{u}(\xi)},$$

with symbol $P(\xi) \geq 0$. Common examples are the heat equation, with $P(u) = -\partial_{xx}^2 u$ and $P(\xi) = \xi^2$, and its centered spatial discretization of mesh size h, with $P(u) = h^{-2}(2u(x) - u(x+h) - u(x-h))$ and $P(\xi) = 2h^{-2}(1 - \cos h\xi)$. In the Fourier domain, the problem reads

$$\partial_t \hat{u} + P(\xi)\hat{u} = 0$$
$$\hat{u}(t = 0, \xi) = \hat{u}_0(\xi), \tag{19}$$

whose solution is given by

$$\hat{u}(t, \xi) = e^{-P(\xi)t}\hat{u}_0(\xi).$$

This section analyzes the convergence of the iterative scheme for the implicit Euler time discretization. In the Fourier domain, the implicit discretization is then given by

$$\hat{U}^{n+1}(\xi) = \frac{1}{1 + P(\xi)\Delta T}\hat{U}^n(\xi).$$

We set $\delta = \delta(\xi) = P(\xi)\Delta T$. Different frequencies satisfy here uncoupled equations. We introduce the function

$$f(\delta) = e^\delta \frac{1}{1 + \delta}.$$

Let us define the sequence $\hat{U}_k^n(\xi)$ as the solution to the iterative scheme (3)—(6) with initial condition $\hat{u}_0(\xi) = 1$ and define $\eta_k^n = \eta_k^n(\xi)$ as

$$\eta_k^n = e^{-\delta n} - \hat{U}_k^n. \tag{20}$$

Following the proof of proposition (1), we verify that

$$\eta_k^{n+1} = e^{-\delta}f(\delta)\eta_k^n + e^{-\delta}(1 - f(\delta))\eta_{k-1}^n$$
$$\eta_0^n = e^{-\delta n} - (e^{-\delta}f(\delta))^n,$$

with $\eta_k^0 = 0$ for $k \geq 0$, assuming that $\hat{u}_0(\xi) = 1$. We will derive an estimate for η_k^n as $n \to \infty$ for small values of k. For this purpose, we introduce

$$\theta_k^n = -\frac{e^{n\delta}}{f^{n-k}(\delta)(1 - f(\delta))^k}\eta_k^n. \tag{21}$$

An easy calculation shows that

$$\theta_k^{n+1} = \theta_k^n + \theta_{k-1}^n$$
$$\theta_0^n = 1 - f^{-n}(\delta).$$

$$(22)$$

We easily verify that $f(\delta) \geq 1$. The terms θ_n^k are all positive. Moreover, we have $\theta_k^n \leq (f-1)f^{-1}\tilde{\theta}_k^n$, where

$$\tilde{\theta}_k^{n+1} = \tilde{\theta}_k^n + \tilde{\theta}_{k-1}^n, \qquad k \geq 0,$$
$$\tilde{\theta}_{-1}^n = 1, \quad n \geq 0,$$
$$\tilde{\theta}_k^0 = 0, \quad k \geq -1.$$

Indeed, we recognize that

$$\tilde{\theta}_k^n = \binom{n}{k+1} = \frac{n!}{(k+1)!(n-k-1)!}$$

hence $\tilde{\theta}_0^n = n$. From this, we deduce the quite accurate estimate

$$0 \leq (-1)^k \eta_k^n \leq \left(e^{-\delta} f(\delta)\right)^n \left(\frac{f(\delta)-1}{f(\delta)}\right)^{k+1} \binom{n}{k+1}.$$

$$(23)$$

It is easy to check that the above estimate implies that of proposition 1 when $\delta = P(\xi)\Delta T$ with λ of order 1 and $n \leq N$. Moreover, we directly verify that $\eta_k^n = 0$ for $k \geq n$ (which means that U_n^k is exact for $k \geq n$). For a fixed λ of order 1, δ converges to 0 with ΔT and we obtain the results of proposition 1. For values of ξ such that $P(\xi)\Delta T$ is of order 1, the theory needs to be improved. There is no hope that such high frequencies are correctly calculated by the iterative scheme. However, we want to make sure that they do not pollute the global solution too much. This is done by obtaining a bound on η_k^n independent of the frequency ξ, or equivalently independent of δ. Let us introduce

$$\alpha_k^n = \sup_{\delta} |\eta_k^n(\delta)|.$$

We now show that for $k \ll n$, we have

$$\alpha_k^n \leq \left(\frac{C(k+1)}{n}\right)^{k+1},$$

$$(24)$$

where C is a constant independent of k and n. We deduce therefore that the iterative scheme converges as $(\Delta T)^{k+1}$ for small values of k *independently* of $\delta = P(\xi)\Delta T$.

In order to bound α_k^n we are led to analyse the function

$$h(\delta) = \frac{1}{1+\delta}\left(1 - \frac{1+\delta}{e^\delta}\right)^{\frac{k+1}{n}}.$$

We verify that $h(0) = 0$ and $h(+\infty) = 0$. Maxima of h are obtained at those points δ verifying

$$\frac{(k+1)\delta}{n} = \frac{e^\delta}{1+\delta} - 1 = g(\delta). \tag{25}$$

We verify that $g''(\delta) > 0$ for $\delta > 0$, from which we deduce the existence of a unique point $\delta_0 > 0$ where $h(\delta_0)$ reaches its maximum. From (25), we obtain that for $(k+1)/n << 1$,

$$\delta_0 = \frac{2(k+1)}{n} + o(\frac{k}{n}),$$

hence, recalling (23)

$$\alpha_k^n \leq \Big(\frac{1}{1+2(k+1)/n}\Big)^n \Big(\frac{1}{2}\Big(\frac{2(k+1)}{n}\Big)^2\Big)^{k+1} \binom{n}{k+1}$$

for $k/n << 1$. Using

$$\binom{n}{k+1} \sim \frac{n^{k+1}}{(k+1)!},$$

and the Stirling formula

$$k! = \Big(\frac{2\pi}{k+1}\Big)^{1/2} e^{-(k+1)} (k+1)^{k+1} (1 + O(k^{-1})),$$

we deduce that

$$\alpha_k^n \leq \Big(\frac{2(k+1)}{e}\frac{1}{n}\Big)^{k+1},$$

for $1 << k << n$. For small values of k, the constant $2/e$ may need to be replaced by another constant. This proves our claim (24).

Let us return to the equation (19). Defining $\eta_k^n(\xi)$ as the error for a frequency ξ after k iterations of the iterative scheme, we have that

$$|\eta_k^n(\xi)| \leq \Big(\frac{C(k+1)}{n}\Big)^{k+1} |\hat{u}_0(\xi)|.$$

Hence, using the Parseval inequality, we deduce the main result of this section:

Proposition 2. *Let $u(t,x)$ be the solution of (18) and $U_k^n(x)$ the solution of the "parareal" scheme (3)—(6) with implicit Euler coarse discretization. Then we have the following estimate*

$$\|u(T,x) - U_k^n(x)\|_{L^2(\mathbb{R})} \leq \Big(\frac{C(k+1)}{n}\Big)^{k+1} \|u_0\|_{L^2(\mathbb{R})}. \tag{26}$$

This proves that the iterative scheme transforms the implicit Euler time discretization of order 1 into a time discretization of order $k+1$ for any linear parabolic problem.

In the actual implementation of the method, a scheme with fine time step δt is used in parallel over each (T^n, T^{n+1}). The gain in computation time obtained by using the iterative scheme is estimated as follows. The cost of the direct scheme, based on the sole use of the fine solver, is proportional to $T/\delta t$. The cost of the iterative scheme is proportional (with the same constant of proportionality) to $(k+1)(T/\Delta T + \Delta T/\delta t)$, with $n = (\Delta T)^{-1}$. For $1 + k$ fixed, the iterative scheme cost is optimized provided $T/\Delta T = \Delta T/\delta t$ since the product $T/\delta t$ is fixed. This implies $\Delta T = \sqrt{T\delta t}$. The accuracy of the iterative scheme will be therefore comparable to the direct simulation provided $k + 1 = 2$ since $\delta t = T^{-1}(\Delta T)^2$.

The maximal gain in time is then found to be

$$G = \frac{1}{4}\left(\frac{T}{\delta t}\right)^{1/2},$$

provided we have at our disposal $(T/\delta t)^{1/2}$ processors. For instance, with $T = 1$ and $\delta t = 10^{-6}$, we ideally obtain a gain of $G = 250$ provided we have 1000 processors and use 2 iterations of the "parareal" scheme.

4   Application to a Nonlinear Parabolic Problem

Our purpose is now to apply the procedure to a nonlinear partial differential equation. In [3], we have considered the following problem

$$\frac{\partial u}{\partial t} - \Delta u + 5u^3 = 5 \cdot \sin(2 \cdot t), \tag{27}$$

over the unit disc Ω, with initial condition $u(t = 0) = \tan(x^2 + y^2 - 1)$, and homogeneous boundary conditions.

We have compared the current implementation with the one of [3] since they do not coincide for nonlinear problems. The generalization to nonlinear problems was done in [3] through a linearization of the coarse time solver. The current scheme is designed to apply to any type of nonlinear equation.

The time interval is $(0, 10)$ and $\Delta T = 0.1, \delta t = 0.0008$. The results presented in Tab. 1 prove, on this example, that the current scheme is more efficient than the first one. The error is measured in the $L^\infty(0, T; L^2(\Omega))$–norm with respect to the exact solution (the latter is computed with a time step much smaller than δt) provided that we have 100 processors at our disposal, the solution with accuracy $2.8\ 10^{-4}$ comes about 14 times faster than if we would use a standard sequential scheme!

The purpose of the last section is to illustrate that this approach can indeed provide much faster solutions for nonlinear, non-differentiable problems.

5   Pricing of an American Put

We consider in this section the pricing of an American option. Up to some change of variables, the American put solves in a simplified setting the fol-

Table 1. $L^\infty(O, T; L^2(\Omega))$ norm of the error for different values of k.

k	1	2	3	4	5	6
current method	$7.4\,10^{-2}$	$1.0\,10^{-2}$	$1.3\,10^{-3}$	$2.8\,10^{-4}$	$2.8\,10^{-4}$	$2.8\,10^{-4}$
method [3]	$1.2\,10^{-2}$	$5.6\,10^{-3}$	$2.7\,10^{-3}$	$1.3\,10^{-3}$	$5.6\,10^{-4}$	$2.8\,10^{-4}$

lowing nonlinear equation [5]

$$\min(\partial_t u - \partial^2_{xx} u, u - g(x)) = 0, \tag{28}$$

with $u(t = 0) = g(x) = \max(e^x - 1, 0)$. Notice that this equation can be written in the framework of the introduction $F(\partial_t u, u, \partial^2_{xx} u) = 0$, where $F(a, b, c) = \min(a - c, b - g(x))$ is a continuous non-differentiable functional.

It is possible to show that the above equation (28) admits a unique weak (viscosity) solution such that u and $\partial_t u$ are continuous. Moreover, there exists a *free boundary*, the optimal exercise boundary $\xi(t)$ such that $u(t, x) = g(x)$ for $x \leq \xi(t)$ and $\partial_t u(t, x) - \partial^2_{xx} u(t, x) = 0$ for $x > \xi(t)$. The derivative $\partial_x u(t, x)$ is moreover continuous everywhere except at $x = \xi(t)$. An approximation of $u(t, x)$ (see below in which sense it is an approximation) is given in Fig. 1.

The discretization chosen to solve (28) approximately is a splitting method, based on solving the parabolic equation on an interval $(t, t + \Delta T)$ and then projecting the solution on the set of functions satisfying $u \geq g$. The implicit scheme is given by

$$\begin{aligned}
\text{Step 1:} \quad & \frac{U^{n+1/2} - U^n}{\Delta T} - \partial^2_{xx} U^{n+1/2} = 0 \\
\text{Step 2:} \quad & U^{n+1} = \max\left(g(x), U^{n+1/2}(x)\right).
\end{aligned} \tag{29}$$

Here again, $u^n(x)$ is an approximation of $u(n\Delta T, x)$.

Step 1 in (29) needs to be further discretized. We replace the space of "positions" $x \in \mathbb{R}$ by the interval $(-3, 5)$ and impose the boundary conditions $u(t, -3) = g(-3)$ and $u(t, 5) = g(5) = 0$. The latter condition is valid as long as $u(t, x)$ is small for x in the vicinity of 5. We see in Figure 1 that such is the case of t up to 1.25 at least but that this assumption no longer holds for $T = 2.5$. Solving the equation for $T \geq 2.5$ requires to solve the parabolic equation on a larger domain. We keep our domain $x \in (-3, 5)$ for all times to simplify. We approximate the second-order derivative $\partial^2_{xx} u$ as

$$\partial^2_{xx} u \approx \frac{u_{m+1} + u_{m-1} - 2u_m}{(\Delta x)^2}, \tag{30}$$

where $u_m \approx u(-3 + m\Delta x)$ and Δx is a constant spatial step.

We are now ready to apply the iterative algorithm (3)—(6) to solve (28). We want to solve (28) for $(t, x) \in (0, 2.5) \times (-3, 5)$ using a time step $\Delta T = 0.05$. The "exact part" (4) is solved using the same splitting algorithm (29)

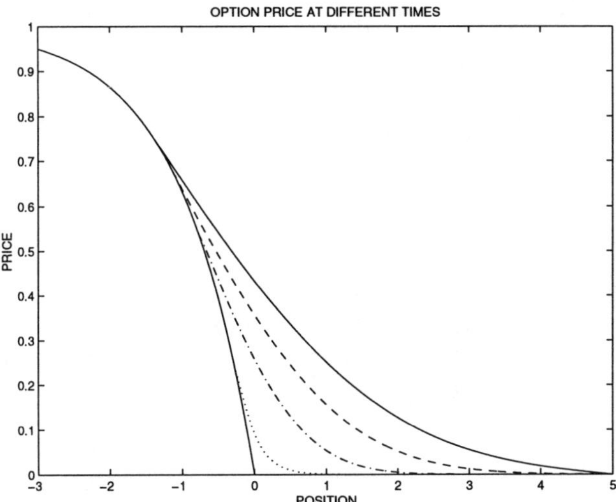

Fig. 1. Option price $u(t,x)$ for several values of time on the interval $(-3,5)$. Represented are the initial condition (solid line), the solution at $t = 0.1$ (dotted line), at $t = 0.5$ (dot-dashed line), at $t = 1.25$ (dashed line) and $t = 2.5$ (solid line).

with a time step $\delta t = 10^{-3}$. The number of points for the spatial discretization is 1500.

The errors on the interval (-3.5) for the four first iterations at final time $T = 2.5$ are give in Fig. 2. A quantitative measure of the $L^\infty(-3,5)$ norm of the errors between the exact solution and $U_k^N(x)$, $\varepsilon_k(x) = u(T,x) - U_k^N(x)$ is presented in Tab. 2. These results confirm the exponential convergence of

Table 2. L^∞ norm of the error $\varepsilon_k(x)$ for different values of k and radius of convergence of the method $\rho_k = \|\varepsilon_k\|_{L^\infty} / \|\varepsilon_{k-1}\|_{L^\infty}$.

k	0	1	2	3	4	5
$\|\varepsilon_k\|_{L^\infty}$	$5.5\,10^{-3}$	$2.3\,10^{-3}$	$5.7\,10^{-4}$	$1.3\,10^{-4}$	$2.9\,10^{-5}$	$6.5\,10^{-6}$
ρ_k	-	0.42	0.24	0.23	0.22	0.23

$U_k^n(x)$ to the solution $u(t,x)$ as $k \to \infty$. However, unlike the convergence of order $O((\Delta T)^k)$ obtained for linear equations in section 3, the numerical results suggest a convergence of the form ρ^k, where ρ is independent of ΔT. There is no theoretical explanation to this fact yet.

The accuracy of solution calculated with the fine discretization $\delta t = 10^{-3}$ is $1.2\,10^{-4}$ in the L^∞ norm (it has been compared with a much finer discretization). We can therefore deduce that the solution of the iterative scheme is as accurate as the fine solution after 3 iterations of the "parareal" algorithm.

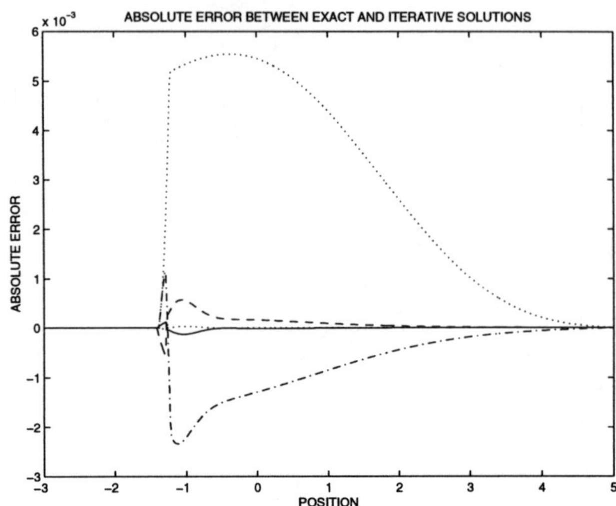

Fig. 2. Errors at time $T = 2.5$ between the "exact solution" obtained by solving (28) with time step $\delta t = 10^{-3}$ and the iterative solutions $U_k^N(x)$ for $k = 0$ (dotted line), $k = 1$ (dot-dashed line), $k = 2$ (dashed line), $k = 3$ (solid line), and $k = 4$ (gray dotted line).

The total gain in computational time is therefore $[T/\delta t]/[(k + 1)(T/\Delta T + \Delta T/\delta t)] = 2500/400 = 6.25$ and requires $\Delta T/\delta t = 50$ processors.

An interesting information in the pricing of American options is the free boundary $\xi(t)$, separating the two regions where it is optimal to exercise the option ($u = g$) and where it is not ($u > g$). We also expect the free boundary $\xi_k(t)$ obtained at iteration k from $U_k^n(x)$ to converge to the exact free boundary $\xi(t)$ obtained from $u(t, x)$. This is confirmed in Fig. (3). Notice that $\xi_k(t)$ is exact at $T^k = k\Delta T$ since the solution $U_k^n(t)$ is exact for $n \leq k$. This can be observed on the right-hand side of Fig. (3). The convergence of the free boundary is comparable to that of the solution (see Tab. 3). The radius of convergence of the method becomes less accurate for large values of k when errors of the order of the spatial discretization Δx are reached.

Table 3. L^∞ norm of the error $\eta_k(t) = \xi(t) - \xi_k(t)$ for different values of k and radius of convergence of the method $\rho_k = \|\eta_k\|_{L^\infty}/\|\eta_{k-1}\|_{L^\infty}$.

k	0	1	2	3	4	5
$\|\eta_k\|_{L^\infty}$	$1.7\,10^{-1}$	$4.8\,10^{-2}$	$1.8\,10^{-2}$	$4.0\,10^{-3}$	$1.3\,10^{-3}$	$8.6\,10^{-4}$
ρ_k	-	0.28	0.37	0.23	0.32	0.66

Acknowledgments. The numerical results in section 4 have been obtained thanks to the software FreeFem+ [4] and have benefited from the

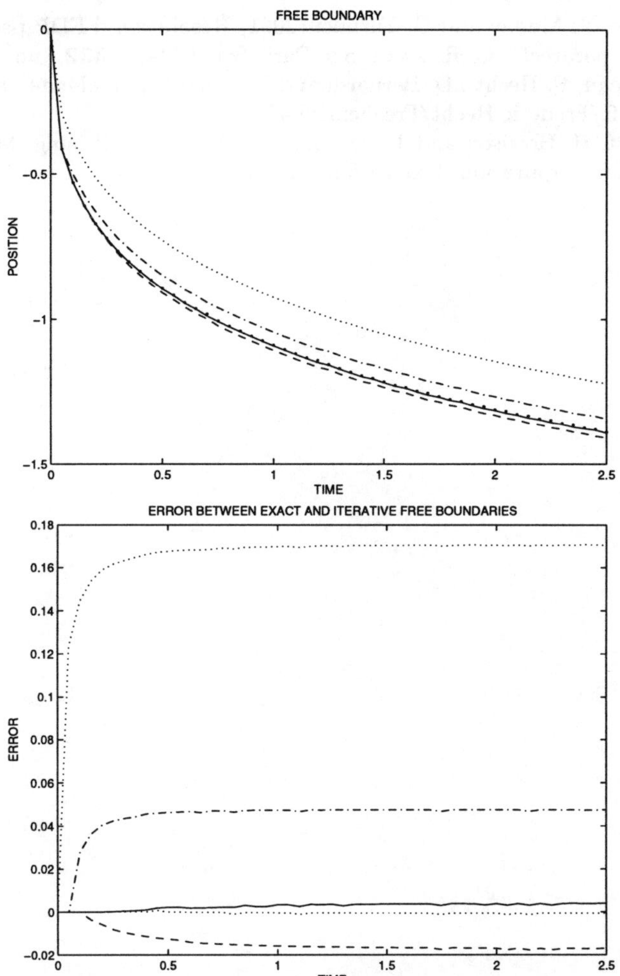

Fig. 3. Right figure: Free boundary $\xi(t)$ (solid line), $\xi_0(t)$ (dotted line), $\xi_1(t)$ (dot-dashed line), $\xi_2(t)$ (dashed line), $\xi_3(t)$ (points). Left figure: Error between the exact free boundary $\xi(t)$ and $\xi_k(t)$ for $k = 0$ (dotted line), $k = 1$ (dot-dashed line), $k = 2$ (dashed line), $k = 3$ (solid line), and $k = 4$ (gray dotted line).

help of Frédéric Hecht and Olivier Pironneau. The first author acknowledges partial support from NSF grant DMS-0072008.

References

1. P. Chartier and B. Philippe (1993) A parallel shooting technique for solving dissipative ODE's, Computing **51**, pp 209–236.
2. J. Erhel J. and S; Rault S. (2000) Algorithme parallèle pour le calcul d'orbites, Techniques et Sciences informatiques, **19**, pp 649–673.

3. J.-L. Lions, Y. Maday and G. Turinici (2001) Résolution d'EDP par un schéma en temps "pararéel", C. R. Acad. Sci. Paris Sér. I Math. **332** , no. 7, 661–668.
4. O. Pironneau, F. Hecht , D. Bernardi and K. Ohtsuka, FreeFem+, http://www-rocq.inria.fr/Frederic.Hecht/FreeFemPlus.htm.
5. P. Wilmott, H. Howison and J. Dewynne (1998) Option Pricing. Mathematical models and computation, Oxford financial press, Oxford.

The Influence of Quadrature Formulas in 2D and 3D Mortar Element Methods

Yvon Maday[1], Francesca Rapetti[1], and Barbara I. Wohlmuth[2]

[1] Laboratoire d'Analyse Numérique, Paris 6 University, Boîte courrier 187,
   75252 Paris cedex 05, France
[2] Mathematisches Institut A, Universität Stuttgart, Pfaffenwaldring 57,
   70 569 Stuttgart, Germany

Abstract. The paper is concerned with the mortar finite element discretization of scalar elliptic equations in three dimensions. The attention is focused on the influence of quadrature formulas on the discretization error. We show numerically that the optimality of the method is preserved if suitable quadrature formulas are used.

1　Introduction

The mortar element method, first proposed in [3], is a non-conforming non-overlapping domain decomposition method. The coupling of different physical models, discretization schemes or non-matching triangulations at the interfaces can be efficiently realized in terms of mortar element methods. To preserve the global optimality of the locally adapted discretizations, the interfaces between the different regions have to be handled appropriately. Due to its high flexibility, this approach has been analysed and implemented in many situations.

The main feature of the mortar element methods is to replace the exact continuity condition at the skeleton of the decomposition with a weak one. An important aspect for the implementation is the realization of the weak coupling across the interfaces. It can be written in terms of a Lagrange multiplier and the jump of the traces. The associated integrals involve discrete functions defined on different non-matching grids and, as a consequence, the computation goes through the intersection of the supports. In two dimensions, for finite element discretizations, the supports can be easily intersected at the common interfaces due to the fact that the latter are one dimensional curves. In three dimensions, finding the intersection of the supports (union of triangles or quadrilaterals) becomes a hard task. The use of quadrature formulas to evaluate these integrals increases considerably the efficiency of the implementation. However, using a standard Galerkin approximation and replacing the exact integration by a quadrature formula based only on the slave or master side does not yield optimal results (see [5] for numerical evidences in two dimensions). The best approximation error requires a quadrature formula based on the slave side and the consistency error requires one on the master side. As a consequence, we are led to work with different test and trial spaces, resulting in a Petrov-Galerkin approach. We focus our attention on the influence of quadrature formulas on the discretization error. Numerical results are presented for a scalar elliptic equation in two and three dimensions. Realizing the difficulty of the proof in three dimensions, we show only numerically that the optimality of the method is preserved if suitable quadrature formulas are used. We are aware that few examples are not sufficient to conclude about this subject but they give a first insight. Asymptotically, the error in the energy norm is $O(h)$ and $O(h^2)$ for the L^2-norm.

2   Problem Formulation and Notation

The key example for the definition and the analysis of the mortar element method is the following second order elliptic boundary value problem

$$-\operatorname{div}(a \operatorname{grad} u) = f \quad \text{in } \Omega,$$
$$u = 0 \quad \text{on } \Gamma_D,$$
$$\frac{\partial u}{\partial n} = 0 \quad \text{on } \Gamma_N,$$
(1)

where a is a sufficiently smooth matrix that is uniformly positive definite in the bounded open set $\Omega \subset \mathbb{R}^3$, Γ_D is a subset of the boundary $\partial\Omega$ of Ω with $\operatorname{meas}(\Gamma_D) > 0$ and $\Gamma_N = \partial\Omega \setminus \Gamma_D$.

Let us introduce the functional space $H^1_{0,D}(\Omega) = \{u \in H^1(\Omega) \,|\, u_{|\Gamma_D} = 0\}$, that is the closure in $H^1(\Omega)$ of all C^∞-functions vanishing on Γ_D. Then, problem (1) admits the following variational formulation:

$$\text{Given } f \in L^2(\Omega), \quad \text{find } u \in H^1_{0,D}(\Omega) \quad \text{such that}$$
$$\int_\Omega a \operatorname{grad} u \cdot \operatorname{grad} v \, d\Omega = \int_\Omega f v \, d\Omega \qquad \forall v \in H^1_{0,D}(\Omega).$$
(2)

Problem (2) admits a unique solution, due to the Lax-Milgram theorem and the Poincaré lemma.

Suppose now that Ω is decomposed into non-overlapping subdomains Ω_k, $k = 1, \dots, \mathcal{K}$, such that

$$\overline{\Omega} = \cup_{k=1}^{\mathcal{K}} \overline{\Omega}_k \quad, \qquad \begin{cases} \Omega_k \cap \Omega_\ell = \emptyset \\ \overline{\Omega}_k \cap \overline{\Omega}_\ell = \Gamma_{k\ell} \end{cases} \quad \forall k \neq \ell.$$
(3)

We introduce the skeleton S of the decomposition

$$S = \cup_{k,\ell} \Gamma_{k\ell}.$$
(4)

$\Gamma_{k\ell}$, Γ_N and Γ_D will be assumed to be the union of polygonal subsets of the boundaries of the subdomains Ω_k: often, such a decomposition is called *geometrically conforming*. To define correctly the trace space, we associate with each interface $\Gamma_{k\ell}$ the *non-mortar (slave) side* which is, by convention, Ω_k whereas Ω_ℓ is the *mortar (master) side*. Let B denote a bounded open subset of $\mathbb{R}^3$. If γ' is a smooth subset of ∂B, the space $H_{00}^{1/2}(\gamma')$ consists of those elements $v \in H^{1/2}(\gamma')$ whose trivial extension $\tilde{v}$ of v by zero to all ∂B belongs to $H^{1/2}(\partial B)$ [6], i.e.

$$H_{00}^{1/2}(\Gamma_{k\ell}) = \{ v \in H^{1/2}(\Gamma_{k\ell}) \,|\, \tilde{v} \in H^{1/2}(\partial \Omega_k) \} \,,$$
$$\|v\|_{H_{00}^{1/2}(\Gamma_{k\ell})} = \|\tilde{v}\|_{1/2,\partial \Omega_k} \,, \tag{5}$$

where the norm of $\tilde{v}$ is evaluated on the side $\partial \Omega_k$. We finally remind that $H_{00}^{1/2}(\Gamma_{k\ell})$ is a proper and continuously embedded subspace of $H^{1/2}(\Gamma_{k\ell})$ [6]. In the following, $(H_{00}^{1/2}(\Gamma_{k\ell}))'$ denotes the dual space of $H_{00}^{1/2}(\Gamma_{k\ell})$. We now return to the characterization of $H_{0,D}^1(\Omega)$. To this purpose, we introduce the space

$$X_* = \{ v \in L^2(\Omega) \,|\, v_{|\Omega_k} \in H^1(\Omega_k) \,, \ k = 1, \ldots, \mathcal{K} \,, \ v_{|\Gamma_D} = 0 \} \,, \tag{6}$$

endowed with the *broken norm*

$$\|v\|_* = \left(\sum_{k=1}^{\mathcal{K}} \|v\|_{1,\Omega_k}^2 \right)^{1/2} , \tag{7}$$

and a proper subspace

$$X_{00} = \{ v \in X_* \,|\, [v]_{|\Gamma_{k\ell}} \in H_{00}^{1/2}(\Gamma_{k\ell}) \,, \quad \forall \Gamma_{k\ell} \subset S \} \tag{8}$$

where $[v]_{|\Gamma_{k\ell}}$ is the restriction of the jump $(v_{|\Omega_k} - v_{|\Omega_\ell})$, for $v \in X_*$, to any interface $\Gamma_{k\ell} \subset S$. Then, the following space

$$V = \{ v \in X_{00} \,|\, <\mu, [v]>_{0,\Gamma_{k\ell}} = 0 \,, \ \forall \mu \in (H_{00}^{1/2}(\Gamma_{k\ell}))' \,, \ \Gamma_{k\ell} \subset S \}$$

is equal to $H_{0,D}^1(\Omega)$. Problem (2) can be written into its equivalent domain decomposition formulation:

$$\text{Given } f \in L^2(\Omega) \,, \quad \text{find } u \in V \quad \text{such that } \forall v \in V$$
$$\textstyle\sum_{k=1}^{\mathcal{K}} \int_{\Omega_k} a \,\mathrm{grad}\, u_k \cdot \mathrm{grad}\, v_k \, d\Omega = \sum_{k=1}^{\mathcal{K}} \int_{\Omega_k} f \, v_k \, d\Omega \,. \tag{9}$$

As a consequence, problem (9) admits a unique solution.

3   Problem Discretization

In each subdomain Ω_k, we choose a family of conforming triangulations $(\mathcal{T}_{k,h})_h$, independently of the ones defined in the neighboring subdomains, i.e. the nodes in $\mathcal{T}_{k,h}$ that belong to $\Gamma_{k\ell}$ do not need to match the nodes of $\mathcal{T}_{\ell,h}$, $k \neq \ell$. We consider triangulations composed of tetrahedra t and we define the local discrete spaces of piecewise linear finite elements on $\mathcal{T}_{k,h}$ such as

$$X_{k,h} = \{v_{k,h} \in H^1(\Omega_k) \,|\, v_{k,h_{|t}} \in \mathbb{P}_1(t)\,, \forall\, t \in \mathcal{T}_{k,h}\,,\, v_{k,h_{|\Gamma_D \cap \partial\Omega_k}} = 0\}\,.$$

We set

$$X_h = \Pi_{k=1}^{K} X_{k,h} = \{v_h \in L^2(\Omega)\,|\, v_{k,h} = v_{h|\Omega_k} \in X_{k,h}\}\,.$$

A particular attention has to be addressed to the trace of elements of X_h on the interface $\Gamma_{k\ell}$ between adjacent subdomains Ω_k and Ω_ℓ. We denote as

$$T_{k\ell,h} = \{w \in L^2(\Gamma_{k\ell})\,|\, \exists\, v_{k,h} \in X_{k,h}\ \text{such that}\ v_{k,h_{|\Gamma_{kl}}} = w\}$$

the space of all continuous piecewise linear functions on $\Gamma_{k\ell}$ on the partition induced by the triangulation $\mathcal{T}_{k,h}$ of $\overline{\Omega}_k$. Note that $T_{k\ell,h} \neq T_{\ell k,h}$ since the triangulations $\mathcal{T}_{k,h}$ and $\mathcal{T}_{\ell,h}$ do not match at the interface $\Gamma_{k\ell}$. In the mortar setting, the value of each discrete function on the mortar side will define *weakly* the values of the discrete function on the non-mortar side. We associate a local space $M_{k\ell,h}$ to each non-mortar side $\Gamma_{k\ell}$ of the skeleton with the following features (that characterizes the mortar element methods among other hybrid formulations):

$$(i)\quad M_{k\ell,h} \subset T_{k\ell,h}\,,$$

$$(ii)\quad \dim M_{k\ell,h} = \dim\left(T_{k\ell,h} \cap H_0^1(\Gamma_{k\ell})\right),\tag{10}$$

$$(iii)\ M_{k\ell,h}\ \text{contains constants on}\ \Gamma_{k\ell}\,.$$

We introduce the space M_h of Lagrange multipliers on the skeleton S as

$$M_h = \Pi_{\Gamma_{k\ell} \subset S}\, M_{k\ell,h}\,.\tag{11}$$

The choice of the Lagrange multiplier space M_h is of key importance for the optimality of the method. If we set

$$u_h = (u_{k,h})_k \quad,\quad m_h = (\mu_{k\ell})_{\Gamma_{k\ell} \subset S}$$

$$a(u_h, v_h) = \sum_{k=1}^{K} \int_{\Omega_k} a\ \text{grad}\ u_{k,h} \cdot \text{grad}\ v_{k,h}\, d\Omega$$

$$b(v_h, m_h) = \sum_{\Gamma_{k\ell} \subset S} \int_{\Gamma_{k\ell}} (v_{k,h} - v_{\ell,h})\, \mu_{k\ell}\, d\Gamma$$

the discrete problem reads

$$\text{Given } f \in L^2(\Omega)\,,\quad \text{find } u_h \in V_h \quad \text{such that } \forall\, v_h \in V_h$$
$$a(u_h, v_h) = \sum_{k=1}^{K} \int_{\Omega_k} f\, v_{k,h}\, d\Omega\tag{12}$$

where V_h is the constrained space

$$V_h = \{v_h \in X_h \mid b(v_h, m_h) = 0, \quad \forall\, m_h \in M_h\}\,.$$

Note that in this way the Lagrange multipliers are not unknowns of the problem since we work with test and trial functions that already satisfy the *weak coupling condition* contained in V_h. The presence of this weak coupling condition prevents V_h from being a subspace of V, i.e. we are using a *non-conforming method* to approximate the solution of problem (9). The second Strang lemma allows to derive the following error bound for such an approximation

$$\|u - u_h\|_{1,*} \le c\left\{ \inf_{v_h \in V_h} \|u - v_h\|_{1,*} + \sup_{w_h \in V_h} \frac{\sum_{\Gamma_{k\ell} \subset S} \int_{\Gamma_{k\ell}} a\,\frac{\partial u}{\partial n}\,[w_{k,h}]\,d\Gamma}{\|w_h\|_{1,*}} \right\}. \quad (13)$$

In the right-hand side of (13), the first term represents the *best approximation error* and the second is the *consistency error*.

In the following sections we address two difficulties: the first is the construction of a basis for the Lagrange multiplier space M_h defined in (11) and the second is the one encountered to satisfy the coupling condition contained in V_h, i.e. $\forall\, \Gamma_{k\ell} \subset S$ and $\forall\, \mu_{k\ell} \in M_{k\ell,h}$,

$$\int_{\Gamma_{k\ell}} (v_{k,h} - v_{\ell,h})\,\mu_{k\ell}\,d\Gamma = 0 \quad, \quad v_{k,h} \in X_{k,h} \quad, \quad v_{\ell,h} \in X_{\ell,h}. \quad (14)$$

3.1   A Basis for $M_{k\ell,h}$

We now detail the candidate proposed in [1], [2], [3] for $M_{k\ell,h}$ satisfying the requirements $(i), (ii), (iii)$ listed in the previous section. To this purpose, we fix some notations. Let $\Gamma = \Gamma_{k\ell}$ and $\mathcal{T}_{k,\Gamma}$ is the triangulation in triangles t that is induced on $\overline{\Gamma}$ by the partitioning $\mathcal{T}_{k,h}$ of the non-mortar side $\overline{\Omega}_k$. Let $\mathcal{V}_{\overline{\Gamma}}$, $\mathcal{V}_\Gamma$ and $\mathcal{V}_{\partial\Gamma}$ denote the set of all nodes of $\mathcal{T}_{k,\Gamma}$, the nodes inside Γ and those on the boundary of Γ, respectively. Note that $\mathcal{V}_{\overline{\Gamma}} = \mathcal{V}_\Gamma \cup \mathcal{V}_{\partial\Gamma}$. The finite element basis functions will be denoted by φ_a, $a \in \mathcal{V}_{\overline{\Gamma}}$. We define

$$\tilde{X}^{\Gamma}_{k,h} = \text{span}\,\{\varphi_a \mid a \in \mathcal{V}_\Gamma\}$$

$$\sigma_a = \text{supp}\,(\varphi_a) = \cup\{t \in \mathcal{T}_{k,\Gamma} \mid a \text{ vertex of } t\}$$

$$\mathcal{N}_a = \{b \in \mathcal{V}_\Gamma \mid b \in \sigma_a\} \quad, \quad \mathcal{N} = \cup_{a \in \mathcal{V}_{\partial\Gamma}} \mathcal{N}_a$$

where $\mathcal{N}_a$ is the set of the internal neighboring nodes of a and $\mathcal{N}$ is the set of those internal nodes which have a neighbor on the boundary of Γ. Now, we set

$$M_{k\ell,h} = \text{span}\,\{\hat{\varphi}_a \mid a \in \mathcal{V}_\Gamma\}$$

where the basis functions $\hat{\varphi}_a$ are defined as follows:

$$\hat{\varphi}_a = \begin{cases} \varphi_a\,, & a \in \mathcal{V}_\Gamma \setminus \mathcal{N}\,, \\[2mm] \varphi_a + \sum_{b \in \mathcal{V}_{\partial\Gamma} \cap \sigma_a} c_{b,a}\,\varphi_b + \sum_{d \in \mathcal{V}_{\partial\Gamma}} c_{d,a}\,\varphi_d\,, & a \in \mathcal{N}\,. \end{cases}$$

To be more precise, we set $c_{b,a} = 1/n_b$ where n_b is the number of nodes in $\mathcal{N}_b$ and the coefficient $c_{d,a} = 1$ if d is the node opposite to a and zero otherwise. We call $d \in \mathcal{V}_{\partial\Gamma}$ *opposite node* to $a \in \mathcal{V}_\Gamma$ if $\partial\sigma_a \cap \partial\sigma_d$ is one edge and $\mathcal{N}_d$ is empty. For more sophisticated choices of $c_{b,a}$ and $c_{d,a}$, we refer to [4]. Note that $\hat{\varphi}_a(b) = \delta_{a,b}$, for all $a, b \in \mathcal{V}_\Gamma$, i.e. the functions $\hat{\varphi}_a$, $a \in \mathcal{V}_\Gamma$, are linearly independent. Requirements (i) and (ii) are satisfied by construction. Observing that, for all $b \in \mathcal{V}_{\partial\Gamma}$ having no opposite node,

$$\sum_{a \in \mathcal{N}_b} c_{b,a} = 1$$

holds and that

$$\sum_{d \in \mathcal{V}_{\partial\Gamma},\, d \text{ opp. to } a \in \mathcal{V}_\Gamma} c_{d,a} = 1\,,$$

we find that (iii) is satisfied. Moreover, we obtain that $\sum_{a \in \mathcal{V}_\Gamma} \hat{\varphi}_a = 1$.

As an example, we present a basis for the Lagrange multiplier space $\mathcal{M}_{k\ell,h}$ where $\mathcal{T}_{k,\Gamma}$ is given in Figure 1. In this case, node 1 is opposite to node 12 so $c_{1,12} = 1$ whereas $c_{1,i} = 0$ for all other nodes $i \neq 12$. Note that all other nodes on Γ have no opposite node. We have

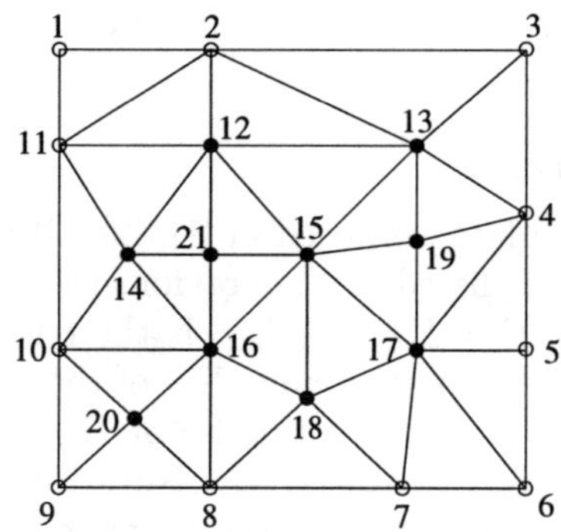

Fig. 1. Example of triangulation on a non-mortar side.

$$\mathcal{V}_{\partial\Gamma} = \{1, \ldots, 11\} \quad,\quad \mathcal{V}_\Gamma = \{12, \ldots, 21\}\,,$$

$$\mathcal{N}_1 = \emptyset\,,\quad \mathcal{N}_2 = \{12, 13\}\,,\quad \mathcal{N}_{13} = \{12, 15, 19\}\,,\quad \mathcal{N} = \mathcal{V}_\Gamma \setminus \{15, 21\}\,,$$

and the basis functions for $\mathcal{M}_{k\ell,h}$ are

$$\hat{\varphi}_{12} = \varphi_{12} + \varphi_1 + \tfrac{1}{2}\varphi_2 + \tfrac{1}{2}\varphi_{11}\,,\quad \hat{\varphi}_{13} = \varphi_{13} + \tfrac{1}{2}\varphi_2 + \varphi_3 + \tfrac{1}{3}\varphi_4\,,$$

$$\hat{\varphi}_{14} = \varphi_{14} + \tfrac{1}{3}\varphi_{10} + \tfrac{1}{2}\varphi_{11}\,,\quad \hat{\varphi}_{15} = \varphi_{15}\,,$$

$$\hat{\varphi}_{16} = \varphi_{16} + \tfrac{1}{3}\varphi_8 + \tfrac{1}{3}\varphi_{10}\,,\quad \hat{\varphi}_{17} = \varphi_{17} + \tfrac{1}{3}\varphi_4 + \varphi_5 + \varphi_6 + \tfrac{1}{2}\varphi_7\,,$$

$$\hat{\varphi}_{18} = \varphi_{18} + \tfrac{1}{2}\varphi_7 + \tfrac{1}{3}\varphi_8\,,\quad \hat{\varphi}_{19} = \varphi_{19} + \tfrac{1}{3}\varphi_4\,,$$

$$\hat{\varphi}_{20} = \varphi_{20} + \tfrac{1}{3}\varphi_8 + \varphi_9 + \tfrac{1}{3}\varphi_{10}\,,\quad \hat{\varphi}_{21} = \varphi_{21}\,.$$

In the following, we denote

$$V_{\overline{S}} = \cup_{\Gamma_{k\ell} \subset S}\, V_{\overline{\Gamma}_{k\ell}} \quad , \quad V_{\partial S} = \cup_{\Gamma_{k\ell} \subset S}\, V_{\partial \Gamma_{k\ell}} \quad , \quad V_{S} = \cup_{\Gamma_{k\ell} \subset S}\, V_{\Gamma_{k\ell}} \,,$$

where the unions are intended without repetitions and $V_{\overline{S}} = V_{S} \cup V_{\partial S}$.

3.2 Evaluation of the Integrals on the Interfaces $\Gamma_{k\ell} \subset S$

Condition (14) is involved in the definition of a basis for the discrete con-
strained space V_h. The computation of the quantity $\int_{\Gamma_{k\ell}} v_{k,h}\, \mu_{k\ell}\, d\Gamma$ raises no
difficulty since the two discrete functions $v_{k,h}$ and $\mu_{k\ell}$ live on the same non-
mortar mesh inherited from Ω_k on $\Gamma_{k\ell}$. On the contrary, the computation of
$\int_{\Gamma_{k\ell}} v_{\ell,h}\, \mu_{k\ell}\, d\Gamma$ involves discrete functions that live on different meshes since
$v_{\ell,h}$ is defined on the master mesh inherited from Ω_ℓ on $\Gamma_{k\ell}$ that may not
match with the one where $\mu_{k\ell}$ is defined, as it is shown in Figure 2. To avoid
intersecting the supports, a task that can be rather difficult and expensive
for interfaces of complicated shape and finite curvature, we resort to a more
efficient technique involving classical numerical quadratures.

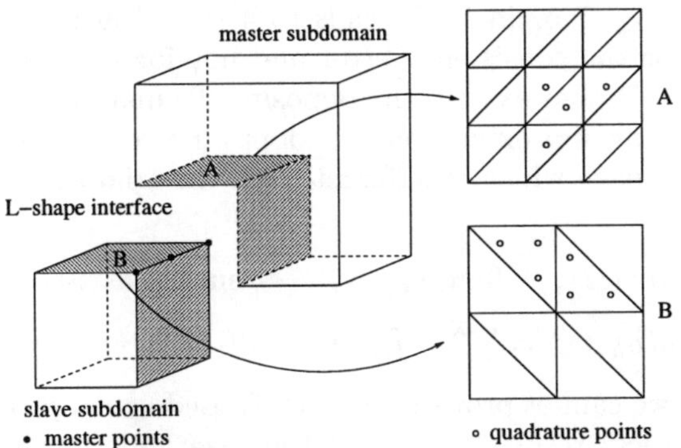

Fig. 2. Example of domain decomposition into two subdomains and non-matching
grids on the skeleton of the decomposition.

In order to simplify the presentation, we focus our attention again on one
interface $\Gamma_{k\ell} \subset S$. We denote by Γ^+ (resp. v^+) the mortar (master) side
of $\Gamma_{k\ell}$ (resp. the value of v on the mortar side) and by Γ^- (resp. v^-) the
non-mortar (slave) side of $\Gamma_{k\ell}$ (resp. the value of v on the non-mortar side).
Numerical quadratures lead to approximate the integrals on the mortar (resp.
on the non-mortar) side as follows

$$\int_{\Gamma^+} w^+ d\Gamma \approx \sum\nolimits_+ w^+ \qquad \left(\text{resp. } \int_{\Gamma^-} w^- d\Gamma \approx \sum\nolimits_- w^-\right)$$

where $\sum_+$ (resp. $\sum_-$) represents the numerical integration on the mortar
(resp. non-mortar) side. Condition (14) can be then rewritten as follows

$$\int_{\Gamma^-} (v^- - v^+)\, \mu^-\, d\Gamma = 0, \quad \forall \Gamma^- \in S, \quad \forall \mu^- \in M_{-,h}\,.$$

Since v^- and μ^- are defined on the slave side, it is natural to apply a quadrature formula on this side to compute $\int_{\Gamma_-} v^- \mu^- \, d\Gamma$. We use then a quadrature formula which is exact, i.e. $\int_{\Gamma_-} v^- \mu^- \, d\Gamma = \sum_- v^- \mu^-$.

To evaluate $\int_{\Gamma_-} v^+ \mu^- \, d\Gamma$, there are two natural possibilities. We define two discrete spaces V_h^{-+} and V_h^{--} by

$$V_h^{-+} = \{ v_h \in X_h \mid \textstyle\sum_- v_h^- \cdot m_h^- = \sum_+ v_h^+ \cdot m_h^- , \quad \forall \, m_h \in M_h \} ,$$

$$V_h^{--} = \{ v_h \in X_h \mid \textstyle\sum_- v_h^- \cdot m_h^- = \sum_- v_h^- \cdot m_h^- , \quad \forall \, m_h \in M_h \} .$$

With these two choices, problem (12) reads respectively

$$\text{given } f \in L^2(\Omega), \quad \text{find } u_h \in V_h^{-+} \quad \text{such that } \forall \, v_h \in V_h^{-+}$$
$$a(u_h, v_h) = \textstyle\sum_{k=1}^{\mathcal{K}} \int_{\Omega^k} f \, v_{k,h} \, d\Omega , \tag{15}$$

$$\text{given } f \in L^2(\Omega), \quad \text{find } u_h \in V_h^{--} \quad \text{such that } \forall \, v_h \in V_h^{--}$$
$$a(u_h, v_h) = \textstyle\sum_{k=1}^{\mathcal{K}} \int_{\Omega^k} f \, v_{k,h} \, d\Omega . \tag{16}$$

In [5], the authors have shown that, in two dimensions, the Galerkin approach to solve (12) in V_h^{-+} or in V_h^{--} leads to a loss of accuracy. In particular, V_h^{-+} is good for the consistency error and not for the best approximation one, whereas V_h^{--} does exactly the opposite. To maintain the optimality of the non-conforming approximation, we adopt a Petrov-Galerkin approach, by choosing a test space which is different from the trial space. Then problem (12) becomes

$$\text{given } f \in L^2(\Omega), \quad \text{find } u_h \in V_h^{--} \quad \text{such that } \forall \, v_h \in V_h^{-+}$$
$$a(u_h, v_h) = \textstyle\sum_{k=1}^{\mathcal{K}} \int_{\Omega^k} f \, v_{k,h} \, d\Omega . \tag{17}$$

Unfortunately, we cannot prove yet the well-posedness of problem (17). However, our numerical results show it is the case and they yield an optimal scheme.

Remark 1. The choice of the test and trial spaces can be motivated by the following interpretation. In the mortar approach, we have to solve a coupled Dirichlet-Neumann problem. On the slave side, a Dirichlet problem has to be considered where the boundary condition is obtained from the trace on the master side. As it can be done in the context of inhomogeneous Dirichlet boundary conditions, we realize the boundary condition in a weak integral form based on quadrature formulas on the mesh which is the mesh on the slave side. Thus the natural choice for the trial space is V_h^{--}. On the other hand, on the master side, we solve a Neumann problem where the inhomogeneous boundary conditions are obtained from the residual on the slave side. In this case, a quadrature formula on the master side is the natural choice. Neumann boundary conditions enter in a weak form on the right-hand side. More precisely, they are seen by the test space. Thus V_h^{-+} is the natural choice for the test space.

4   Matrix Form of the Discrete Problem

To underline the difference with a Galerkin (symmetric) approach, we write the three discrete problems (15), (16) and (17) in matrix form. We start by considering on each subdomain the unconstrained stiffness matrix associated with homogeneous Neumann type conditions on any interface $\Gamma \subset S$ (the system is the same in the three cases). In the next step, the local systems are coupled by means of the mortar condition and this will make the difference between the Galerkin and the Petrov-Galerkin approaches.

Let u_h be the solution of the discrete problem: we expand each of its components $u_{k,h}$ in terms of the basis functions of the finite dimensional space $X_{k,h}$ defined in the subdomains. We then have m^k coefficients u_s^k, $s = 1, \ldots, m^k$, where m^k is the dimension of $X_{k,h}$. Note that, due to the coupling condition, only $m^k - n_{k\ell}$ coefficients are real unknowns of the problem in $\overline{\Omega}_k$, where $n_{k\ell}$ is the number of nodes on the non-mortar interfaces $\Gamma_{k\ell}$ of $\partial\Omega_k$.

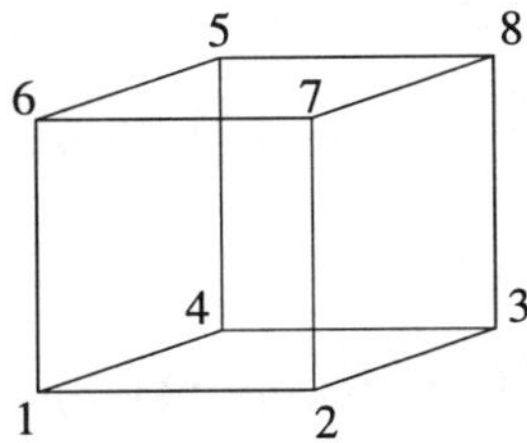

Fig. 3. Let us consider a subdomain Ω_k and let us suppose that $\overline{\Gamma}_{1456} = \partial\Omega_D \cap \partial\Omega_k$, for example. We assume that Γ_{1276} and Γ_{5678} are mortar interfaces whereas Γ_{1234}, Γ_{2387} and Γ_{4385} are non-mortar interfaces. The real unknowns in $\overline{\Omega}_k$ are the values associated to nodes in $\overline{\Omega}_k \setminus (\Gamma_{1234} \cup \Gamma_{2387} \cup \Gamma_{4385} \cup \overline{\Gamma}_{1456})$.

For each interface $\Gamma_{k\ell} \subset S \cap \partial\Omega_k$, we have

$$C_1\, u^{k,-}_{\mathcal{V}_{\partial\Omega_k \cap \Gamma_{k\ell}}} = D_1\, u^{\ell,+}_{\mathcal{V}_{\partial\Omega_\ell \cap \Gamma_{k\ell}}} + D_2\, u^{\ell,+}_{\mathcal{V}_{\partial\Omega_\ell \cap \partial\Gamma_{k\ell}}} - C_2\, u^{k,-}_{\mathcal{V}_{\partial\Omega_k \cap \partial\Gamma_{k\ell}}}$$

where

$$C_1(i,j) = \int_{\Gamma_{k\ell}} \varphi_j^- \, \hat{\varphi}_i^- \, d\Gamma, \quad \forall\, i \in \mathcal{V}_{\partial\Omega_k \cap \Gamma_{k\ell}}, \quad \forall\, j \in \mathcal{V}_{\partial\Omega_k \cap \Gamma_{k\ell}},$$

$$C_2(i,j) = \int_{\Gamma_{k\ell}} \varphi_j^- \, \hat{\varphi}_i^- \, d\Gamma, \quad \forall\, i \in \mathcal{V}_{\partial\Omega_k \cap \Gamma_{k\ell}}, \quad \forall\, j \in \mathcal{V}_{\partial\Omega_k \cap \partial\Gamma_{k\ell}},$$

$$D_1(i,j) = \int_{\Gamma_{k\ell}} \varphi_j^+ \, \hat{\varphi}_i^- \, d\Gamma, \quad \forall\, i \in \mathcal{V}_{\partial\Omega_k \cap \Gamma_{k\ell}}, \quad \forall\, j \in \mathcal{V}_{\partial\Omega_\ell \cap \Gamma_{k\ell}},$$

$$D_2(i,j) = \int_{\Gamma_{k\ell}} \varphi_j^+ \, \hat{\varphi}_i^- \, d\Gamma, \quad \forall\, i \in \mathcal{V}_{\partial\Omega_k \cap \Gamma_{k\ell}}, \quad \forall\, j \in \mathcal{V}_{\partial\Omega_\ell \cap \partial\Gamma_{k\ell}}.$$

Now, we denote

$$C_1^-(i,j) = \sum_{-} \varphi_j^- \, \hat{\varphi}_i^- , \quad \forall\, i \in \mathcal{V}_{\partial\Omega_k \cap \Gamma_{k\ell}}, \quad \forall\, j \in \mathcal{V}_{\partial\Omega_k \cap \Gamma_{k\ell}},$$

$$C_2^-(i,j) = \sum_{-} \varphi_j^- \, \hat{\varphi}_i^- , \quad \forall\, i \in \mathcal{V}_{\partial\Omega_k \cap \Gamma_{k\ell}}, \quad \forall\, j \in \mathcal{V}_{\partial\Omega_k \cap \partial\Gamma_{k\ell}}.$$

and

$$D_1^-(i,j) = \sum_- \varphi_j^+ \, \hat{\varphi}_i^- \,, \quad \forall i \in \mathcal{V}_{\partial\Omega_k \cap \Gamma_{k\ell}}, \quad \forall j \in \mathcal{V}_{\partial\Omega_\ell \cap \Gamma_{k\ell}},$$

$$D_2^-(i,j) = \sum_- \varphi_j^+ \, \hat{\varphi}_i^- \,, \quad \forall i \in \mathcal{V}_{\partial\Omega_k \cap \Gamma_{k\ell}}, \quad \forall j \in \mathcal{V}_{\partial\Omega_\ell \cap \partial\Gamma_{k\ell}},$$

$$D_1^+(i,j) = \sum_+ \varphi_j^+ \, \hat{\varphi}_i^- \,, \quad \forall i \in \mathcal{V}_{\partial\Omega_k \cap \Gamma_{k\ell}}, \quad \forall j \in \mathcal{V}_{\partial\Omega_\ell \cap \Gamma_{k\ell}},$$

$$D_2^+(i,j) = \sum_+ \varphi_j^+ \, \hat{\varphi}_i^- \,, \quad \forall i \in \mathcal{V}_{\partial\Omega_k \cap \Gamma_{k\ell}}, \quad \forall j \in \mathcal{V}_{\partial\Omega_\ell \cap \partial\Gamma_{k\ell}}.$$

Note that when the value of a function v^+ is needed in a point x^- of a grid $\mathcal{T}^-$ which is different from the grid $\mathcal{T}^+$ where the function v^+ lives, we project x^- on $\mathcal{T}^+$ and we compute v^+ in the projection point. We denote, in a block layout,

$$Q_{k\ell}^+ = [(C_1^-)^{-1} D_1^+ \,, (C_1^-)^{-1} D_2^+ \,, -(C_1^-)^{-1} C_2^-],$$

$$Q_{k\ell}^- = [(C_1^-)^{-1} D_1^- \,, (C_1^-)^{-1} D_2^- \,, -(C_1^-)^{-1} C_2^-].$$

Then the interface condition for $v_h \in V_h^{-\,+}$ on the interface $\Gamma_{k\ell}$ reads

$$\mathbf{v}_{\mathcal{V}_{\partial\Omega_k \cap \Gamma_{k\ell}}}^{k,-} = Q_{k\ell}^+ [\mathbf{v}_{\mathcal{V}_{\partial\Omega_\ell \cap \Gamma_{k\ell}}}^{\ell,+}, \mathbf{v}_{\mathcal{V}_{\partial\Omega_\ell \cap \partial\Gamma_{k\ell}}}^{\ell,+}, \mathbf{v}_{\mathcal{V}_{\partial\Gamma_{k\ell} \cap \partial\Gamma_{k\ell}}}^{k,-}], \tag{18}$$

and for $v_h \in V_h^{-\,-}$ on the same interface $\Gamma_{k\ell}$ reads

$$\mathbf{v}_{\mathcal{V}_{\partial\Omega_k \cap \Gamma_{k\ell}}}^{k,-} = Q_{k\ell}^- [\mathbf{v}_{\mathcal{V}_{\partial\Omega_\ell \cap \Gamma_{k\ell}}}^{\ell,+}, \mathbf{v}_{\mathcal{V}_{\partial\Omega_\ell \cap \partial\Gamma_{k\ell}}}^{\ell,+}, \mathbf{v}_{\mathcal{V}_{\partial\Gamma_{k\ell} \cap \partial\Gamma_{k\ell}}}^{k,-}]. \tag{19}$$

In the non-symmetric approach we have the relation (19) for the trial functions and the relation (18) for the test functions.

Remark 2. We note that in the implementation, the matrix $(C_1^-)^{-1}$ will be never assembled. To compute the action of $(C_1^-)^{-1}$ on a vector, one has to solve a mass matrix system per interface. If dual Lagrange multipliers [7] are used, the algorithm is even more efficient since the matrix C_1^- is diagonal and its inversion is for free.

If A denote the block stiffness matrix, the matrix form is

$$(Q^+)^t A \, Q^+ \, \mathbf{U} = (Q^+)^t \, \mathbf{F}\,, \qquad \text{for problem (15)},$$

$$(Q^-)^t A \, Q^- \, \mathbf{U} = (Q^-)^t \, \mathbf{F}\,, \qquad \text{for problem (16)},$$

$$(Q^+)^t A \, Q^- \, \mathbf{U} = (Q^+)^t \, \mathbf{F}\,, \qquad \text{for problem (17)}.$$

Note that Q^+ and Q^- are block matrices built from $Q_{k\ell}^+$ and $Q_{k\ell}^-$. Within the Galerkin approaches (15) and (16), the final system has a symmetric and positive definite matrix. The system can be easily solved by a preconditioned Conjugate Gradient method. We note that there is a unique coupling matrix to compute but the results with these approaches are not optimal. Within the Petrov-Galerkin approach, the final system has a non-symmetric matrix. In this case, we use a Bi-Conjugate Gradient method. We note that there are two coupling matrices to compute but the results within this approach are optimal.

Thanks to the mortar element method philosophy, in the adopted iterative procedure, residuals can be computed in parallel as observed in [1]. In particular, since we do not want to deal with the assembled matrix $Q^t A Q$, we have to work with the "objects" $Q_{k\ell}$, A_k ($k, \ell = 1, \ldots, \mathcal{K}$) separately. For a vector $\mathbf{v} \in \mathbb{R}^n$ with $n = \sum_{k=1}^{\mathcal{K}} m^k - \sum_{\Gamma_{k\ell} \subset S} n_{k\ell}$, where $n_{k\ell}$ is the number of nodes in the interior of the non-mortar side $\Gamma_{k\ell}$, the matrix-vector product $Q^t A Q \mathbf{v}$, where Q is Q^+ or Q^- according to the system we solve, can be done in three steps:

($step_1$) we compute $\mathbf{q} = Q\mathbf{v}$ with $\mathbf{q} \in \mathbb{R}^s$ with $s = \sum_{k=1}^{\mathcal{K}} m^k$, separately on each non-mortar interface $\Gamma_{k\ell}$;

($step_2$) we compute $\mathbf{p} = A\mathbf{q}$ separately in each subdomain Ω_k;

($step_3$) we compute $\mathbf{w} = Q^t \mathbf{p}$ with $\mathbf{w} \in \mathbb{R}^n$, separately on each non-mortar interface $\Gamma_{k\ell}$.

In this three-step product, the most expensive step is the second one due to the size of the subdomain matrices; the products involving the coupling matrix are not expensive at all (the number of nodes, and consequently of the unknowns, on S is inferior to that in the adjacent volumes and the Lagrange multiplier supports on S are rather small). As a consequence, the iterative solution of the system in the three cases is accomplished with a sensible gain in time and without involving additional memory space.

5   Numerical Results in 3D

We consider the configuration presented in Figure 4 (left), where

$$\Omega = \Omega_1 \cup \Omega_2 = (0,2) \times (0,1) \times (0,1) \; ; \; \Gamma = \{(1,y,z) \,|\, y, z \in (0,1)\} \,,$$

$$\Gamma_D = ABCD \cup EFGH \qquad\qquad ; \; \Gamma_N = \partial\Omega \setminus \Gamma_D \,.$$

Since $\partial\Gamma \cap \partial\Gamma_N \neq \emptyset$, we have a decomposition with "cross-points", i.e. the unknowns that are associated to all the points on $\partial\Gamma \cap \partial\Gamma_N$, even if they belong to the slave side, they are real degrees of freedom.

The right hand side f and the boundary conditions for $-\Delta u = f$ are chosen such that the exact solution is given by

$$u(x,y,z) = \cos(\pi y) \, \cos(\pi z) \, [\, 2\,x - x^2 + \sin(\pi x)\,]$$

and in Figure 4 (right) we present its behavior on the interface Γ. The normal derivative on the interface of the analytical solution is given by

$$\partial_n u(y,z) = -\pi \, \cos(\pi y) \, \cos(\pi z)$$

where $n = (1,0,0)^t$ and its value will be used later to evaluate the consistency error on the slave part. We note that the discrete Lagrange multiplier

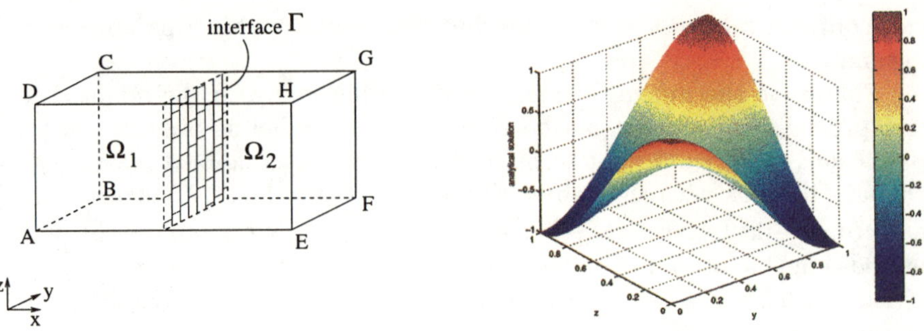

Fig. 4. Computational configuration and analytical solution on the interface.

approximates the flux of the solution across the interface Γ. We use low order quadrature formulas. On the reference triangle of vertices $(0,0), (1,0), (0,1)$, we take the quadrature points $(\frac{2}{3}, \frac{1}{6}), (\frac{1}{6}, \frac{1}{6}), (\frac{1}{6}, \frac{2}{3})$ with weights equal to $\frac{1}{3}$.

Table 1. Error in L^2 norm with two cubes. A Petrov-Galerkin approach is adopted.

Nodes	Error in Ω_1	Error in Ω_2	Global error
125/64	0.10649	0.04576	0.11591
343/125	0.06009	0.03183	0.06801
1000/512	0.02954	0.01418	0.03276
2197/1331	0.01750	0.08101	0.01928

The global error in the L^2 norm is presented in Figure 5 together with the asymptotic order $O(h^2)$ and the detail is given in Table 1. As it can be remarked, the computed error is in good agreement with the theoretical one (the two slopes are parallel).

Table 2. Error in H^1 norm with two cubes. A Petrov-Galerkin approach is adopted.

Nodes	Error in Ω_1	Error in Ω_2	Global error
125/64	1.10769	0.61424	1.26660
343/125	0.81070	0.50506	0.95516
1000/512	0.57225	0.32739	0.65928
2197/1331	0.44235	0.24082	0.50366

The global error in the H^1 norm is presented in Figure 6 together with the asymptotic order $O(h)$ and the detail is given in Table 2. Again, the computed error is in good agreement with the theoretical one.

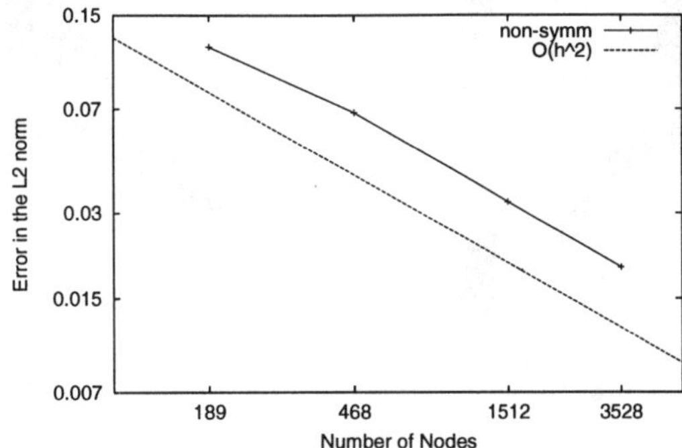

Fig. 5. Global error in L^2 norm with two cubes adopting a Petrov-Galerkin approach.

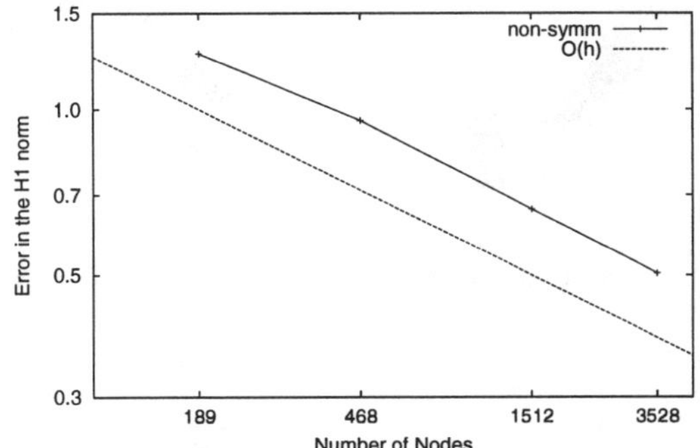

Fig. 6. Global error in H^1 norm with two cubes adopting a Petrov-Galerkin approach.

In Figures 7 and 8, we present the numerical solution on the interface Γ computed on the master and slave sides discretized with non-matching grids when a Petrov-Galerkin approach is considered. Note that already with few mesh nodes per subdomain, the computed solution shown in Figure 7 is close to the analytical one presented in Figure 4 (right); as long as we increase the number of mesh nodes, the two computed functions given in Figure 8 become closer to the analytical one.

Now we check, on the interface Γ, that the Galerkin approach for problem (12) gives a bad approximation error when $V_h = V_h^{-+}$ and a bad consistency error when $V_h = V_h^{--}$. For the best approximation error on the interface, of asymptotic order $O(h^{3/2})$, we compare the Petrov-Galerkin approach and the Galerkin one with $V_h = V_h^{--}$ in Figure 9 and $V_h = V_h^{-+}$ in Figure 10, respectively. As predicted, the Galerkin approach with $V_h = V_h^{-+}$ is not optimal for what concerns this error.

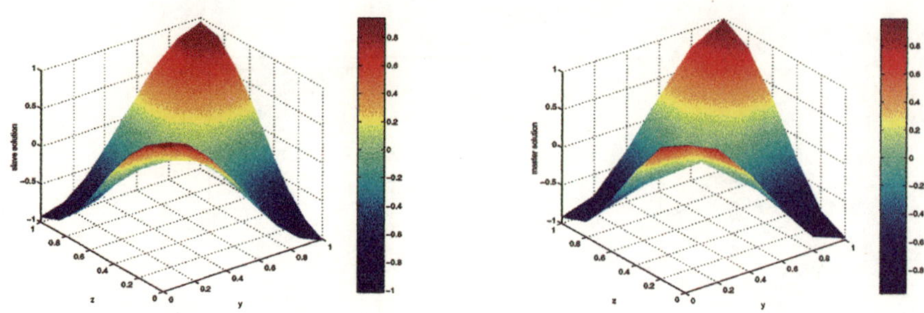

Fig. 7. Slave and master solutions with resp. 49 and 25 mesh points on Γ.

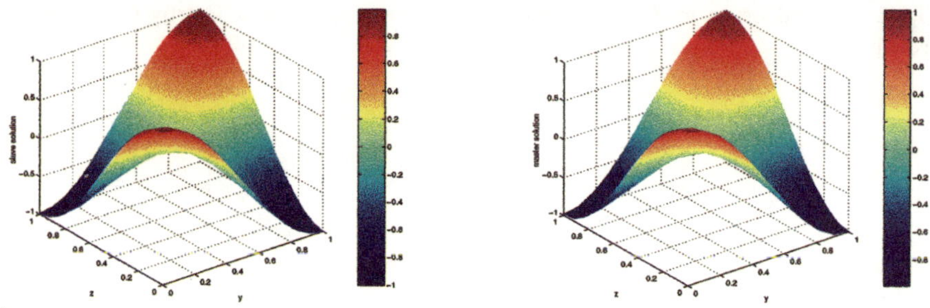

Fig. 8. Slave and master solutions with resp. 149 and 121 mesh points on Γ.

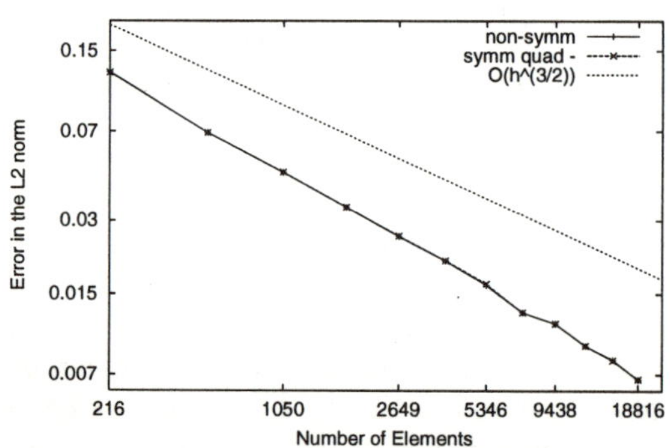

Fig. 9. Best approximation error on Γ for the Galerkin approach with $V_h = V_h^{--}$ and the Petrov-Galerkin one.

For the consistency error on the interface, of asymptotic order $O(h^{1/2})$, we compare the Petrov-Galerkin approach and the Galerkin one with $V_h = V_h^{--}$ in Figure 11 and $V_h = V_h^{-+}$ in Figure 12, respectively. As predicted, the Galerkin approach with $V_h = V_h^{--}$ is not optimal for this error.

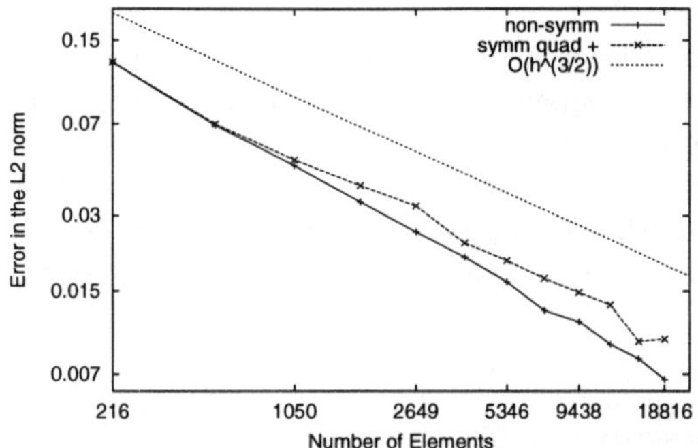

Fig. 10. Best approximation error on Γ for the Galerkin approach with $V_h = V_h^{-+}$ and the Petrov-Galerkin one.

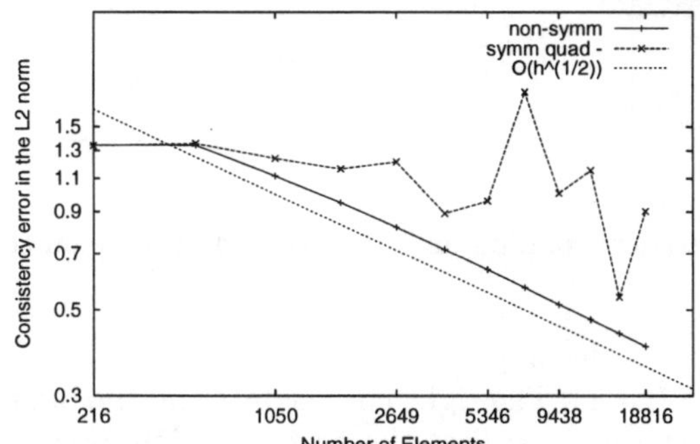

Fig. 11. Consistency error on Γ for the Galerkin approach with $V_h = V_h^{--}$ and the Petrov-Galerkin one.

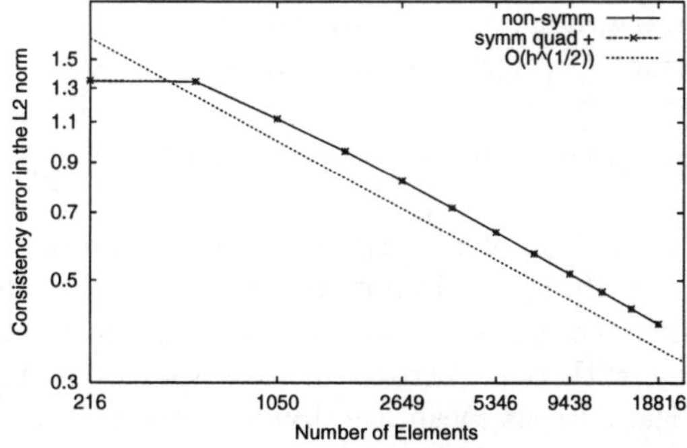

Fig. 12. Consistency error on Γ for the Galerkin approach with $V_h = V_h^{-+}$ and the Petrov-Galerkin one.

6   Numerical Results in 2D

In this section, we present some numerical results in two dimensions obtained by using the mortar element method with Lagrange multiplier spaces based on a dual basis [7]. In such a case, a biorthogonality relation between the nodal basis functions of these spaces and the finite element trace spaces holds. A main advantage of these Lagrange multiplier spaces is that the locality of the support of the nodal basis functions of the constrained space can be preserved.

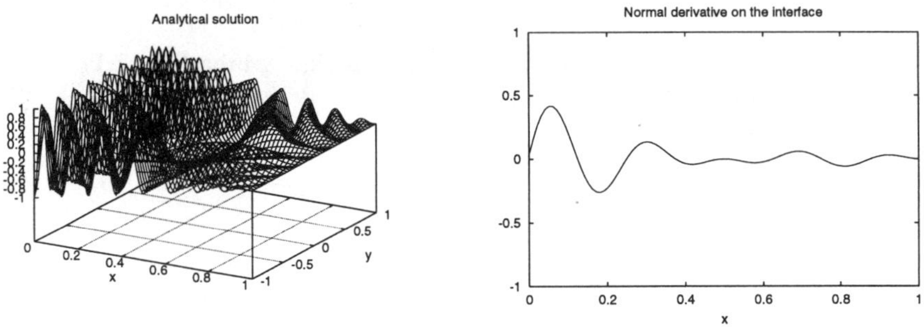

Fig. 13. Analytical solution and its normal derivative on the interface.

The unit square is decomposed into two rectangles $\Omega_1 := (0,1) \times (0,0.5)$ and $\Omega_2 := (0,1) \times (0.5,1)$. The right hand side f and the boundary conditions for $-\Delta u = f$ are chosen such that the exact solution, presented in Figure 13 (left), is given by

$$u(x,y) = (1-x)\,\cos(50(x - \frac{1}{2})y)\,.$$

Then, the discrete Lagrange multiplier is an approximation of the flux across the common interface $\Gamma = \{(x,0.5)\,|\,x \in (0,1)\,\}$ that is

$$\partial_n u(x) = -50(1-x)(x - \frac{1}{2})\sin(25(x - \frac{1}{2}))$$

which is given as a function of x in Figure 13 (right). Here, we have fixed the normal direction $n = (0,1)^t$. In Figure 14 we present the adopted decomposition together with the coarse mesh and the numerical solution computed on the finest mesh with our Petrov-Galerkin approach. At each level, the quadrilaterals of the current mesh are divided by 4 up to a level where the global number of elements is 688128.

As in the 3D case, we use low order quadrature formulas. On the reference interval $[0,1]$, we take the three Gaussian quadrature points 0.11270167, 0.5, 0.88729833 and the corresponding weights 5/18, 4/9, 5/18. In the following

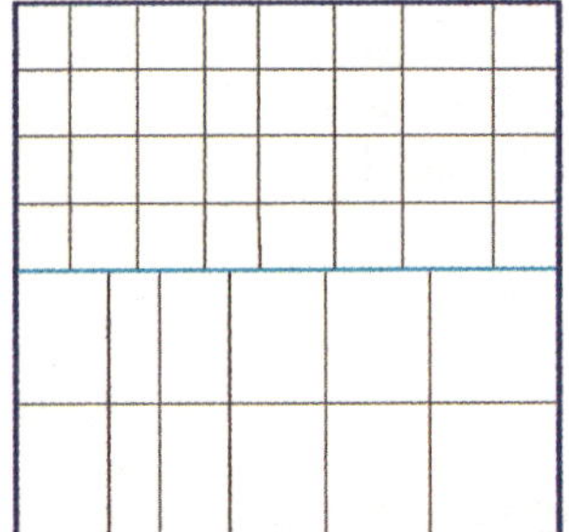 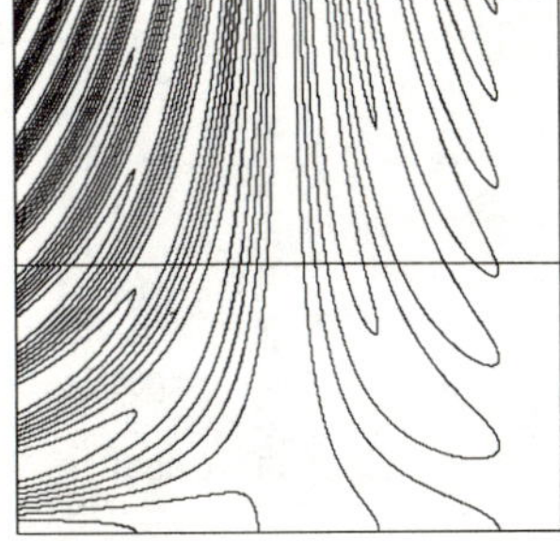

Fig. 14. Coarse mesh of quadrilaterals in the two-domain configuration and numerical solution computed on the finest grid composed of 688128 elements.

figures, the dotted line indicates the asymptotical order of the mortar element method. The solid line is obtained in the non-symmetric case and the dotted line with stars is obtained in the symmetric case if the quadrature formula is applied only on the master side. The three figures are connected with the error in the energy norm, L^2 norm and the mesh dependent Lagrange multiplier norm, the latter given by $(\sum_{e \in \Gamma} h_e \|\mu_\Gamma - \frac{\partial u}{\partial n}\|_{0,e}^2)^{1/2}$, where e denotes an edge on the slave side of Γ. The non-symmetric Petrov-Galerkin approach yields in the energy and L^2 norm asymptotically the same order as predicted by the theory for exact integration. This is not the case for the symmetric approach with $V_h = V_h^{-+}$.

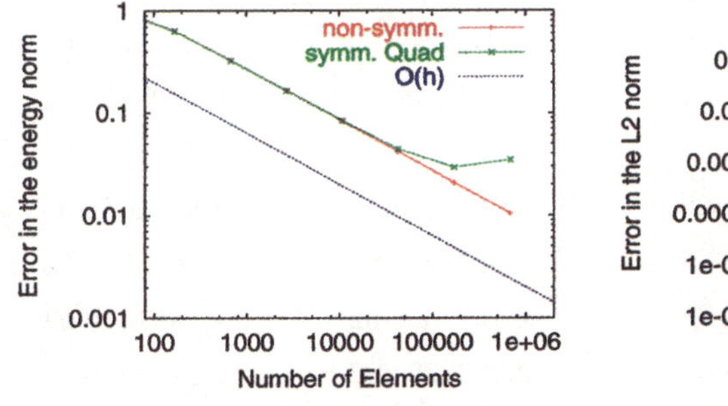 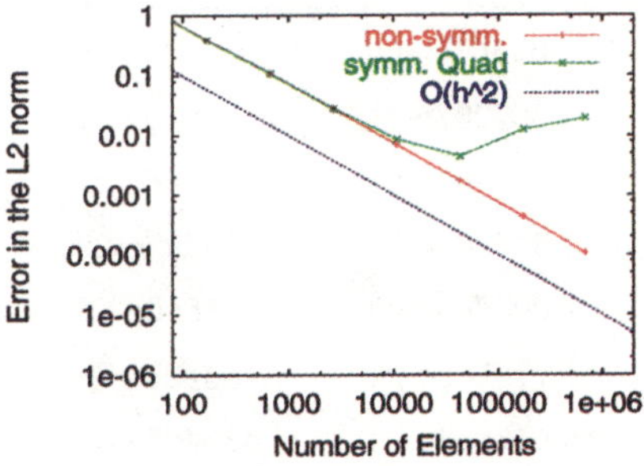

Fig. 15. Relative error in energy (left) and L^2 (right) norm for the Galerkin approach with $V_h = V_h^{-+}$ and the Petrov-Galerkin one.

In Figure 16, we show the computed errors in the mesh dependent Lagrange multiplier norm. From the point of theory, we expect to find an $O(h)$ behavior. However, we observe asymptotically $O(h^{3/2})$. This corresponds to the best approximation error of the Lagrange multiplier space and can be shown under the assumption that the error in the energy norm is asymptotically equally distributed.

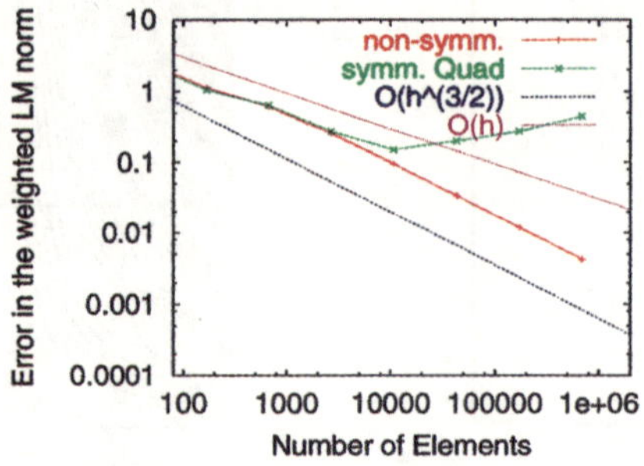

Fig. 16. Error in the mesh dependent Lagrange multiplier norm for the Galerkin approach with $V_h = V_h^{-+}$ and the Petrov-Galerkin one.

In Table 3, we report the error in the energy and L^2 norm within the exact integration and Petrov-Galerkin approaches. We remark that the values in columns 2 and 4 are very close to the one in columns 1 and 3, stating once again the optimality of the non-symmetric approach. There is no significant difference in the obtained accuracy. The Petrov-Galerkin approach which is considerably cheaper gives the same qualitative and quantitative results as the symmetric mortar based on exact integration.

Table 3. Relative error for exact integration and Petrov-Galerkin approach

ex. int. L^2 norm	P.-G. L^2 norm	ex. int. energy norm	P.-G. energy norm
1.564561e+00	1.590183e+00	1.021536e+00	1.022896e+00
3.867007e-01	3.891088e-01	6.427550e-01	6.430641e-01
1.087453e-01	1.087666e-01	3.283115e-01	3.283159e-01
2.813904e-02	2.813497e-02	1.667178e-01	1.667166e-01
7.097746e-03	7.133550e-03	8.364393e-02	8.364368e-02
1.778188e-03	1.788415e-03	4.185135e-02	4.185134e-02
4.447664e-04	4.485467e-04	2.092929e-02	2.092929e-02
1.112059e-04	1.114683e-04	1.046497e-02	1.046496e-02

Finally, Table 4, we compare the error in the weighted Lagrange multiplier norm for the four different approaches. Note that the one based on exact integration is the most expensive one. The results for the symmetric approach with $V_h = V_h^{-+}$ are not good and this is confirmed by the values displayed in column 2 of Table 4. On the other hand, the values in column 3 of Table 4 for the symmetric approach with $V_h = V_h^{--}$ are rather good but less than the ones in column 4 of the same table, that are related to the Petrov-Galerkin approach. As a consequence, it can be remarked that applying the quadrature rule only on the slave side gives surprisingly good results for this example. More numerical tests have to be carried out to illustrate the difference between the symmetric approach with $V_h = V_h^{--}$ and the Petrov-Galerkin approach. It seems that the Lagrange multiplier norm is the most sensitive norm to measure the influence of the consistency error.

Table 4. Error in the weighted Lagrange multiplier norm.

Exact integration	Galerkin (V_h^{-+})	Galerkin (V_h^{--})	Petrov-Galerkin
2.572471e+00	2.536110e+00	2.713026e+00	2.692954e+00
1.024982e+00	1.053929e+00	1.179669e+00	1.126765e+00
6.150985e-01	6.347660e-01	6.142472e-01	6.178719e-01
2.615208e-01	2.760115e-01	2.618566e-01	2.615416e-01
9.573677e-02	1.511278e-01	9.606547e-02	9.581336e-02
3.388482e-02	2.023391e-01	3.417012e-02	3.390462e-02
1.193008e-02	2.779688e-01	1.237064e-02	1.193274e-02
4.204846e-03	4.476061e-01	4.804128e-03	4.204771e-03

References

1. Ben Belgacem F., Maday Y. (1994) A Spectral Element Methodology Tuned to Parallel Implementations, CMAME, **116**, 59-67.
2. Ben Belgacem F., Maday Y. (1997) The mortar element method for three-dimensional finite elements, M^2AN, **31**, 289-302.
3. Bernardi C., Maday Y., Patera A. (1994) A new nonconforming approach to domain decomposition: the mortar element method, in Nonlinear Partial Differential Equations and Their Applications, H. Brezis and J.L. Lions, eds. Pitman, 13–51.
4. Braess D., Dahmen W. (1998) Stability estimates of the mortar finite element method for 3-dimensional problems, East-West J. Num. Math., **6**, 249-263.
5. Cazabeau L., Lacour C., Maday Y. (1997) Numerical quadratures and mortar methods, Computational Science for the 21st Century, John Wiley and Sons, 119–128.
6. Lions J.L., Magenes E. (1972) Non-homogeneous Boundary Value Problems and Applications I, Springer-Verlag, New York.
7. Wohlmuth I.B. (2001) Discretization methods and iterative solvers based on domain decomposition, Lecture Notes in Computational Sciences and Engineering, **17**, Springer.

Portable Efficient Solvers for Adaptive Finite Element Simulations of Elastostatics in Two and Three Dimensions

Andrew C. Bauer[1], Swapan Sanjanwala[1], and Abani K. Patra[1]

University at Buffalo, State University of New York, Amherst NY 14260

Abstract. Adaptive finite element methods (FEM), generate linear equation systems that require dynamic and irregular patterns of data storage, access and computation, making their parallelization very difficult. Moreover, constantly evolving computer architectures often require new algorithms altogether. We describe here several solvers for solving such systems efficiently in two and three dimensions on multiple parallel architectures.

1 Introduction

In recent years, very efficient parallel iterative solvers have been developed using domain decomposition ideas. Solver libraries based on these developments are also widely available (see for e.g. Balay et. al. [1]). However, in our work developing parallel adaptive hp finite element schemes [3,2] we have encountered two major difficulties. Firstly, for adaptive hp FEM schemes which give rise to linear systems that are irregularly sparse, and, poorly conditioned, there exist few solution algorithms that pose no restrictions on the mesh and partitioning and are easy to implement. A second difficulty is encountered in the design of a solver for *efficient* use across a large class of parallel architectures. The choice of solution method for best efficiency is usually architecture, discretization and problem dependent. Thus, one must either change algorithms (code) based on architecture or design a solver that is able to adjust to the architecture. Repeated code changes of this type are neither easy nor affordable. Clearly, efficient solvers meeting the criteria of portability across parallel architectures need to be developed.

The efficiency of domain decomposition based solvers on classical distributed memory parallel machines (homogeneous processors with equal memory and processing capabilities) for both h version and p version finite element discretizations has been shown by several researchers including Bramble, Pasciak, Schatz, Dryja, Widlund, Mandel, Pavarino [4,6,7,9,10,13,14]. In all cases polylogarithmic bounds for the condition number in terms of the mesh parameters have been obtained. For the hp-version, Ainsworth [11], Guo and Cao [15], and Oden, Patra and Feng [12] have introduced different domain decomposition based preconditioners in two and three dimensions. However, none of the three dimensional versions introduced to date, have been backed up with implementation in a large scale parallel code. Moreover, most of the preconditioners proposed to date require the existence of an *actual* coarse finite element grid in which the finer adaptive hp mesh is embedded. The elements of the coarse grid so defined constitute the partitioning for parallel computing. The task of obtaining such a proper partitioning of the mesh in which each partition is a finite element is difficult for geometries and partitionings used in many applications. Furthermore, these algorithms are not very efficient when implemented on the heterogeneous parallel architectures (clusters of machines with different processing powers and memories) that are increasingly popular.

We describe in this paper the development and implementation of solvers designed specifically for adaptive hp methods. We will also describe here an approach to designing solvers that are efficient across multiple parallel architectures.

2 Model Problem and Finite Element Approximation

2.1 Elastostatics

$$\sigma_{ij,j} + f_i = 0 \text{ in } \Omega \subset \mathbb{R}^n, \ i,j = 1,..,n, n = 2,3, \tag{1}$$

$$\sigma_{ij} = 2\mu\varepsilon_{ij} + \lambda\varepsilon_{kk}, \ \varepsilon_{ij} = \frac{1}{2}(u_{i,j} + u_{j,i}),$$

$$u = \ g \text{ on } \partial\Omega_D, \quad \sigma.n = t \text{ on } \partial\Omega_N,$$

where σ_{ij} is the Cauchy stress, f_i is the body force, ε_{ij} is the corresponding strain, μ, λ are the Lamè parameters, $u = \{u_i\}$ is the displacement field, and $\partial\Omega_D$ is the part of the boundary on which Dirichlet boundary conditions are applied and $\partial\Omega_N$ is the part of the boundary on which Neumann boundary conditions are applied. The standard method of multiplying with an appropriate test function and integrating by parts may be applied to obtain the weak form:
Find $u \in \mathbf{V}$ such that

$$\mathbf{B}(u,v) = \mathbf{L}(v) \quad \forall \, v \in \mathbf{V}, \tag{2}$$

where $\mathbf{V} = \{v : v \in H^1(\Omega)^n, v = 0 \text{ on } \partial\Omega_D\}$, and $\mathbf{B}(u,v) = \int_\Omega \sigma_{ij}:\varepsilon_{ij} \, d\Omega$ is the bilinear form on $\mathbf{V}$, and $\mathbf{L}(v) = \int_\Omega fv \, d\Omega + \int_{\partial\Omega_N} tv \, ds$ is a continuous linear functional in $\mathbf{V}$.

2.2 Finite Element Spaces and Decomposition

We now introduce approximation spaces $\mathbf{V^{hp}} \subset \mathbf{V}$ that describe the finite element discretization and a partitioning of the space for parallel computing.

S1. Ω is a connected domain affine equivalent to a union of polygons or polyhedra, and can be represented as the union of N_D subdomains Ω_I such that

$$\overline{\Omega}_I \cap \overline{\Omega}_J = \emptyset \quad I \neq J, \text{ and } \overline{\Omega} = \cup_{I=1}^{N_D} \overline{\Omega}_I$$
$$\text{further let } H_I = diam(\Omega_I), H = \max_I H_I.$$

Each of these subdomains is further partitioned into a finer mesh of quadrilaterals or bricks (the finite elements) $\omega_K^I, K = 1, 2, ..., N_I,$

$$\overline{\Omega}_I = \cup_{K=1}^{N_I} \omega_K^I \ , \quad \omega_K^I \cap \omega_L^I = \emptyset, K \neq L$$

and

$$h_K^I = diam(\omega_K^I) \ , \quad h_I = \min_K h_K^I \ , \quad h = \min_I h_I.$$

This defines a two level hierarchy of partitionings $--$ a *coarse partitioning* and a *fine partitioning* both of which are assumed to be quasi-uniform of size $O(H_I)$ and $O(h_I)$ for each I (see Figure 1).

S2. Each element ω_K^I is the affine image of a master element $\overline{\widehat{\omega}} = [-1, 1]^n$, $n = 2, 3$. The master element has 9-nodal points in two dimensions: four vertices, $\widehat{\mathbf{a}}_1, \widehat{\mathbf{a}}_2, \widehat{\mathbf{a}}_3, \widehat{\mathbf{a}}_4$, four edge nodes $\widehat{\mathbf{e}}_1, \widehat{\mathbf{e}}_2, \widehat{\mathbf{e}}_3, \widehat{\mathbf{e}}_4$, and a centroid node, $\widehat{\mathbf{c}}_0$. In three dimensions the master element has 27-nodes: eight vertices $\widehat{\mathbf{a}}_1, .., \widehat{\mathbf{a}}_8$, twelve edges $\widehat{\mathbf{e}}_1, .., \widehat{\mathbf{e}}_{12}$, six face nodes $\widehat{\mathbf{f}}_1, .., \widehat{\mathbf{f}}_6$, and a centroid node $\widehat{\mathbf{c}}_0$ (see Figure 2). Corresponding polynomial shape functions in three dimensions are:

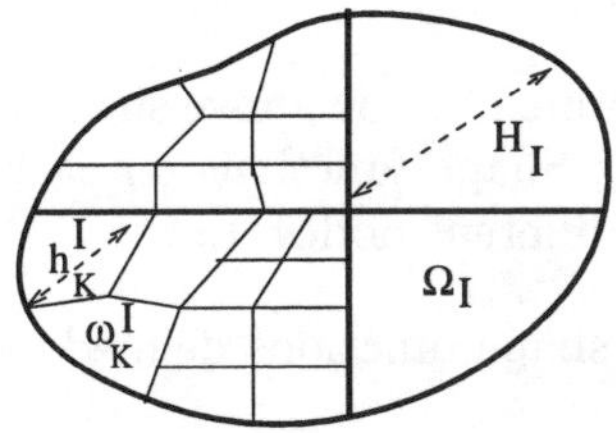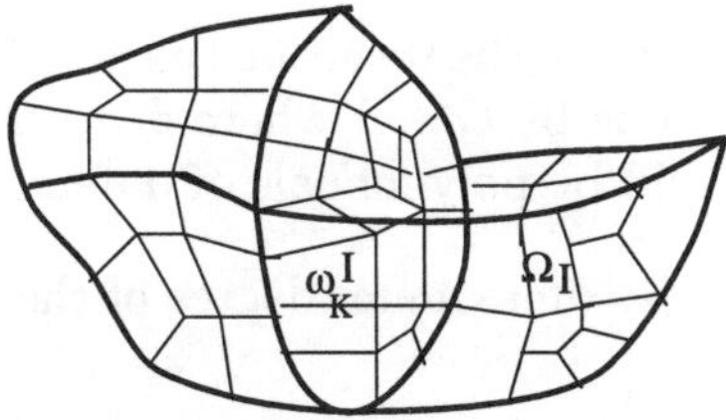

Fig. 1. Partitioning into sub-domains and meshes

- *Vertex Functions:* These are the standard trilinear functions

$$\widehat{\psi}_i(\xi, \eta, \zeta) = \frac{1}{2^n}(1 \pm \xi)(1 \pm \eta)(1 \pm \zeta) \qquad i = 1, 2, .., 8 \quad (\xi, \eta, \zeta) \in [-1, 1]^3$$

$$\widehat{\psi}_i(\widehat{\mathbf{a}}_j) = \delta_{ij} \qquad 1 \leq i, j \leq 8.$$

- *Edge Functions:* A polynomial function of degree $\widehat{p}_e, e = 1, 2, 3, .., 12,$ vanishing at the vertices, is assigned to the edge nodes, e_i e.g.

$$\widehat{\gamma}_i^{\widehat{p}_1}(\xi, \eta, \zeta) = \frac{-1}{4}(1 + \eta)(1 + \zeta)\rho_i(\xi) \ i = 2, ..., \widehat{p}_1;$$

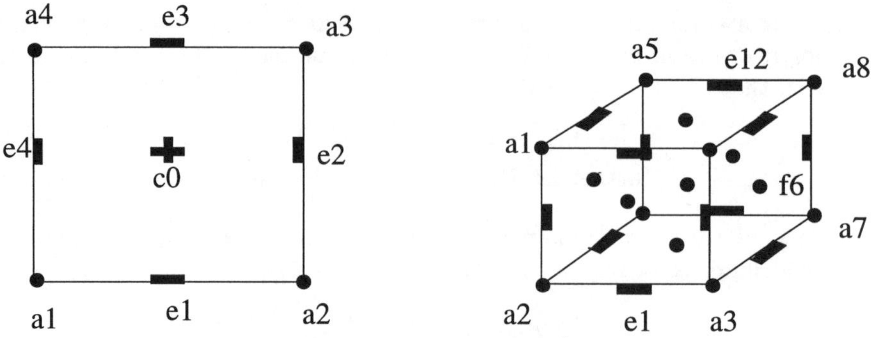

Fig. 2. The master elements

here

$$\rho_i(\xi) = \sqrt{\frac{2\widehat{p}_1 - 1}{2}} \int_{-1}^{\xi} P_{\widehat{p}_1 - 1}(s)ds,$$

with $P_{\widehat{p}_1 - 1}$ the Legendre polynomial of degree $(\widehat{p}_1 - 1)$.

- *Face Functions:* Corresponding to node $\widehat{\mathbf{f}}_{\mathbf{i}}$, e.g.

$$\widehat{f}_{ijk}(\xi, \eta, \zeta) = \rho_i(\xi)\rho_j(\eta)(1 - \zeta) \quad 2 \leq i, j \leq \widehat{p}_f.$$

- *Interior(Bubble) Functions:* Corresponding to node $\widehat{\mathbf{c}}_{\mathbf{0}}$, introduce the functions

$$\widehat{b}_{ijk}(\xi, \eta, \zeta) = \rho_i(\xi)\rho_j(\eta)\rho_k(\zeta) \quad 2 \leq i, j \leq \widehat{p}_b.$$

Denote the resulting space of functions spanned by the above shape functions by $Q^{\widehat{p}}(\widehat{\omega})$ where $\widehat{p} = \max\{\widehat{p}_e, \widehat{p}_f, \widehat{p}_b\}$. Shape functions on each ω_K^I will be polynomials of different degree at different nodes. Let

$$p_K^I = \text{maximum degree of the polynomial shape functions defined on } \omega_K^I,$$

and

$$p_I = \max_K p_K^I, \qquad p = \max_I p_I.$$

If $F_K^I : \widehat{\omega} \to \omega_K^I$ is an affine invertible map from $\widehat{\omega}$ to element ω_K^I, then corresponding shape functions for ω_K^I are of the form

$$\psi_{Ki}^I = \widehat{\psi}_i \circ (F_K^I)^{-1},$$
$$\gamma_{Ki}^{I_{p_j}} = \widehat{\zeta}_i^{\widehat{p}_j} \circ (F_K^I)^{-1}, \text{ etc.},$$

and the restriction u_K^I of $u_{hp} \in \mathbf{V}^{\mathbf{hp}}$ to ω_K^I is of the form,

$$u_K^I = \widehat{u}_K \circ (F_K^I)^{-1}.$$

S3. Interelement constraints are imposed so that, globally, $\mathbf{V}^{\mathbf{hp}} \subset \mathbf{C}^{\mathbf{0}}(\overline{\mathbf{\Omega}})$.

S4. We now introduce a subdivision of the *vertex, edge* and *face* functions. We will denote *vertex* functions based on nodes $\mathbf{a}_i$ such that $\mathbf{a}_i \in \Omega_I \cup \Omega_J, I \neq J$ as *node* functions. *Vertex* functions other than the *node* functions (i.e. *vertex* functions in the interior of the subdomains) will continue to be denoted as vertex functions. Similarly, *edge* functions with support on the subdomain interface will be denoted *side* functions and *face* functions with support on the subdomain interface will be called *interface* functions.

The result of ($S4$.) is a decomposition of the finite element space

$$\mathbf{V^{hp}} = \mathbf{\Phi_N} + \mathbf{\Phi_S} + \mathbf{\Phi_I} + \mathbf{\Phi_V} + \mathbf{\Phi_E} + \mathbf{\Phi_F} + \mathbf{\Phi_B}.$$

This decomposition of the finite element space will induce a decomposition of the set of unknowns, the stiffness matrix and the right hand side. Thus, (2) may be rewritten in matrix form as:

$$B_{(u,v)} = \begin{bmatrix} NN & NS & NI & NV & NE & NF & NB \\ SN & SS & SI & SV & SE & SF & SB \\ IN & IS & II & IV & IE & IF & IB \\ VN & VS & VI & VV & VE & VF & VB \\ EN & ES & EI & EV & EE & EF & EB \\ FN & FS & FI & IV & IE & FF & FB \\ BN & BS & BI & BV & BE & BF & BB \end{bmatrix} \begin{Bmatrix} u_N \\ u_S \\ u_I \\ u_V \\ u_E \\ u_F \\ u_B \end{Bmatrix} = \begin{Bmatrix} f_N \\ f_S \\ f_I \\ f_V \\ f_E \\ f_F \\ f_B \end{Bmatrix}. \quad (3)$$

Note that the blocks are loosely organized in increasing order of the locality of support. The BB block, ordered last, has support only in one element, while the FF block, ordered next, has support only in two neighboring elements in the same partition. The EE and VV have support on larger sets of elements in the same partition while the $NN, SS,$ and II have support spanning more than one partition.

S5. Based on the partitioning of $\Omega = \cup_{I=1}^{N_D} \Omega_I$ we identify a subset of the set of *nodes* and *sides* as a set of *cross-points* (see Figure 3). These are functions related to the *nodes* and *sides* belonging to more than *two* subdomains. Thus, $N = \{N^c, N^s\}$ and $S = \{S^c, S^s\}$.

3   Solution Algorithms

Linear equation systems like (3) may be solved by either a direct or an iterative method. The direct methods are usually variants of the Gaussian elimination process in which the matrix is factorized into a triangular form followed by a backsubstitution. Parallelization of such methods on computers with distributed memory computers with high costs of communication is complex and usually not very efficient because of the high volume of communication required. Another drawback of such systems is the large "fill-in"

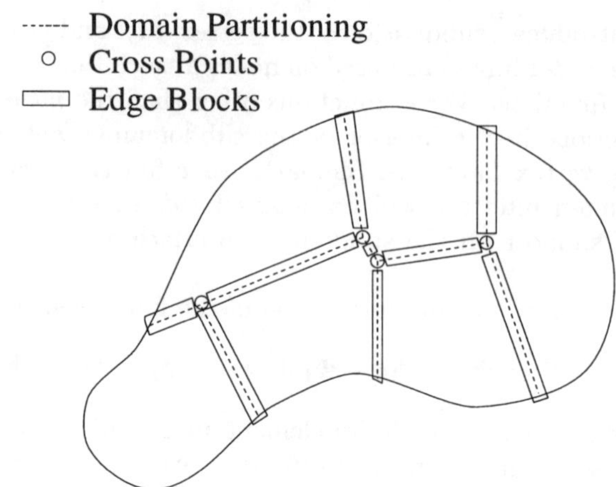

Fig. 3. Cross-points and edges of a partitioned domain

(additional storage) generated during the solution process. Iterative solvers, where the main cost is incurred in a matrix–vector multiplication, are often preferred as lower cost alternatives. The matrix–vector product is intrinsically easier to parallelize and requires very little communication overhead. Further, they require significantly less memory/disk storage since no additional storage is required in the solution process. However, the weakness of such solvers is their lack of robustness (no guarantee of convergence can be provided for all systems) and often slow convergence for poorly conditioned systems. Nevertheless, with *proper preconditioning techniques* these solvers can be made efficient and quite robust. Based on the properties noted above and numerical experience implementing a wide variety of both types of solvers we make the following observations:

1: The cost of communication relative to computation, availability and access speeds of memory, and numerical conditioning of the system determine the relative efficiency of direct and iterative solver algorithms for large sparse systems on a particular computer.

2: On computers with low costs of communication relative to computation, sufficient memory and/or if the problem is poorly conditioned direct solvers are preferable.

3: Iterative solvers are preferable on computers with high costs of communication, small memory relative to problem size and if the problem or a preconditioned variant is well conditioned.

Thus, for optimal performance one must choose the correct algorithm for the machine and the problem.

3.1 Iterative Substructuring

Clearly, appropriate combinations of the direct and iterative solution methodologies that exploit the structure of the equation system at hand and the abilities of the computer will provide the most efficient solution methodologies. Solvers of the type espoused by [6,12–14] appear to provide the best framework for such a combination. The essential idea in these solvers is to order the linear systems in a manner such that there are many dense blocks in the sparse matrix. In the first step of the solution process these dense blocks are eliminated from the systems using a direct solver methodology (often termed substructuring or partial orthogonalization). Subsequently the remaining smaller sub-system is solved using an iterative solver of the Krylov space type with an appropriate preconditioner.

We may view the elimination process as a transformation of the finite element subspaces. For instance, we may eliminate the BB block from (3) by modifying $\{\Phi_N, \Phi_S, \Phi_I, \Phi_V, \Phi_E, \Phi_F\}$ to $\{\widetilde{\Phi}_N, \widetilde{\Phi}_S, \widetilde{\Phi}_I, \widetilde{\Phi}_V, \widetilde{\Phi}_E, \widetilde{\Phi}_F\}$ such that:

$$\mathbf{B}(u_\alpha, u_\beta) = 0, \quad \alpha = \{N, S, I, V, E, F\}, \quad \beta = \{B\}. \tag{4}$$

This process may be further expanded in a recursive process including additional components in the subset β. The next possibility is:

$$\mathbf{B}(u_\alpha, u_\beta) = 0, \quad \alpha = \{N, S, I, V, E\}, \quad \beta = \{F, B\}. \tag{5}$$

Another choice of α and β is:

$$\mathbf{B}(u_\alpha, u_\beta) = 0, \quad \alpha = \{N, S, I\}, \quad \beta = \{V, E, F, B\}. \tag{6}$$

This choice essentially eliminates all the interior unknowns on a processor.

This process corresponds to a direct solution (LU factorization and back-substitution) of the β subset of unknowns. In the iterative substructuring procedure the modified set of equations corresponding to the set α are then solved using an iterative procedure e.g. any of the Krylov space schemes. The primary issue that has then to be tackled is the preconditioning of this modified set of equations. Further, note that depending on the choices of the set α and β we end up with different mixes of direct and iterative solution methodologies. This will be key to our construction of portable solver algorithms.

3.2 Preconditioners for Iterative Substructuring

We outline here a series of preconditioners $C(u, v)$ for the reduced systems $\mathbf{B}(u_\alpha, v_\alpha) = \mathbf{L}(v_\alpha)$ outlined above.

1. **Cross-Point:** The cross-point preconditioner $C(u, v)$ has a block from all dof/shape functions along the subdomain interface that belong to more than two subdomains and a diagonal preconditioner for the rest of the

dof/shape functions. Figure 3 shows the cross-points for a partitioned domain.

$$C_{(u,v)} = \begin{cases} \mathbf{B}_{(u,v)} \ \forall u, v \in N^c, S^c \\ \mathbf{B}_{(u,u)} \ \forall u \in N^s, S^s, F. \end{cases} \tag{7}$$

2. **NN Block:** The NN block preconditioner is comprised of all of the linear shape functions on the interface and a diagonal preconditioner for the higher order shape functions. We note that the NN block can be substantial in size and requires specialized parallel solution techniques. In our application we have implemented this using the PLAPACK[16] suite of parallel direct solvers.

$$C_{(u,v)} = \begin{cases} \mathbf{B}_{(u,v)} \ \forall u, v \in N \\ \mathbf{B}_{(u,u)} \ \forall u \in \alpha \backslash N \end{cases} \tag{8}$$

3. **Edge Block:** The blocks of the edge block preconditioner are constructed from dof/shape functions which are shared by only two subdomains and the diagonals for the rest of the dof/shape functions. This is shown in Figure 3. The edge blocks consist of all dof/shape functions along the edge.

$$C_{(u,v)} = \begin{cases} \mathbf{B}_{(u,v)} \ \forall u, v \in N^s, S^s, F \\ \mathbf{B}_{(u,u)} \ \forall u \in N^c, S^c. \end{cases} \tag{9}$$

4. **Cross-Point and Edge Block:** The cross-point and edge block preconditioner is just a combination of the cross-point preconditioner and the edge block preconditioner applied in sequence.

Analysis of these preconditioners to show dependence of the condition number on mesh and partitioning parameters can be performed by extending the approach of Guo and Cao [15] and will be described in a forthcoming paper.

4   Portable Hybrid Solvers

The key idea is to introduce flexibility into the solution algorithm so that it can take advantage of the efficiencies of different solution methodologies depending on the environment i.e. machine, discretization and problem. The solver code must therefore combine iterative and direct solution methodologies and provide mechanisms for automatically combining them.

The mix is achieved by organizing unknowns into a multi-level hierarchy (Table 1, Figure 5). Higher levels in the hierarchy are closely coupled and will be usually assigned to the same block of memory. Such coupling can be revealed either from the underlying grid structure or an analysis of the graph of the grid/matrix (see for instance Plassman and Jones [19]). In our work here we will use the grid information to obtain the hierarchy. Such closely coupled blocks can be solved using a direct solution methodology (substructuring). The lower levels are then solved using an iterative solver. The iterative solver is accelerated using a preconditioner that is a direct solution at the lowest levels of the hierarchy. The number of levels, and the apportioning of levels among the substructuring, iterative solution and preconditioning can be determined for optimal efficiency at run-time for the combination of machine and problem. This decision process will be based on simple tests on memory access, computation speed, communication costs and corresponding problem specific parameters like maximal block sizes etc.

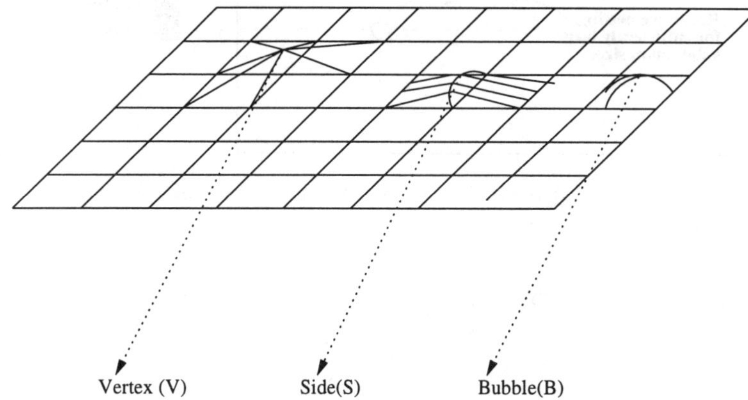

Fig. 4. Support of vertex, edge and bubble shape functions in an FEM mesh

The first step in designing the solver is to decompose the finite element space into a multi-level hierarchy using the natural hierarchy of higher order finite elements and nested partitionings. At the element level the bubble functions are non-zero only on the interior of one element and form the beginning of the hierarchy i.e. $supp(B) = int(\Omega_K)$. The edge and vertex functions have support on patches of two and four elements respectively and constitute the next level of the hierarchy. Typical 2D elements and their interactions are shown in Figure 4.

Now consider a nested partitioning of the domain as show in Figure 5. Here the partitioned subdomain ($\Omega_{I,I=1,np}$ where np is the number of processors) is further decomposed into nested subdomains (Ω_{IJ}) ($\Omega = \sum_{I=0}^{np} \sum_{J=0}^{nn} \Omega_{IJ}$). The number of nestings, nn, can be different for different subdomains. Observe that Ω_3 is further partitioned into two nested domains Ω_{3L} and Ω_{3R}. Because of this, the unknowns in the subdomain can be classified into four major groups. The bubble degrees of freedom of the elements, the unknowns associated with interior vertices and edges, unknowns which are shared by more than one nested subdomain on the level 1 interface and finally unknowns which are shared by more than one processor on the level 0 interface.

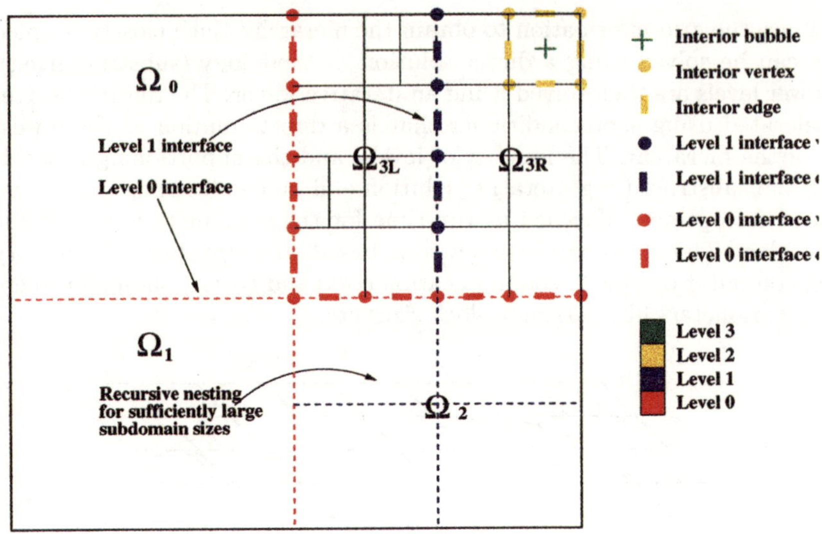

Fig. 5. Multi-level decomposition of degrees of freedom

Accordingly, the hierarchy of the unknowns of processor 3 is shown in Table 1.

Table 1. Multi-level hierarchy of unknowns associated with adaptive hp finite element methods

3.	u_B	Level 3 →	Bubble dof local to each element
2 (a).	u_{V_L} u_{E_L}	Level 2 →	Vertex, edge unknowns local to sub-partition L
2 (b).	u_{V_R} u_{E_R}	Level 2 →	Vertex, edge unknowns local to sub-partition R
1.	u_{N_1} u_{S_1}	Level 1 →	Vertex, edge unknowns shared across intraprocessor partitions
0.	u_{N_0} u_{S_0}	Level 0 →	Vertex, edge unknowns shared across processors

In the absence of such detailed mesh information, similar structure/hierarchy information for the linear systems can be generated by using either the graph of the input matrix or the graph of the grid [19].

4.1   Multi-level Substructuring and Preconditioning

Multi-level Substructuring The substructuring method outlined in 3.1 is subsequently applied to the above described hierarchy. We begin with the highest level, the bubble unknowns, and go downwards until we switch to the iterative solver at one of the lower levels, i.e. we solve (4-6) increasing the set of the unknowns β one level at a time using the hierarchy defined above. Note that the substructuring process can be stopped at any level and the rest of the problem solved using an iterative solver and an appropriate preconditioner.

Preconditioners Once the decision to switch from the direct to the iterative solver is made, a suitable preconditioner can be constructed using the same hierarchical strategy but now we start from the lowest level (i.e. level 0) and keep expanding the set to be used for preconditioning. The simplest choice can be described as:

$$\mathbf{C}_{(i,j)} = \begin{cases} \mathbf{B}_{(u_\alpha, u_\alpha)} & \alpha = \{N_0\} \\ diag(\mathbf{B}_{(u_\beta, u_\beta)}) & \beta = \{S_0, N_1, S_1, ...\} \end{cases}. \tag{10}$$

Note, that for poorly conditioned systems if the iteration counts go up the preconditioner can be expanded to include more levels. This has the effect of increasing the proportion of direct solver relative to the iterative solver.

4.2   Parallel Implementation

All the substructuring processes can be performed in parallel. Moreover, the matrix vector products in the Preconditioned Conjugate Gradient (PCG) algorithm is easily parallelized by computing each processor's contribution to the product separately and assembling the product. To save memory, in our implementation the transformation matrix and vector needed to recover the element level bubble unknowns (level 3 unknowns) will not be stored. Instead, we will reconstruct these matrices as needed. At lower levels matrices are saved, since the computation of these is expensive. Again, memory savings are achieved by reusing the storage for the substructurings among the sub-partitions of a processor.

Tuning of this solver can be achieved by adjusting the levels to be substructured and the levels to be used in the preconditioner. In the limit the whole problem is solved by the direct method (nested dissection) or iteratively. Moreover, the memory requirements of the whole solution algorithm can be regulated by controlling the number of nestings (nn) since the multiple nestings of a processor use the same storage for the interior condensation procedure. The two parameters necessary to carry out this tuning are

- number of higher levels to substructure, and,
- number of lower levels to use in the preconditioner.

Figure 6 shows a simple graphic to illustrate such a hybrid solver that can be tuned between its two extremes. Based on preprocessing data of an analyzer, an optimizer module could automatically carry out the performance tuning to give the optimal mix for the actual supercomputer. This technique ensures portability without explicit user intervention.

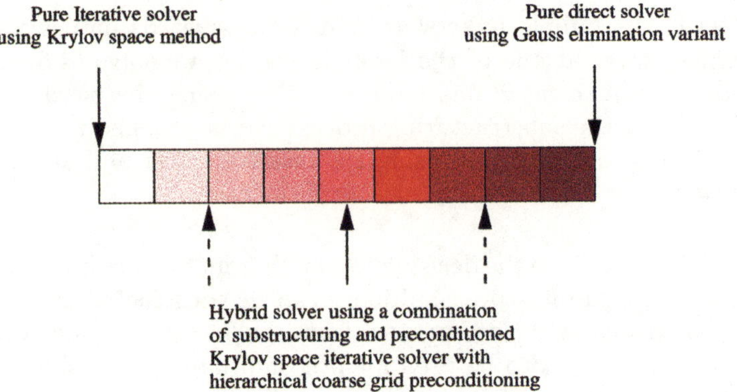

Fig. 6. Hybrid solvers

4.3   Tuned Solution Strategy

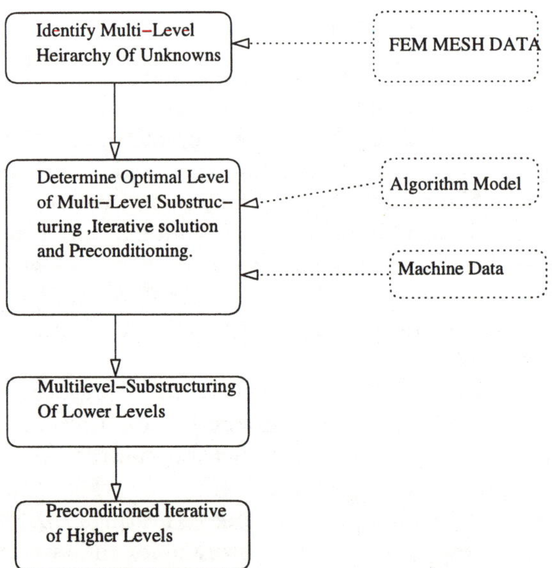

Fig. 7. Adaptive solution algorithm

Figure 7 shows the overall solution strategy. In the first preprocessing phase we identify the hierarchy of unknowns using mesh data. Then using an algorithm model (described next) and the machine data we determine the best mix of direct and iterative solvers by determining the number of levels to substructure and the number of levels to use in the iterative solver.

4.4   Performance Models

The need for good performance models for computational efficiency is emphasized by the fact that the success of this solver in minimizing solution times hinges on its ability to make accurate run time decisions on the proper solution strategy for a problem and computer. Any performance model must incorporate diverse machine characteristics and possibly the physics of the problem for accuracy.

The solution time can be expressed as a sum of computation and communication times. The model incorporates cache effects in the computation using a simple analytical approach and neglects it during communication. The total cost Γ is decomposed into the sum of computation and communication costs,

$$\Gamma = \Gamma_{comp} + \Gamma_{comm},$$

where Γ_{comp} is the computation cost and Γ_{comm} is the communication cost.

These can now be separately modeled. We have also adopted the approach that we can measure all costs in terms of time consumed in any operation. Such an approach does not account for the memory limitations. This is clearly a constrained minimization problem. However, for the purposes of our study here where we only have a few discrete variables, namely the number of levels to substructure/precondition, a formal optimization procedure is deemed unnecessary.

Computation Cost   For instance a simple model of the computational cost Γ_{comp} is

$$\Gamma_{comp} = \quad \Gamma_{substructuring} + \Gamma_{matrix-vector} + \Gamma_{preconditioning}$$

$$= \tau\left[\frac{1}{P}\sum_{j=1}^{n_{sub}}\sum_{k=1}^{m^j}(n_k)^3 + \frac{n_{iter}}{P}(N - \sum_{j=1}^{n_{sub}}\sum_{k=1}^{m^j}n_k)^2 + \right.$$

$$\sum_{j=1}^{n_{prec}}\sum_{k=1}^{m^j}(n_k)^3. \tag{11}$$

where n_{sub} is the number of levels substructured, m^j is the number of blocks at level j, n_k is the size of each block at level j, N is the total number of unknowns, n_{iter} is the number of iterations of the iterative solver and P is the number of processors. τ is the expected time for a floating point operation.

Communication Cost

$$\Gamma_{comm} = n_{iter} \times \log_2 P((N - \sum_{j=1}^{n_{sub}}\sum_{k=1}^{m^j}n_k) * \nu + \delta). \tag{12}$$

where δ is the message startup cost and ν is the cost of sending a single real number. We note that it will be difficult to predict the number of iterations

n_{iter} which is obviously dependent on n_{prec} and the problem at hand but insensitive to the computer architecture. However, if n_{prec} is appropriate the number is bounded below a hundred.

Note that this model assumes that the matrix-vector product is parallelized with 100 percent efficiency, substructuring costs are approximately $O(n^3)$ for each block of size n and the factorization of the preconditioner is not parallelized.

Determining Parameters The model described above depends on parameters that characterize a particular computer architecture. Values for these parameters β, τ and δ are selected on the basis of sample computations. For our implementation we have used matrix-vector multiplications of different problem sizes on varying number of processors to determine the parameters. Note that such data is also available directly from vendors but vendor data is usually based on peak performance numbers that may not be achieved in practice for the operating conditions and code optimizations available.

5   Numerical Results

We provide here a sample of numerical results obtained with the solvers designed here. Firstly, for the conventional iterative substructuring solver we illustrate the dependence of the number of iterations on the mesh parameters h and p for problems in both two and three space dimensions. Secondly, for the multi-level tunable solver design we illustrate the architecture dependence of the solver's efficiency and the ability of the model to pick the correct parameters for automatically tuning the solver for optimal performance regardless of platform.

5.1   Iterative Substructuring Solver Results

In the first set of results presented in Figure 8 we illustrate the behavior of the iterative substructuring solver for the preconditioners defined earlier. We note that except for the cross-point preconditioner all the others control the iteration count well as we refine the grid and/or increase the numbers of processors. In the next set of results shown in Figure 9 we illustrate the dependence on p enrichment. Here we notice that the iteration count is well controlled for all the preconditioners.

In the next set of results (Figure 10), we illustrate the performance of the solver for problems in three dimensions. We observe now that the cross-point and the edge block show increasing iteration counts with respect to h refinement. Finally, in the last set of results illustrating performance of this solver we show the iteration counts as a function of p refinement in Figure 11. We observe that the NN block and cross-point and edge block preconditioners control the number of iterations quite well.

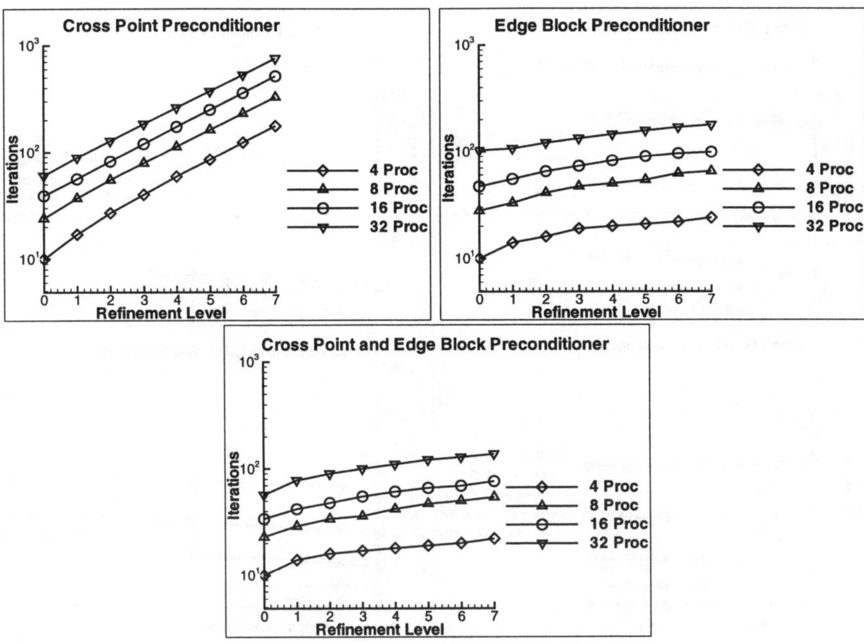

Fig. 8. Iterative substructuring solve: iteration count vs. h refinement level in two dimensions

5.2   Portable Solver Results

The primary objective of these tests is to establish

- the architecture, physics and problem size dependent nature of the optimal solution scheme,
- the ability of the model to predict the appropriate mix of direct and iterative methodologies.

5.3   Architectural Adaptability

Levels of Substructuring In the first series of tests, we investigate the dependence of the optimal number of levels of substructuring for different architectures. Figure 12 shows the optimum number of levels required for the same problem on three different architectures. Since the underlying architectures differ in the communication and computation cost, the number of levels required for optimal solution time is different. The IBM architectures have faster processors and hence lesser computation time. Here it is worthwhile to substructure larger blocks i.e. use a lesser number of levels. The trend is also noted in Figure 12 which plots the optimum levels for different problem sizes.

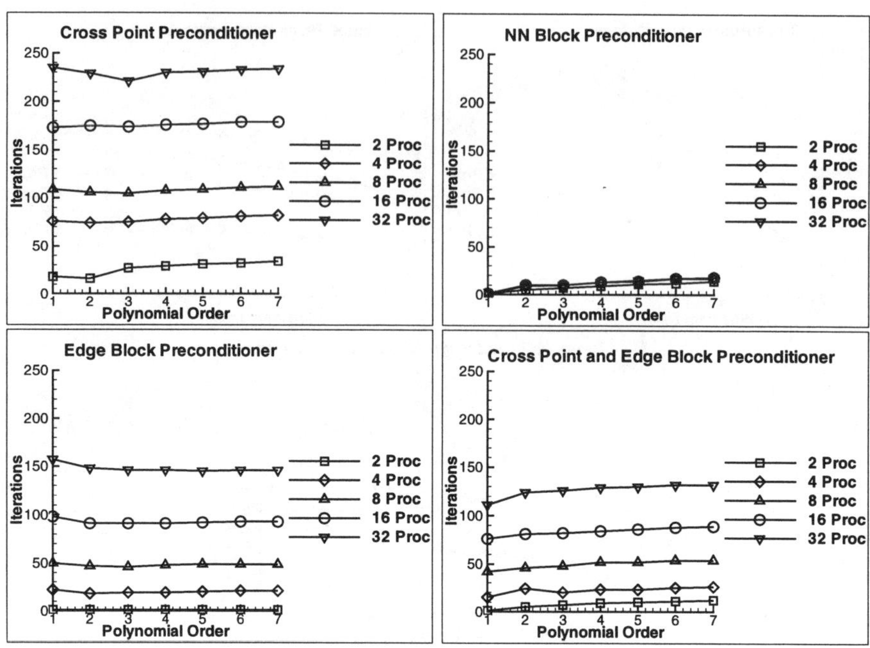

Fig. 9. Iterative substructuring solve: iteration count vs. p refinement level in two dimensions

Levels of Preconditioning In the next set of tests, we study the effect of including different levels in the preconditioner defined in Figure 10. Note that preconditioning is a direct solution methodology and costs are of $O(n^3)$ where n is the block size. It is thus necessary to perform this in parallel. Figure 13 shows the solution time for various levels of preconditioning. We note that the number of levels of preconditioning is never reduced below that corresponding to the linear unknowns on the real interface (unknowns shared across processors). This level of preconditioning is denoted by level 0. Additional levels are obtained from the hierarchy of the nested partitioning. The results are for four processors. Due to higher computation and communication speeds increasing the degree of preconditioning decreases the solution time for the IBM, while for the SGI it shows a marginal increase in time. The problem size also affects the times. For smaller problems it usually is worthwhile to have higher levels of preconditioning.

In the next set of tests we will investigate the speedups on different architectures for a fixed problem size. Figure 14 shows the best solution time on an IBM SP, SGI Origin and IBM Blue Horizon for different numbers of processors. While we have obtained very good (super-linear) speedups, it re-

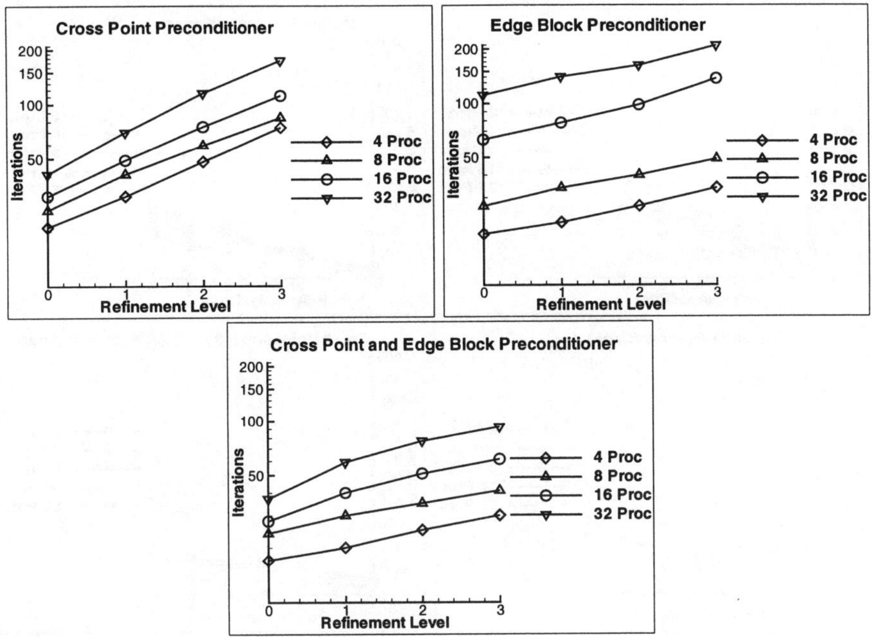

Fig. 10. Iterative substructuring solve: iteration count vs. h refinement level for three dimensions

quired different solution strategies for each combination of architecture and number of processors.

We note that these are not the fastest speeds when compared to times obtained by other popular solution strategies like a fully iterative technique using very efficient sparse storage schemes. But success of those schemes is highly dependent on the physics of the problem and tuning to a specific architecture. Our model problem is not sufficiently complex to exercise those limitations. However, it does illustrate the architecture dependent nature of performance and the need to change algorithms for differing architectures.

5.4   Model Prediction

In the next set of tests, we will evaluate the ability of the models introduced in the previous chapter to predict the optimal number of levels for different combinations of problem size and architecture. Model parameters were estimated using sample matrix vector products of sizes ranging from 2,000-10,000 unknowns per processor and measuring the time for computation and communication. The parameter values for the architectures are listed in Table 2.

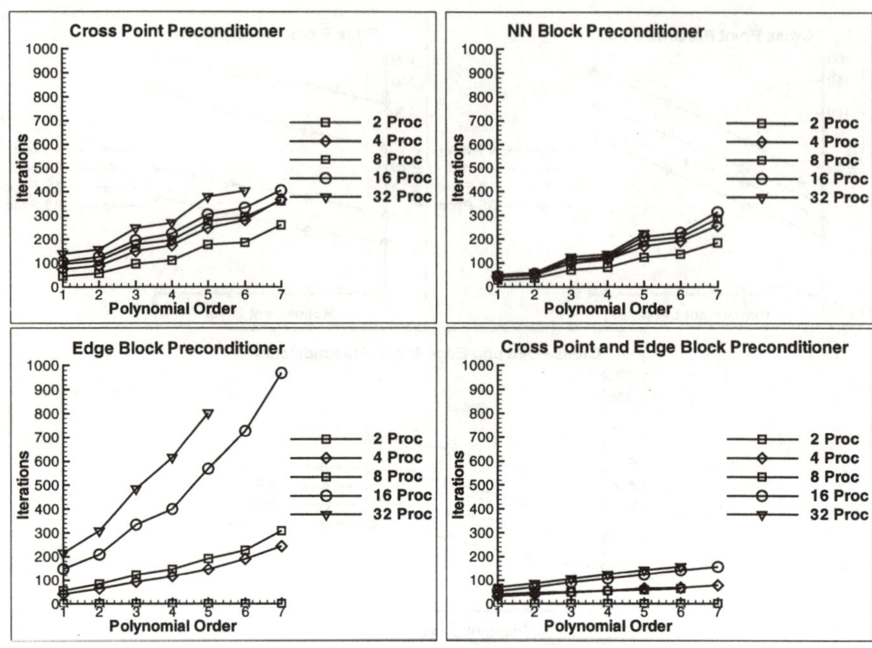

Fig. 11. Iterative substructuring solve: iteration count vs. p refinement level for three dimensions

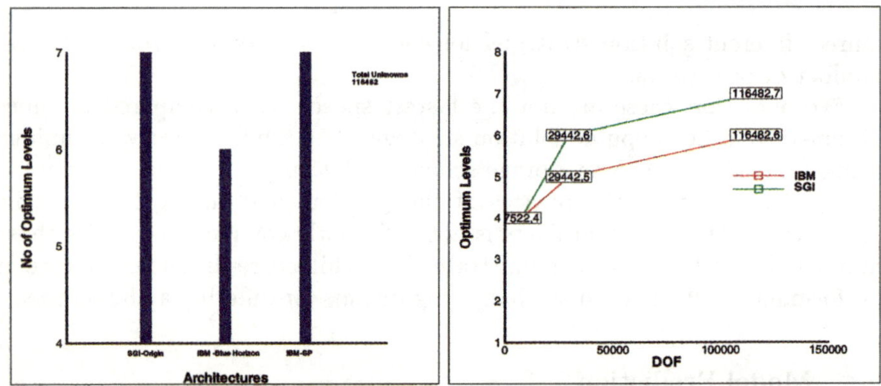

Fig. 12. Right: Optimum number of levels of substructuring for different architectures for a fixed problem. Left: Optimum number of levels of substructuring for different problem sizes on different architectures. The first number in the box is the problem size and second number in the box is the optimum number of levels

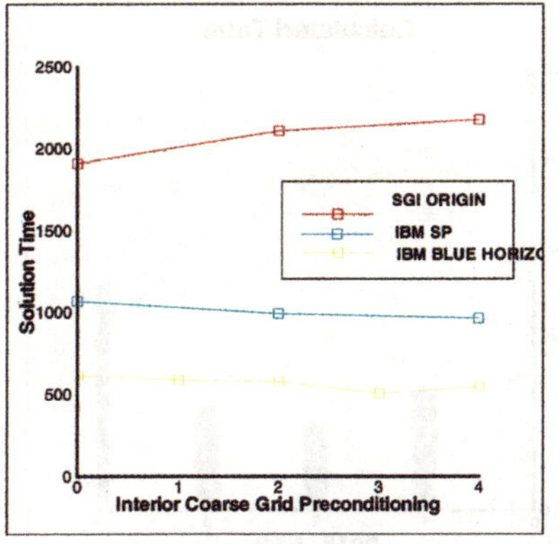

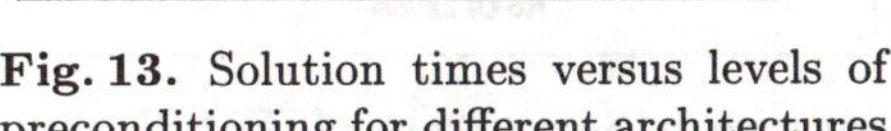

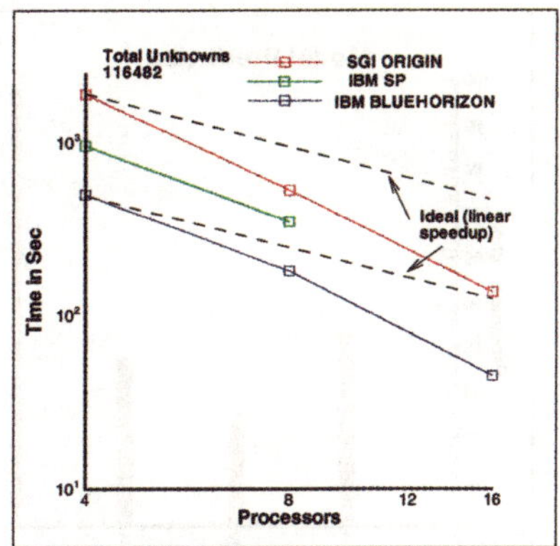

Fig. 13. Solution times versus levels of preconditioning for different architectures

Fig. 14. Speedups for different architectures using the best schemes

Table 2. Parameter values for different architectures

	Architecture	Avg. Computation/Communication $\left(\frac{\nu}{\tau}\right)$
1	SGI ORIGIN 2000 (R10K Processors)	146
2	IBM SP (Power 3 processors) (processors in the same cluster)	36
3	IBM SP (Power 3 processors) (processors in different cluster)	134

Figures 15-16 show the model predictions and actual solution times on two different architectures, namely the IBM and the SGI. The model correctly predicts the number of levels for optimum performance. Note that the cache effects in the interprocess communication and TLB misses [17] have not been accounted for in the model. However, even this simple model appears to be adequate in predicting the relative performance as a function of the number of levels of substructuring. The prediction of actual solution time is not very accurate (it appears to be scaled).

6   Conclusions and Future Work

We describe here several efficient solvers for the solution of systems of equations arising from adaptive *hp* finite elements in two and three dimensions on multiple parallel architectures. Our primary solution methodology is an iterative substructuring scheme with coarse grid preconditioning. We also outline the development of a new self-tuning solver using extensions of the basic iterative scheme into a multi-level scheme where the number of levels to substructure/use in the preconditioner are chosen based on a machine dependent model of the computational cost. Numerical results illustrate the behavior of both solvers. Analysis of the base iterative substructuring scheme

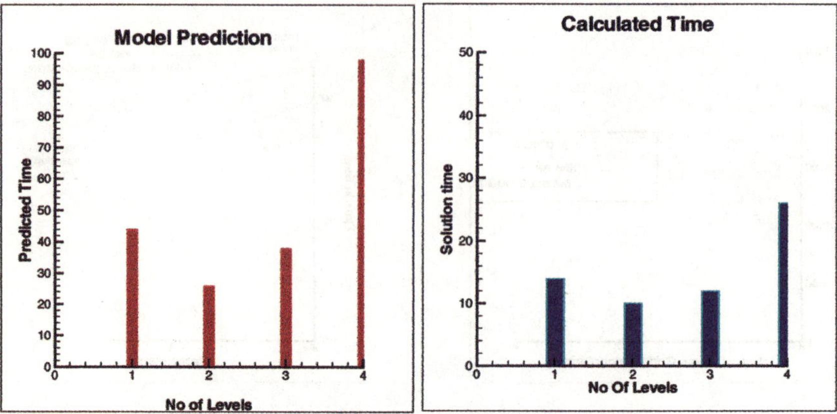

Fig. 15. Model prediction and performance on 8 processors of IBM SP

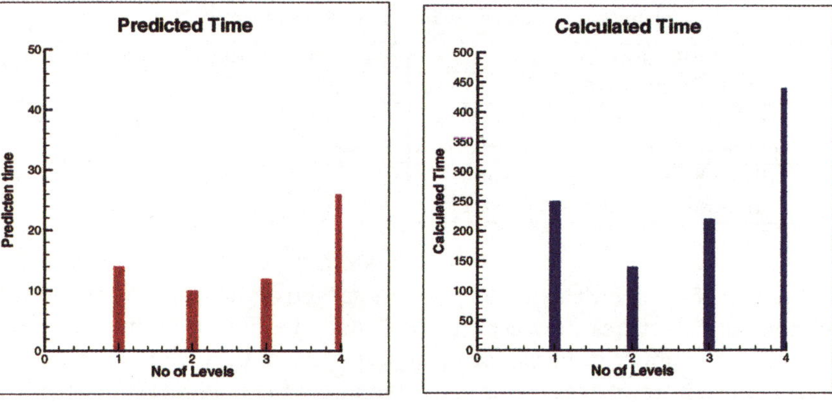

Fig. 16. Model prediction and performance on 8 processors of SGI Origin

and extension of the self-tuning solver to three space dimensions will be the subject of a forthcoming paper.

Acknowledgments: The financial support of the National Science Foundation through Grant ASC9702947 is acknowledged. Computer time was provided by the National Partnership for Advanced Computing Infrastructure, San Diego and Center for Computational Research, University at Buffalo.

References

1. Satish B., Gropp, W. D. et. al. (2000) PETSc 2.0 Users Manual, ANL-95/11 - Revision 2.0.28. Argonne National Laboratory, Chicago
2. Laszloffy A., Long J., Patra A. (2000) Simple Data Management Schemes and Scheduling Schemes For Managing the Irregularities in Parallel Adaptive *hp* Finite Element Simulations. Parallel Computing. **26**, 1765-1788

3. Patra A., Oden J. T. (1997) Computational techniques for adaptive hp finite element methods. Finite Elements in Analysis and Design. **25**, 27-39

4. Bramble J., Pasciak J., Schatz A. (1986) The Construction of Preconditioners for Elliptic Problems by Substructuring I. Math. Comp. **47(175)**, 103-134

5. Bramble J. H., Ewing R. H. et. al. (1992) Domain Decomposition for Problems with Partial Refinement. SIAM J. Sci. Sta. Comp. **13(1)**, 397-410

6. Widlund O. B. (1988) Iterative Substructuring Methods: Algorithms and Theory for Elliptic Problems in the Plane. In Glowinski R., Golub G. H. et. al., editors, First International Symposium on Domain Decomposition Methods for Partial Differential Equations. SIAM, Philadelphia

7. Dryja M. (1989) An Additive Schwarz Algorithm for Two- and Three Dimensional Finite Element Elliptic Problems. In Chan T., Glowinski R. et al., editors, Domain Decomposition Methods. SIAM, Philadelphia

8. Babuska I., Craig A. et. al. (1991) Efficient preconditioning for the p version finite element method in two dimensions. SIAM J. Numer. Anal. **28(3)**, 624-661

9. Mandel J. (1990) Iterative solvers by substructuring for the p-version finite element method. Comput. Meth. Appl. Mech. Engg. **80**, 117-128

10. Mandel J. (1990) Two level domain decomposition preconditioning for the p-version finite element method in three dimensions. Int. J. Numer. Meth. in Engg. **29**, 1095-1108

11. Ainsworth M. (1996) A Preconditioner Based on Domain Decomposition of h-p Finite Element Approximation on Quasi-Uniform Meshes. SIAM J. Numer. Anal. **33(4)**, 1358-1377

12. Oden J. T., Patra A., Feng Y. S. (1997) Domain decomposition solvers for adaptive hp finite element methods. SIAM J. Numer. Anal. **34(6)**, 2090-2118

13. Pavarino L. F., Widlund O. B. (1999) Iterative Substructuring Methods For Spectral Element Discretizations of Elliptic Systems I. Compressible Linear Elasticity. SIAM J. Numer. Anal. **37(6)**, 353-374

14. Pavarino L. F., Widlund O. B. (1999) Iterative Substructuring Methods For Spectral Element Discretizations of Elliptic Systems II. Mixed Methods For Linear Elasticity and Stokes Flow. SIAM J. Numer. Anal. **37(6)**, 375-402

15. Guo B. Q., Cao W. (1998) Domain Decomposition Method for the h-p Version Finite Element Method. Comp. Meth. Appl. Mech. Engrg. **157**, 425-440

16. van de Geijn R., Gunnels J. PLAPACK Users Guide. http://www.cs.utexas.edu/users/plapack/

17. Loshin D. (1998) Efficient Memory Programming. Mcgraw-Hill, New York

18. Patra A., Oden J. T. (1995) Problem Decomposition Strategies for Adaptive hp Finite Element Methods. Computing Systems in Engineering. **6(2)**, 97-109

19. Jones M. T., Plassmann P. E. (1993) The Efficient Parallel Iterative Solution of Large Sparse Linear Systems. In George A., Gilbert J., Liu J. W. H., eds. Graph Theory and Sparse Matrix Computation. **56** of IMA Volumes in Mathematics and Its Applications, Springer-Verlag, 229-245

Editorial Policy

§1. Volumes in the following four categories will be published in LNCSE:

i)   Research monographs
ii)   Lecture and seminar notes
iii) Conference proceedings

Those considering a book which might be suitable for the series are strongly advised to contact the publisher or the series editors at an early stage.

§2. Categories i) and ii). These categories will be emphasized by Lecture Notes in Computational Science and Engineering. **Submissions by interdisciplinary teams of authors are encouraged.** The goal is to report new developments – quickly, informally, and in a way that will make them accessible to non-specialists. In the evaluation of submissions timeliness of the work is an important criterion. Texts should be well-rounded, well-written and reasonably self-contained. In most cases the work will contain results of others as well as those of the author(s). In each case the author(s) should provide sufficient motivation, examples, and applications. In this respect, Ph.D. theses will usually be deemed unsuitable for the Lecture Notes series. Proposals for volumes in these categories should be submitted either to one of the series editors or to Springer-Verlag, Heidelberg, and will be refereed. A provisional judgment on the acceptability of a project can be based on partial information about the work: a detailed outline describing the contents of each chapter, the estimated length, a bibliography, and one or two sample chapters – or a first draft. A final decision whether to accept will rest on an evaluation of the completed work which should include

– at least 100 pages of text;
– a table of contents;
– an informative introduction perhaps with some historical remarks which should be
  accessible to readers unfamiliar with the topic treated;
– a subject index.

§3. Category iii). Conference proceedings will be considered for publication provided that they are both of exceptional interest and devoted to a single topic. One (or more) expert participants will act as the scientific editor(s) of the volume. They select the papers which are suitable for inclusion and have them individually refereed as for a journal. Papers not closely related to the central topic are to be excluded. Organizers should contact Lecture Notes in Computational Science and Engineering at the planning stage.

In exceptional cases some other multi-author-volumes may be considered in this category.

§4. Format. Only works in English are considered. They should be submitted in camera-ready form according to Springer-Verlag's specifications. Electronic material can be included if appropriate. Please contact the publisher. Technical instructions and/or TEX macros are available via http://www.springer.de/author/tex/help-tex.html; the name of the macro package is "LNCSE – LaTEX2e class for Lecture Notes in Computational Science and Engineering". The macros can also be sent on request.

General Remarks

Lecture Notes are printed by photo-offset from the master-copy delivered in camera-ready form by the authors. For this purpose Springer-Verlag provides technical instructions for the preparation of manuscripts. See also *Editorial Policy*.

Careful preparation of manuscripts will help keep production time short and ensure a satisfactory appearance of the finished book. The actual production of a Lecture Notes volume normally takes approximately 12 weeks.

The following terms and conditions hold:

Categories i), ii), and iii):
Authors receive 50 free copies of their book. No royalty is paid. Commitment to publish is made by letter of intent rather than by signing a formal contract. Springer-Verlag secures the copyright for each volume.

For conference proceedings, editors receive a total of 50 free copies of their volume for distribution to the contributing authors.

All categories:
Authors are entitled to purchase further copies of their book and other Springer mathematics books for their personal use, at a discount of 33,3 % directly from Springer-Verlag.

Addresses:

Professor Timothy J. Barth
NASA Ames Research Center
NAS Division
Moffett Field, CA 94035, USA
e-mail: barth@nas.nasa.gov

Professor Michael Griebel
Institut für Angewandte Mathematik
der Universität Bonn
Wegelerstr. 6
D-53115 Bonn, Germany
e-mail: griebel@iam.uni-bonn.de

Professor David E. Keyes
Computer Science Department
Old Dominion University
Norfolk, VA 23529–0162, USA
e-mail: keyes@cs.odu.edu

Professor Risto M. Nieminen
Laboratory of Physics
Helsinki University of Technology
02150 Espoo, Finland
e-mail: rni@fyslab.hut.fi

Professor Dirk Roose
Department of Computer Science
Katholieke Universiteit Leuven
Celestijnenlaan 200A
3001 Leuven-Heverlee, Belgium
e-mail: dirk.roose@cs.kuleuven.ac.be

Professor Tamar Schlick
Department of Chemistry and
Courant Institute of Mathematical
Sciences
New York University
and Howard Hughes Medical Institute
251 Mercer Street, Rm 509
New York, NY 10012-1548, USA
e-mail: schlick@nyu.edu

Springer-Verlag, Mathematics Editorial IV
Tiergartenstrasse 17
D-69121 Heidelberg, Germany
Tel.: *49 (6221) 487-185
e-mail: peters@springer.de
http://www.springer.de/math/
peters.html

Lecture Notes
in Computational Science
and Engineering

Vol. 1   D. Funaro, *Spectral Elements for Transport-Dominated Equations.*   1997.
X, 211 pp. Softcover. ISBN 3-540-62649-2

Vol. 2   H. P. Langtangen, *Computational Partial Differential Equations.* Numerical Methods and Diffpack Programming.   1999.   XXIII, 682 pp.   Hardcover.
ISBN 3-540-65274-4

Vol. 3   W. Hackbusch, G. Wittum (eds.), *Multigrid Methods V.* Proceedings of
the Fifth European Multigrid Conference held in Stuttgart, Germany, October
1-4, 1996. 1998. VIII, 334 pp. Softcover. ISBN 3-540-63133-X

Vol. 4   P. Deuflhard, J. Hermans, B. Leimkuhler, A. E. Mark, S. Reich, R. D. Skeel
(eds.), *Computational Molecular Dynamics: Challenges, Methods, Ideas.* Proceedings of the 2nd International Symposium on Algorithms for Macromolecular Modelling, Berlin, May 21-24, 1997. 1998. XI, 489 pp. Softcover. ISBN 3-540-63242-5

Vol. 5   D. Kröner, M. Ohlberger, C. Rohde (eds.), *An Introduction to Recent Developments in Theory and Numerics for Conservation Laws.* Proceedings of the International School on Theory and Numerics for Conservation Laws, Freiburg / Littenweiler, October 20-24, 1997. 1998. VII, 285 pp. Softcover. ISBN 3-540-65081-4

Vol. 6   S. Turek, *Efficient Solvers for Incompressible Flow Problems.* An Algorithmic and Computational Approach. 1999. XVII, 352 pp, with CD-ROM. Hardcover.
ISBN 3-540-65433-X

Vol. 7   R. von Schwerin, *Multi Body System SIMulation.* Numerical Methods,
Algorithms, and Software. 1999. XX, 338 pp. Softcover. ISBN 3-540-65662-6

Vol. 8   H.-J. Bungartz, F. Durst, C. Zenger (eds.), *High Performance Scientific and
Engineering Computing.* Proceedings of the International FORTWIHR Conference
on HPSEC, Munich, March 16-18, 1998. 1999. X, 471 pp. Softcover. 3-540-65730-4

Vol. 9   T. J. Barth, H. Deconinck (eds.), *High-Order Methods for Computational
Physics.* 1999. VII, 582 pp. Hardcover. 3-540-65893-9

Vol. 10   H. P. Langtangen, A. M. Bruaset, E. Quak (eds.), *Advances in Software
Tools for Scientific Computing.* 2000. X, 357 pp. Softcover. 3-540-66557-9

Vol. 11   B. Cockburn, G. E. Karniadakis, C.-W. Shu (eds.), *Discontinuous Galerkin
Methods.* Theory, Computation and Applications. 2000. XI, 470 pp. Hardcover.
3-540-66787-3

Vol. 12   U. van Rienen, *Numerical Methods in Computational Electrodynamics. Linear Systems in Practical Applications.* 2000. XIII, 375 pp. Softcover. 3-540-67629-5

Vol. 13   B. Engquist, L. Johnsson, M. Hammill, F. Short (eds.), *Simulation and Visualization on the Grid.* Parallelldatorcentrum Seventh Annual Conference, Stockholm, December 1999, Proceedings. 2000. XIII, 301 pp. Softcover. 3-540-67264-8

Vol. 14   E. Dick, K. Riemslagh, J. Vierendeels (eds.), *Multigrid Methods VI.* Proceedings of the Sixth European Multigrid Conference Held in Gent, Belgium, September 27-30, 1999. 2000. IX, 293 pp. Softcover. 3-540-67157-9

Vol. 15   A. Frommer, T. Lippert, B. Medeke, K. Schilling (eds.), *Numerical Challenges in Lattice Quantum Chromodynamics.* Joint Interdisciplinary Workshop of John von Neumann Institute for Computing, Jülich and Institute of Applied Computer Science, Wuppertal University, August 1999. 2000. VIII, 184 pp. Softcover. 3-540-67732-1

Vol. 16   J. Lang, *Adaptive Multilevel Solution of Nonlinear Parabolic PDE Systems. Theory, Algorithm, and Applications.* 2001. XII, 157 pp. Softcover. 3-540-67900-6

Vol. 17   B. I. Wohlmuth, *Discretization Methods and Iterative Solvers Based on Domain Decomposition.* 2001. X, 197 pp. Softcover. 3-540-41083-X

Vol. 18   U. van Rienen, M. Günther, D. Hecht (eds.), *Scientific Computing in Electrical Engineering.* Proceedings of the 3rd International Workshop, August 20-23, 2000, Warnemünde, Germany. 2001. XII, 428 pp. Softcover. 3-540-42173-4

Vol. 19   I. Babuška, P. G. Ciarlet, T. Miyoshi (eds.), *Mathematical Modeling and Numerical Simulation in Continuum Mechanics.* Proceedings of the International Symposium on Mathematical Modeling and Numerical Simulation in Continuum Mechanics, September 29 - October 3, 2000, Yamaguchi, Japan. 2002. VIII, 301 pp. Softcover. 3-540-42399-0

Vol. 20   T. J. Barth, T. Chan, R. Haimes (eds.), *Multiscale and Multiresolution Methods. Theory and Applications.* 2002. X, 389 pp. Softcover. 3-540-42420-2

Vol. 21   M. Breuer, F. Durst, C. Zenger (eds.), *High Performance Scientific and Engineering Computing.* Proceedings of the 3rd International FORTWIHR Conference on HPSEC, Erlangen, March 12-14, 2001. 2002. XIII, 408 pp. Softcover. 3-540-42946-8

Vol. 22   K. Urban, *Wavelets in Numerical Simulation. Problem Adapted Construction and Applications.* 2002. XV, 181 pp. Softcover. 3-540-43055-5

Vol. 23   L. F. Pavarino, A. Toselli (eds.), *Recent Developments in Domain Decomposition Methods.* 2002. XII, 243 pp. Softcover. 3-540-43413-5

For further information on these books please have a look at our mathematics catalogue at the following URL: http://www.springer.de/math/index.html

Production: Druckhaus Beltz, Hemsbach